MYCORRHIZAS IN ECOSYSTEMS

Mycorrhizas in Ecosystems

Edited by

D.J. Read

Department of Animal and Plant Sciences
University of Sheffield
Sheffield S10 2TN, UK

D.H. Lewis

Department of Animal and Plant Sciences
University of Sheffield
Sheffield S10 2TN, UK

A.H. Fitter

Department of Biology
University of York
Heslington, York YO1 5DD, UK

I.J. Alexander

Department of Plant and Soil Science
University of Aberdeen
Aberdeen AB9 1UE, UK

CAB INTERNATIONAL

CABI is a trading name of CAB International

CABI Head Office
Nosworthy Way
Wallingford
Oxfordshire OX10 8DE
UK

CABI North American Office
875 Massachusetts Avenue
7th Floor
Cambridge, MA 02139
USA

Tel: +44 (0)1491 832111
Fax: +44 (0)1491 833508
Email: cabi@cabi.org
Web site: www.cabi.org

Tel: +1 617 395 4056
Fax: +1 617 354 6875
Email: cabi-nao@cabi.org

1005107257

A catalogue record for this book is available from the British Library, London, UK

ISBN-13: 978-0-85198-786-6
ISBN-10: 0-85198-786-9

First published 1992
Reprinted 1994
Transferred to print on demand 2006

Printed and bound in the UK by CPI Antony Rowe, Eastbourne.

Contents

Preface

This volume is dedicated to two pioneers of the study of mycorrhizal function, Professor Jack Harley and Professor Gösta Lindeberg, both of whom did so much to add to our awareness of the importance of the symbiosis.

In nature, the roots of most plants are infected by fungi to form mycorrhizas which play a central role in the capture of nutrients from the soil. Most of our knowledge of the biology of the mycorrhizal symbiosis has been derived from studies carried out under controlled conditions in the laboratory or glasshouse. There is an increasing awareness of the need to extend these studies to the more natural situations in which the symbiosis evolved and in which it normally functions. This volume brings together a series of papers that place major emphasis upon mycorrhizal function in nature. They were originally presented as invited oral contributions to the third European Symposium on Mycorrhizas which was held at the University of Sheffield from 19 to 23 August 1991.

All of the major types of mycorrhiza are considered. Factors determining their distribution and effectiveness in different ecosystems from the boreal forest to the tropics are examined in depth and attempts to improve productivity of ecosystems by inoculation with more effective fungal symbionts are described. Important papers describe the application of the latest techniques for the study of intact hose–fungus systems and discuss the results obtained with reference to mycorrhizal function in the field. These contributions include considerations of the capture of ions from soil and the factors which influence their transfer by way of fungal hyphae to the root and into the plant cell. Recent views on the functional relationship between fungal associations of roots (mycorrhiza) and those of shoots (mycophylla) are also explored.

The volume will be of interest to all those who wish to obtain a more complete understanding of the functioning of natural ecosystems.

The assistance of the following people is gratefully acknowledged: Terry Croft, Mark Cosgrove, Rob Bradley, Bob Bartlett, Glyn Woods, David Smart, Nuala Ruttle, Jonathan Leake, Simon Kerley, Gary Bending, Helen Jones, Marco Martins, Ranjan Jayaratne, Chris Scanlon, Angie Doncaster, Sue Carter, Angela Sellars,

Jane Bird, Charles Lane, Andy Langdale, Allison Hartill, Chris Williams, Diana Lewis, Catherine Kemp, and Richard Young. Particular thanks go to Jayne Young who made all of the local arrangements for the symposium and for this volume.

Financial and other support was generously provided by the New Phytologist Trust, The British Mycological Society, The Lord Mayor and Sheffield City Council, The University of Sheffield, Shell Research Limited and British Airways.

Part One

Status and Function of Vesicular-Arbuscular (VA) Mycorrhiza in Ecosystems

1 Nutrient Dynamics at the Soil–Root Interface (Rhizosphere)

H. Marschner

Institute of Plant Nutrition, University of Hohenheim, P.O.B. 70 05 62, 7000 Stuttgart 70, Germany

Introduction

The soil–root interface, the rhizosphere, is characterized by gradients, which occur both in a radial and longitudinal direction along an individual root. Gradients may exist for mineral nutrients, pH, redox potential and reducing processes, root exudates and microbial activity. These gradients, which are determined by soil and plant factors, strongly affect the acquisition of mineral nutrients and play a key role in the adaptation of plants to adverse soil chemical conditions, as occur for example in acid soils (Marschner, 1991). In this chapter a few examples are given for such gradients.

Nutrient Concentration

Concentrations and dynamics of mineral nutrients in the rhizosphere are not only determined by the capacity of the various zones along the root axis to absorb mineral nutrients and water (Marschner, 1991) but also by the rate of delivery of nutrients via mass flow and diffusion from the bulk soil to the rhizoplane. For example, delivery of Ca by mass flow may exceed uptake rates, thus leading to accumulation of Ca in the rhizosphere and at the rhizoplane. In contrast, for P and K, usually present in the soil solution at only low concentrations, depletion in the rhizosphere soil is a typical feature and may, for example, lead to release of K from the non-exchangeable fraction and even induce enhanced 'weathering' of clay minerals.

pH

Rhizosphere pH may differ from that of the bulk soil pH by more than two units, depending on the form of nitrogen supply (NH_4^+; NO_3^-; symbiotic

3

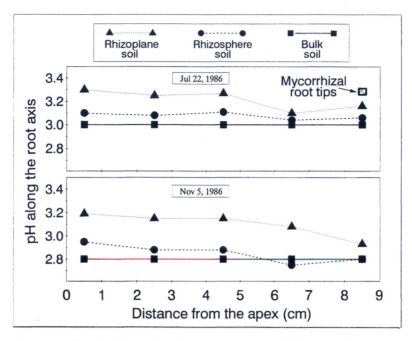

Fig. 1.1. Bulk soil pH and pH at the rhizoplane and in the rhizosphere soil along the axis of long-roots of 60-year-old Norway spruce (Marschner *et al.*, 1991).

N_2-fixation), plant genotype, the nutritional status of the plants and the pH buffering capacity of the soil (Marschner *et al.*, 1986). In most instances, these pH changes are brought about by differences in net excretion of H^+ owing to an imbalance in cation–anion uptake. In this respect the uptake of NH_4^+ and NO_3^- is of particular importance as these ions are required by the plants in large amounts. As a result of the much higher mobility of NO_3^- in the soil and its more rapid delivery to roots by mass flow and diffusion, uptake of NO_3^- often exceeds that of NH_4^+ with a corresponding increase in rhizosphere and rhizoplane pH, even in strongly acid soils (Fig. 1.1).

By contrast, rhizosphere acidification is a typical feature of legumes depending on N_2 fixation or of non-graminaceous species under Fe-deficiency. There is increasing evidence that rhizosphere acidification is also often caused by nutrient deficiency-induced enhanced excretion of organic acids (see below). These root-induced changes in rhizosphere pH have important effects, for example, on the solubility and thus acquisition of mineral nutrients, P and the micronutrients Zn, Fe and Mn in particular.

Redox Potential and Reducing Processes

In plants adapted to submerged conditions (e.g. lowland rice, reed) the redox potential of the rhizosphere is increased and oxidizing conditions prevail in the rhizosphere by the release of O_2 derived from the shoot ('internal ventilation'). This oxidation is essential to decrease phytotoxic concentrations of volatile organic acids and of Mn^{2+} and Fe^{2+} present in the bulk soil solution under submerged conditions. In contrast, in well-aerated calcareous soils, availability of Fe, and also often Mn, is very low leading to deficiency (e.g. 'Fe-chlorosis'). Non-graminaceous species respond to Fe-deficiency by inducing activity of a plasma membrane-bound reductase in the rhizodermal cells of apical root zones leading to enhanced reduction rates of Fe^{III}, often combined with enhanced net excretion of H^+ and phenolics (Marschner *et al.*, 1986). As a side-effect of this response, both Mn reduction and uptake are increased.

Root Exudates

The amounts of organic carbon that roots release into the rhizosphere may reach as much as 40% of the total dry matter production of a plant (Lynch and Whipps, 1990). Various forms of stress such as mechanical impedance, anaerobiosis, and mineral nutrient deficiency (e.g. of P, K, Zn and Fe) may strongly increase this carbon release (Marschner, 1991). The release of low molecular weight root exudates is of particular importance in relation to the nutrient dynamics in the rhizosphere. Some of these exudated solutes are directly involved in mobilization of sparingly soluble mineral nutrients, others more indirectly via affecting the microbial activity in the rhizosphere. For example, tryptophan may serve as source

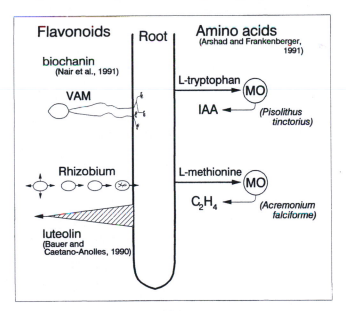

Fig. 1.2. Possible role of certain low molecular weight root exudates as sources for hormone production or as 'signals' for microorganisms in the rhizosphere. VAM = vesicular-arbuscular mycorrhiza; MO = microorganism.

for auxin production by ectomycorrhizal fungi such as *Pisolithus tinctorius*, and methionine as a source for ethylene production by soil fungi like *Acremonium falciforme*. Increased concentrations of these phytohormones in the rhizosphere may strongly affect growth, morphology and physiology of the roots and, thus, indirectly acquistion of mineral nutrients. Also, certain flavonoids in root exudates, such as luteolin and biochanin, may respectively act as 'signals' for *Rhizobium* and as a stimulating factor for hyphal growth of vesicular-arbuscular (VA) mycorrhizal fungi (Fig. 1.2).

Other root exudates such as organic acids and phytosiderophores affect nutrient dynamics in the rhizosphere directly. Nutrient deficiency may enhance release of such compounds which in turn enable the acquisition of the nutrients. In response to P-deficiency members of the Proteaceae as well as annual species like *Lupinus albus* (a non-mycorrhizal species) form 'proteoid roots' which strongly acidify the rhizosphere (Dinkelaker *et al.*, 1989). This acidification is brought about by excretion of citric acid (Table 1.1) which mobilizes sparingly soluble phosphates either by chelation (Fe, Al) or formation of sparingly soluble Ca citrate in the rhizosphere. This localized acidification of the rhizosphere by citric acid leads to mobilization not only of P but also of Fe, Zn and Mn (Table 1.1) and consequently higher concentrations of these nutrients in the plants. In rape, acidification of the rhizosphere induced by P-deficiency is confined to apical root zones and is brought about by excretion of both malic and citric acids (Hoffland *et al.*, 1989).

When grown in a P-deficient calcareous soil, the amounts of citric acid released into the rhizosphere by *Lupinus albus* may reach up to 23% of net photosynthesis (Dinkelaker *et al.*, 1989). This appears to be a high cost for P-acquisition. However, in view of the benefits (mobilization not only of P but also other growth-limiting nutrients) and the costs of alternatives (allocation of 10–20% of the net photosynthates to the fungus in VA mycorrhizal associations), this strategy of 'proteoid-root' plants seems to be quite efficient as it enables the plants, in contrast to the mycorrhizal associations, rapidly to turn on and off the response mechanisms and, thus, regulate the carbon costs effectively.

A particular type of root exudation exists in graminaceous species (Fig. 1.3). In response to Fe-deficiency, the release of non-proteinogenic amino acids, the

Table 1.1. Soil pH and concentrations of citrate and micronutrients in bulk and rhizosphere soil of *Lupinus albus* grown in a P-deficient soil (23% $CaCO_3$) (Dinkelaker *et al.*, 1989).

	Bulk soil	Rhizosphere soil (proteoid roots)
pH (H_2O)	7.5	4.8
Citrate ($\mu g\ g^{-1}$ soil)	n.d.*	47.7
DTPA† extract ($\mu mol\ kg^{-1}$)		
Fe	34.0	251.0
Mn	44.0	222.0
Zn	2.8	16.8

*not detectable; †DTPA = diaethylenetriamine pentacetic acid.

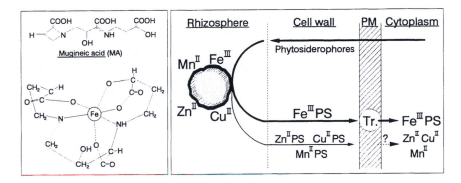

Fig. 1.3. Molecular structure of the phytosiderophore (PS), mugineic acid, and of the corresponding Fe^{III} chelate (based on Nomoto *et al.*, 1987) and a proposed mechanism of PS release, mobilization in the rhizosphere and uptake of Fe^{III} and other metal cations. Tr = translocator for Fe^{III} PS; PM = plasma membrane.

so-called phytosiderophores, is markedly increased (Takagi *et al.*, 1984; Römheld and Marschner, 1990). Phytosiderophores form Fe^{III} chelates of high stability and are absorbed selectively by a transport system located in the plasma membrane of root cells. Although this phytosiderophore system in graminaceous species is similar to the siderophore system in microorganisms, it does not 'recognize' non-plant siderophores. Accordingly, the uptake rates of Fe supplied as Fe^{III} phytosiderophores are orders of magnitude higher than those of Fe^{III}-siderophores or synthetic Fe^{III} chelates (Table 1.2).

Acid phosphatases are a special type of root exudate which hydrolyse organically bound P in the rhizosphere. Acid phosphatase activity is much higher at the rhizoplane than in the bulk soil and, in many plant species, increased under P deficiency. The importance of acid phosphatases for P acquisition under field conditions has been demonstrated for Norway spruce growing in acid soil where depletion of P in the rhizosphere and rhizoplane soil was confined to the P_{org} fraction whereas the P_{inorg} fraction remained unchanged (Häussling and

Table 1.2. Fe mobilization from a calcareous soil by Fe^{III} chelators and uptake rates of Fe^{III} supplied as $^{59}Fe^{III}$ chelates to −Fe barley plants (Römheld and Marschner, 1990).

Chelator (10^{-5}M)	Mobilization (nmol g^{-1} soil 12 h^{-1})	Uptake rate (nmol g^{-1} root DW 4 h^{-1})
Phytosiderophore (HMA)*	23.6	3456.00
Microbial (Desferal)	19.2	1.21
Synthetic (DTPA)†	2.0	0.51

* HMA = 3-hydroxy-mugineic acid; † DTPA = diethylenetriamine pentacetic acid.

Marschner, 1989). As no distinction was made in this study between mycorrhizal and non-mycorrhizal roots, the phosphatase activity may have been derived from fungi, roots or both. Current studies on nutrient dynamics in the rhizosphere of different plant species, including Norway spruce, involve non-destructive test methods which allow semi-quantitative measurements of acid phosphatase activity at the rhizoplane, in the rhizosphere and in the bulk soil of mycorrhizal and non-mycorrhizal plants.

Rhizosphere Microorganisms

For the nutrient dynamics in the rhizosphere and nutrient acquisition, two groups of microorganisms are of particular importance, those non-infecting species confined to the rhizosphere itself and the infecting ones which form mycorrhiza.

Non-infecting rhizosphere microorganisms

Since roots act as source of organic carbon, the population density of microorganisms, especially bacteria, is considerably higher in the rhizosphere than in the bulk soil. Non-infecting rhizosphere microorganisms may affect mineral nutrition of plants either directly or indirectly. Direct effects can be brought about by mobilization (e.g. reduction of Mn), mineralization of N_{org} and nitrification, or production of chelators such as phenolics and siderophores. On the other hand, rhizosphere microorganisms may impair the mineral nutrition of plants, for example by oxidation of Mn, competition for P, denitrification or utilization

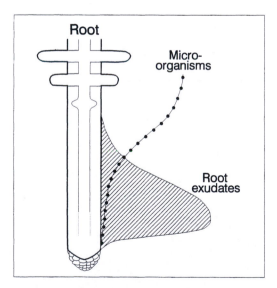

Fig. 1.4. Schematic presentation of spatial separation of root exudates (phytosiderophores, organic acids) and microbial activity in the rhizosphere of soil-grown plants.

of root exudates as carbon sources. The last aspect is particularly important as it may considerably decrease the efficiency of those mechanisms in which enhanced release of organic acid or phytosiderophores plays a key role under conditions of deficiency.

High decomposition rates of root exudates can readily be demonstrated in plants growing in non-axenic nutrient solutions. However, along the axis of roots in soil-grown plants gradients exist in both root exudation, for example, of phytosiderophores, and microbial activity (Römheld, 1991). Microbial activity is usually low in apical zones and steeply increases towards basal zones, whereas exudation of phytosiderophores is high in apical zones and, additionally, is confined to only a few hours during the day. This spatial separation of root exudation and microbial activity (Fig. 1.4) is probably of crucial importance for the efficiency of phytosiderophores or organic acids (e.g. citric acid in proteoid root zones) for mobilization of sparingly soluble mineral nutrients in the rhizosphere.

N_2-fixing (diazotrophic) bacteria are also present in the rhizosphere. In C4 species, particularly when growing on N-deficient soils in the tropics, these bacteria may contribute substantially to the N nutrition of plants. In temperate climates, however, stimulation of growth by diazotrophic bacteria is most probably mainly the result of hormonal effects. Certain strains of *Azospirillum brasilense* for example, produce auxin and thereby strongly enhance root-hair formation and lateral root initiation (Martin *et al.*, 1989). The corresponding increase in root surface area may considerably enhance P-acquisition and thus plant growth on P-deficient soils.

Mycorrhizas

Mycorrhizal infection may change root activity, root growth and exudation and, via the external mycelium, extend the soil volume available for acquisition of mineral nutrients and/or water. In principle, three possibilities exist.

1. Root growth and activity are not affected and mycorrhizal hyphae add surface area and increase the spatial availability of mineral nutrients.
2. Mycorrhizal hyphae supply nutrients such as P or N which are growth-limiting factors in non-mycorrhizal plants. If it is assumed that photosynthetic capacity (source) is unlimited, not only shoot but also root growth will be enhanced and thus acquisition of nutrients (and water) by the roots.
3. Mycorrhizas are either ineffective in delivering nutrients, or nutrients are not growth-limiting in non-mycorrhizal plants; under conditions of source limitation, mycorrhizas depress root growth by sink competition. Harmful effects on shoot growth are then to be expected if the external mycelium cannot fully compensate for the root's functions in uptake of nutrients and water.

In view of these various possibilities of mycorrhizal effects on plant growth, it is necessary for a better understanding of the interactions and for predictions of mycorrhizal effects to quantify the capacity of the external mycelium for nutrient and water uptake. It is also important to characterize the conditions at the hyphae–soil interface (hyphosphere) for both VA and ectomycorrhizas in a similar manner as for the root–soil interface. In order to achieve this characterization, special techniques have to be applied to allow a spatial separation of root and hyphal compartments (e.g. Finlay and Read, 1986, for ectomycorrhizas; George *et al.*, Chapter 5 this volume, for VA mycorrhizas).

VA mycorrhizal hyphae may extend several centimetres from the root surface and thereby enlarge the depletion zone for P accordingly. The external hyphae of VAM have the capacity to deliver a high proportion of the host plant demand not only of P but also Zn and Cu (George *et al.*, Chapter 5 this volume). In contrast, the capacity for uptake and delivery of K, Ca and Mn seems to be very low or negligible. This may lead to a decrease in concentrations of K and Mn in the host plant when VA mycorrhizal infection decreases root surface area (and thus K-acquisition) or root exudation (and thus activity of Mn-reducing rhizosphere microorganisms) (George *et al.*, Chapter 5 this volume). There is good evidence for a substantial capacity of external hyphae to absorb also NH_4-N and deliver it to the host plant. In the hyphosphere as in the rhizosphere, uptake of NH_4-N is associated with substrate acidification. Although hyphae interconnect roots of different species, including both legumes and non-legumes, hyphal transfer of fixed N from legumes to non-legumes seems to be very small (Bethlenfalvay *et al.*, 1991).

Compared with VA mycorrhizas in annual species, our knowledge on the role of ectomycorrhizas in the acquisition of mineral nutrients by forest trees is poor (Vogt *et al.*, 1991). There are various reasons for this, such as the large number of fungal species which differ in morphology (e.g. extension of the external

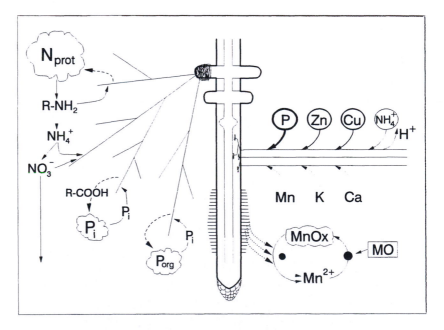

Fig. 1.5. Schematic presentation of some established components (processes) of the nutrient dynamics in the 'hyphosphere' of ecto- and VA-mycorrhizal roots. N_{prot} = protein nitrogen; MO = microorganism.

mycelium) and the succession of species from seedlings to older trees. With regard to such acquisition, there are at least two principal differences from VA systems (Fig. 1.5): (1) ectomycorrhizal hyphae may completely enclose the lateral roots and thus prevent direct acquisition of mineral nutrients by these roots, and (2) the external mycelium at least of some species mobilizes mineral nutrients from sources which may not be available to host roots, for example by releasing proteases and, thus, utilizing protein-N (Leake and Read, 1990), and by excretion of organic acids and phenolics.

This additional capacity of ectomycorrhizal hyphae may be of particular importance for the mineral nutrition of trees under two circumstances: (1) nitrogen-limited sites where the additional carbon input into mycorrhizal structures and turnover is paid off by the better accessibility to the most growth-limiting nutrient (Högberg, 1989) and (2) on nutrient–poor sites where the mineral soil is poorly accessible to roots (e.g. because of acidity or hypoxia) and where most of the roots are confined to the humus layer. Here, the contact between roots and substrate is poor and the external mycelia act as connecting structures between plant and substrate. On nutrient rich sites in general and on sites with high atmospheric input of bound N in particular, however, it is questionable whether ectomycorrhizas play an important role in the mineral nutrition of trees (Marschner *et al.*, 1991).

Conclusion

Depending on plant species, nutritional status of plants and soil conditions, rhizospheres of roots can change considerably. Localized releases of exudates (e.g. organic acids, phytosiderophores) as root responses to deficiency are examples which enhance the efficiency of acquisitions of mineral nutrients on nutrient-poor sites. In mycorrhizal plants, the external mycelium of the fungus may take over some of the roots' functions. Whether the additional carbon demand for the fungus is paid off depends on circumstances (e.g. nutrient poor/rich sites, which nutrient is limiting). For a given amount of carbon investment into plant parts below ground for acquisition of mineral nutrients, root-induced changes in the rhizo-sphere may increase flexibility and specificity of the plant response whereas mycorrhizas may ensure a higher long-term stability, in which protection against root pathogens may also be an important component.

References

Arshad, M. and Frankenberger Jr., W.T. (1991) Microbial production of plant hormones. *Plant and Soil* 133, 1–8.

Bauer, W.D. and Caetano-Anolles, G. (1990) Chemotaxis, induced gene expression and competitiveness in the rhizosphere. *Plant and Soil* 129, 45–52.

Bethlenfalvay, G.J.H., Reyes-Solis, M.G., Camel, S.B. and Ferrera-Cerrato, R. (1991) Nutrient transfer between the root zones of soybean and maize plants connected by a common mycorrhizal mycelium. *Physiologia Plantarum* 82, 423–432.

Dinkelaker, B., Römheld, V. and Marschner, H. (1989) Citric acid secretion and precipitation of calcium citrate in the rhizosphere of white lupin (*Lupinus albus* L.). *Plant, Cell and Environment* 12, 285–292.

Finlay, R.D. and Read, D.J. (1986) The structure and function of the vegetative mycelium of ectomycorrhizal plants. II. The uptake and distribution of phosphorus by mycelial strands interconnecting host plants. *New Phytologist* 103, 157–165.

Häussling, M. and Marschner, H. (1989) Organic and inorganic soil phosphates and acid phosphatase activity in the rhizosphere of 80-year-old Norway spruce [*Picea abies* (L.) Karst.] trees. *Biology and Fertility of Soils* 8, 128–133.

Hoffland, E., Findenegg, G.R. and Nelemans, J.A. (1989) Solubilization of rock phosphate by rape. II. Local root exudation of organic acids as a response to P-starvation. *Plant and Soil* 113, 161–165.

Högberg, P. (1989) Growth and nitrogen inflow rates in mycorrhizal and non-mycorrhizal seedlings of *Pinus sylvestris*. *Forest Ecology and Management* 28, 7–17.

Leake, J.R. and Read, D.J. (1990) Proteinase activity in mycorrhizal fungi. I. The effect of extracellular pH on the production and activity of proteinase by ericoid endophytes from soils of contrasted pH. *New Phytologist* 115, 243–250.

Lynch, J.M. and Whipps, J.M. (1990) Substrate flow in the rhizosphere. *Plant and Soil* 129, 1–10.

Marschner, H. (1991) Mechanism of adaptation of plants to acid soils. *Plant and Soil* 134, 1–20.

Marschner, H., Römheld, V., Horst, W.J. and Martin, P. (1986) Root-induced changes in the rhizosphere: Importance for the mineral nutrition of plants. *Zeitschrift für Pflanzenernährung und Bodenkunde* 149, 441–456.

Marschner, H., Häussling, M. and George, E. (1991) Ammonium and nitrate uptake rates and rhizosphere-pH in non-mycorrhizal roots of Norway spruce [*Picea abies* (L.) Karst.]. *Trees* 5, 14–21.

Martin, P., Glatzle, A., Kolb, W., Omay, H. and Schmidt, W. (1989) N_2-fixing bacteria in the rhizosphere: quantification and hormonal effects on root development. *Zeitschrift für Pflanzenernährung und Bodenkunde* 152, 237–245.

Nair, M.G., Safir, G.R. and Siqueira, J.O. (1991) Isolation and identification of vesicular-arbuscular mycorrhiza stimulatory compounds from clover (*Trifolium repens*) roots. *Applied and Environmental Microbiology* 57, 434–439.

Nomoto, K., Sugiura, Y. and Takagi, S. (1987) Mugineic acids, studies on phytosiderophores. In: G., Winkelmann, D. van der Helm and J.B. Neilands (eds), *Iron Transport in Microbes, Plants and Animals*. VCH Verlagsgesellschaft, Weinheim, Germany, pp. 401–425.

Römheld, V. (1991) The role of phytosiderophores in acquisition of iron and other micronutrients in graminaceous species: an ecological approach. *Plant and Soil* 130, 127–134.

Römheld, V. and Marschner, H. (1990) Genotypical differences among graminaceous species in release of phytosiderophores and uptake of iron phytosiderophores. *Plant and Soil* 123, 147–153.

Takagi, S., Nomoto, K. and Takemoto, T. (1984) Physiological aspect of mugineic acid, a possible phytosiderophore of graminaceous plants. *Journal of Plant Nutrition* 7, 469–477.

Vogt, K.A., Publicover, D.A. and Vogt, D.J. (1991) A critique of the role of ectomycorrhizas in forest ecology. *Agriculture, Ecosystems and Environment* 35, 171–190.

2 Mycophyllas and Mycorrhizas: Comparisons and Contrasts

K. Clay

Department of Biology, Indiana University, Bloomington, IN 47405, USA

Introduction

In a treatise on beneficial plant diseases, Charles Wilson (1977) wrote, 'Considering the varied beneficial fungus–root associations (mycorrhizae) that have evolved it is reasonable to suspect that beneficial fungus–leaf associations ('mycophyllae') may have also evolved. To my knowledge, none have been described.' The same year, Bacon *et al.* (1977) published their account of the association between a fungal endophyte and toxic tall fescue grass (*Festuca arundinacea*), initiating a large body of research on mutualistic symbioses between fungal endophytes and plant leaves. In the intervening 15 years, many additional mycophyllas have been identified. These discoveries raise the questions of whether mycophyllas are aerial analogues of mycorrhizas and whether they have similar underlying mechanisms. The purpose of this chapter is to compare and contrast these two plant–fungal symbioses with respect to several basic attributes.

I limit my discussion of mycophyllas primarily to interactions between clavicipitaceous endophytes and their host grasses because most published data are from these interactions and they have more functional similarities to mycorrhizas than other described mycophyllas. A large number of endophytic fungi from leaves of dicotyledons and gymnosperms have been isolated and described (Carroll, 1988) but, unlike the clavicipitaceous endophytes of grasses, these typically are localized and intracellular. Neighbouring leaves or needles on the same branch may support very different fungal floras (Riesen and Sieber, 1985; Petrini, 1986). However, much of the following discussion pertains equally well to all mycophyllas, not just those involving grasses.

The endophyte of *F. arundinacea* is typical of many from grasses (Bacon and Siegel, 1988). The fungus occurs in the aleurone layer of the seed where it remains dormant until the seed germinates. Within a few days of germination, it grows out of the seed into the base of the young seedling via hyphae which are intercellular and are most dense at the base of the tiller in the leaf sheath. Newly produced tillers are colonized by growth of the fungus from older, neighbouring tillers, maintaining a systemic infection. At the initiation of flowering, the endophyte proliferates in the developing inflorescence and grows into developing ovules. The seed thus represents the reproductive and dispersal unit of both plant and fungus.

Above- and Below-ground Environments

As their names suggest, mycophyllas and mycorrhizas are two symbioses with aerial and subterranean plant tissues respectively. Mycorrhizal fungi are ideally located for the acquisition of nutrients by serving as an interface between the roots of the host plant and the surrounding soil. To some extent, everything that enters or leaves a plant root may pass through the fungus, which itself penetrates some plant root cells (Harley and Smith, 1983). The soil, which provides a medium for the growth of hyphae between roots of the same plant species and between species (Allen and Allen, 1990), is buffered against extreme changes in temperature, moisture and nutrient concentrations. Mycorrhizal fungi occur in a biologically diverse environment of plant roots, decaying organic matter, soil micro- and macro-organisms including other mycorrhizal fungi, and the mineral soil. Interactions with the diverse soil flora and fauna could reduce the opportunity for evolutionary specialization with just the host plant.

In contrast, clavicipitaceous endophytes occur intercellularly in the leaves, stems and reproductive organs of their host plants. The only contact the fungi have with the outside environment occurs if they form a stroma on the surface of leaves or inflorescences. However, many endophytes never sporulate and remain completely internal within the plant for their entire life cycle. Plant leaves are subjected to relatively rapid changes in light, temperature, pH, moisture and concentrations of CO_2 and other gases, compared with the more uniform rhizosphere environment. The location of the fungi within leaves ensures their close proximity to the site of carbon fixation and access to sugars. As a result, endophytes get 'first crack' at products of photosynthesis and may be able to control the flow of carbohydrates out of the leaves into the roots and mycorrhizal fungi (Thrower and Lewis, 1973).

The aerial environment is a poor substrate for hyphal growth between plants by endophytic fungi, so that interplant connections by one fungus very rarely occur. As a result, the dichotomy between infected and uninfected plants is sharper and their competitive interactions may be more intense, compared to interconnected mycorrhizal plants (Allen and Allen, 1990). Further, the phyllosphere represents a biotically less diverse environment than the rhizosphere. With the exception of herbivores and pathogens or possibly other endophytes, clavicipitaceous endophytes rarely encounter other organisms besides their host. Thus, endophytes are more likely to be specifically coevolved with a single host species than mycorrhizal fungi.

Species Composition

Fungi associated with plant roots in mycorrhizal symbioses are mostly species of Basidiomycetes or Zygomycetes. Some Ascomycetes form mycorrhizas but they are much less common except in the Ericales (Harley and Smith, 1983). Mycor-rhizal species form the majority of all terrestrial plants and some plant families (Ericaceae, Orchidaceae) form very distinctive mycorrhizas (Read, 1983; Warcup, 1991).

In contrast, the large majority of described mycophyllas involve Ascomycetes or Deuteromycetes (Petrini, 1986), including species of Clavicipitaceae forming systemic infections in grasses. Ascomycetous fungi also form the majority of lichen symbioses (Hawksworth, 1988), making them the primary fungal symbiont with photosynthetic tissues or cells. All plants that have thus far been examined have had fungal endophytes living inside healthy leaves (Petrini, 1986), suggesting that mycophyllas are at least as common as mycorrhizas. All grasses are not infected by clavicipitaceous endophytes but, in some species, infection is ubiquitous (Clay and Leuchtmann, 1989).

A large proportion of conidial fungi produce secondary compounds with antibiotic or toxic effects on other organisms. Basidiomycetes and Zygomycetes also produce toxic secondary compounds but account for relatively fewer cases of animal poisonings (Purchase, 1974). Ascomycetes may have been predisposed to form mycophyllas given their biochemical capabilities to deter competitors. Endophytes of grasses, like the related *Claviceps*, produce a range of alkaloid mycotoxins (Clay, 1988).

Clavicipitaceous endophytes primarily parasitize grasses although some sedges and rushes are also infected (Clay, 1990b). The fungi, which can be readily isolated from healthy, surface-sterilized grass leaves or seeds, form one of the three major radiations in the family Clavicipitaceae along with *Claviceps*, an ovarian para-site that has a similar graminaceous host range, and *Cordyceps* which is a parasite of other fungi or insects (Diehl, 1950). Sporulating endophytes, which form stromata on host leaves and stems, are grouped into four genera totalling about 20 species. Species of *Balansia* and *Myriogenospora* infect primarily warm-season grasses with the C4 photosynthetic pathway while *Atkinsonella* and *Epichloe*, and *Acremonium* anamorphs, infect cool-season grasses with the C3 photosynthetic pathway. Anamorphic endophytes are classified in the genus *Acremonium* sect. *albo-lanosa* (White and Morgan-Jones, 1987; Clay, 1990b).

Costs and Benefits

An accounting of the costs and benefits, or currency, of an interaction between partners in symbiotic interactions is critical for understanding the interaction (Janzen, 1985). Fungi symbiotic with roots or leaves both receive sugars from their hosts, providing the basis for fungal nutrition. Thus, in both mycophyllas and mycorrhizas, the host plants 'pay' the fungi with sugars. Endophytic hyphae in grasses are most concentrated in and around intercalary meristems and developing inflorescences (Clay, 1990b), two locations that are strong sinks for photosynthate. Many of these endophytes are maternally transmitted through the seed (Clay, 1990b). Other endophytic fungi, unrelated to the clavicipitaceous endophytes of grasses, are also seed-transmitted in a similar way. Examples have been described in perennial ryegrass, *Helianthemum* and *Casuarina* but there is no evidence that the fungi significantly affect the ecology or physiology of their host plants (Bose,

1947; Boursnell, 1950, Latch *et al.*, 1984). As a result they receive an additional benefit from host plants in the form of dispersal but it is not clear whether this entails significant energetic costs to the plant. In contrast, there is evidence that endophyte-infected seeds are avoided by predators, an effect which would be to the benefit of the host (Cheplick and Clay, 1988, Madej and Clay, 1991). Despite early reports to the contrary (Rayner, 1927), there is no evidence that mycorrhizal fungi are ever seed-transmitted.

In contrast to seed-borne endophytes, others abort or inhibit developing inflorescences of hosts, leading to partial or total loss of the hosts' reproductive capacity so that an additional cost that these mycophyllous grasses incur is a reduction in seed production (Clay, 1991a). However, vegetative growth by stolons or rhizomes is unaffected and can actually be stimulated by infection (Bradshaw, 1959; Clay, 1991a). The difference between seed-transmission and inflorescence abortion can be variable; Sampson (1933) found that *E. typhina* infecting red fescue could cause abortion of some inflorescences yet be seed-borne in other inflorescences on the same plant. No direct effects of mycorrhizas upon the reproductive systems of their host plants have been recorded.

What do the fungi pay their host plants in return? In the case of mycorrhizal fungi, the plant benefits primarily from the increased uptake of mineral nutrients from the soil (Harley and Smith, 1983). A large number of experiments in controlled environments have demonstrated that mycorrhizal plants produce more biomass than non-mycorrhizal plants in nutrient-limited situations. This is not to say that improved mineral nutrition is the only benefit accruing to mycorrhizal plants but this is certainly the major effect.

The primary benefit received by mycophyllous plants appears to be protection against herbivores and other pests (Clay, 1991b). The fungi produce a wide variety of alkaloids which are present in host tissues. These compounds can be highly toxic to animals that feed on the plant or act as deterrents to those that do not. Field and laboratory experiments have repeatedly demonstrated enhanced resistance to a wide variety of insects, nematodes, vertebrate herbivores and pathogens (Clay, 1991b). Alkaloid concentrations in infected tall fescue were positively correlated with fertilization regime in field plots, suggesting that soil fertility could affect host plant resistance to pests (Lyons *et al.*, 1986). Mycorrhizal and mycophyllous fungi could indirectly interact as a result of plant mineral nutrition.

Acquisition of nutrients and protection from pests are not the only benefits for host plants in mycorrhizal and mycophyllous symbioses, respectively. Mycorrhizal fungi can enhance resistance of host plants to insects and nematodes (Clay, 1987a) as well as other fungi. In both types of symbiosis, the fungi can produce plant hormones that affect many plant physiological processes, directly or indirectly affecting photosynthetic rate, drought tolerance and stomatal behaviour (Harley and Smith, 1983; Arachevalata *et al.*, 1989). Fungal effects on pest damage, resource acquisition and hormonal balance can in turn have indirect effects on the competitive ability of plants in mixed communities (Allen and Allen, 1990; Clay, 1990a). Successional communities exhibit an increase in the proportion of mycorrhizal and mycophyllous species over time, indicating ecological benefits of fungal symbiosis. However, given the differing abilities of endophytic and mycorrhizal fungi to form inter-plant and inter-specific connections, competitive inequalities may be magnified by endophytic fungal infection and minimized by mycorrhizal fungal infection.

Variation in Interactions

The distinction between mutualism and antagonism in symbiotic associations is rarely fixed. In both mycorrhizal and mycophyllous interactions, the net balance of cost and benefit can vary with the host's identity, with the stage of the life cycle and with the presence or absence of resources and pests. Tobacco stunt disease appears to result from infection by mycorrhizal fungi (Jones and Hendrix, 1987). Mycorrhizas form between *Rhizoctonia* spp. and orchids (Warcup, 1991) but some *Rhizoctonia* spp. are pathogens infecting a wide variety of plants. Many leaf endophytes are closely related to pathogenic fungi so it is not surprising to observe pathogenic effects at certain times (Carroll, 1988). A sterilizing endophyte, like *Epichloe typhina*, is pathogenic on annual grasses but can be mutualistic when it infects rhizomatous perennial grasses (Clay, 1990b).

Studies with grasses and their clavicipitaceous endophytes suggest that several factors can alter the balance between costs and benefits, and therefore the degree of mutualism. Infected seedlings of tall fescue produced significantly more biomass than uninfected seedlings when grown at high nutrient concentrations in green-house experiments but this relationship was reversed when they were grown at low soil nutrient concentrations (Cheplick *et al.*, 1989). The same general pattern held true for adult plants but the detriment to infected plants at low nutrients was less. These data suggest that the endophyte can act as a drain on young plants growing in stressful conditions. Further, it illustrates the opposite trend from mycorrhizal plants where benefits from infection are positively correlated with nutrient limitation (Harley and Smith, 1983).

As indicated previously, the degree of benefit and harm to host grasses also corresponds to the sexuality of the fungi. Sexual endophytes generally reduce or eliminate seed production by hosts while seed-borne endophytes obviously do not.

There is also some evidence that the benefit of endophyte-infection in grasses is greatest where pest pressure is greatest. In environments lacking pests, the advantage for endophyte-infected plants is minimal or even reversed (Clay, 1991b). This represents a conditional mutualism, as in the case of many protective mutualisms (Thompson, 1982), where the status of the interaction between symbionts depends on the presence or absence of a third species. One might predict that the frequency of endophytic infection in grasses is greatest where pest pressure is also greatest. Endophyte-infected *Sporobolus poretii* and *Melica decumbens* have been reported to increase in grassland communities relative to other species under heavily grazed conditions (Shaw, 1873; Bacon *et al.*, 1986). Some data from greenhouse studies suggest that extreme pest pressure by leaf-feeding insects or nematodes could favour mycorrhizal plants (Clay, 1987b).

Seed Transmission vs. Spores

Mycorrhizal and mycophyllous fungi find new hosts in different ways. Both types of fungi typically produce spores capable of infecting new plants. The exceptions are completely seed-borne endophytes that do not sporulate. Dispersal of spores is accomplished by the wind or water or soil movement and, even in some instances, by animal vectors that bring the spores directly to their target.

Mycorrhizal fungi can colonize new plants in the soil by vegetative growth of hyphae from one plant to another (Allen and Allen, 1990). Mycophyllous fungi, which occur in the above-ground environment, are not capable of horizontal spread by vegetative growth between plants. They can, however, spread systemically by vegetative growth within a host plant.

Clavicipitaceous endophytes exhibit a remarkable adaptation for reproduction and dispersal not paralleled by mycorrhizal fungi. Species of *Atkinsonella* and *Epichloe*, as well as their anamorphic derivatives, are often seed-transmitted from generation to generation (Clay, 1990b). The fungus does not colonize pollen and so is not paternally transmitted. Infected seeds are completely viable and, in two grasses, the rates of seed germination and seedling growth were significantly enhanced by endophyte infection (Clay, 1987b). Infected seeds are also rich in alkaloids, making them less attractive to seed predators (Cheplick and Clay, 1988; Madej and Clay, 1991).

Seed-transmission of a symbiont eliminates the uncertainty of finding a new host; the fungus and plant are dispersed as a unit. Successful reproduction by the fungus depends directly on seed germination and seedling establishment. The greater certainty of the identity of the future host (an offspring of the previous host) relaxes selection for a broad host range that would be more important where the finding a host of any particular species is less certain.

An interesting form of co-dispersal occurs in those species like *Cyperus virens* where infected plants produce viviparous plantlets on inflorescences instead of seeds (Clay, 1991a). The fungus is present in the plantlets which are capable of dispersal, establishment and growth. Other viviparous grasses also possess endophytes (Clay, 1991a). In these cases, both the genotype of the host plant and the fungus remain constant generation after generation, allowing the potential of extreme specificity.

Clavicipitaceous endophytes of grasses and sedges can directly alter the hosts' reproductive systems, which control to a large extent the amounts and distribution of genetic variability in the host population. Thus, they can more directly affect the evolutionary and coevolutionary potential of their hosts than mycorrhizal fungi, which have no direct effects on their hosts' reproductive systems. This ability depends largely on the location of the fungi in aerial plant tissues, including the reproductive structures.

Specialization vs. Generalization

The abilities of mycorrhizal or mycophyllous fungi to spread vegetatively between plants and to be seed-transmitted, and their locations around roots and inside leaves may all influence the specificity of the fungi from the level of a particular host genotype to an entire plant community. Specificity has been determined for both types of symbioses by inoculations of plants with particular strains or isolates of fungi, as well as by compiling known host associations from the field. The latter approach may be flawed if cryptic host-specific strains exist within a well-defined taxon. The increasing utilization of genetic markers in studies of host ranges will clarify the extent to which morphologically identical fungal strains may differ in host range.

Although it can be argued that, in general, mycophyllas are more specific than mycorrhizas, the comparison is somewhat unfair given that the clavicipitaceous endophytes of grasses are already restricted to a single plant family. However, there are many examples of endophytes apparently being specific to a single host species. Only a single example of this level of specificity has been identified in mycorrhizas (Smits, 1983; cited in Molina *et al.*, 1992).

Specificity in mycorrhizal associations has been subject to a great deal of research ranging from cataloguing field patterns of host specificity to experiments with laboratory-based microcosms, to study of host-related patterns of variation in amounts of DNA in fungi. Rather than review a large body of data covered elsewhere in this volume, it is sufficient to say that the view that mycorrhizal fungi all show broad host ranges is too sweeping. In a recent review, Molina *et al.* (1992) provide examples of mycorrhizal fungi falling into three general host range classes. Narrow host-range fungi are restricted to a single host genus while intermediate host-range species are restricted primarily to a single family. In contrast, mycorrhizal fungi with broad host ranges can infect species in different families, orders, or classes. They provide extensive lists of narrow and intermediate host-range fungi. The specificity of many or most mycorrhizal fungi is at the level of family or higher. In those cases where fungi are restricted to a single genus, that host genus typically has many species and is geographically widespread (e.g., *Pinus, Quercus*).

A number of studies on the specificity of clavicipitaceous endophytes for particular host species or genera now allow comparisons with similar work from mycorrhizal associations. Specificity has been investigated in three ways. The first is the compilation of hosts infected by a given endophytic species (see Diehl, 1950). The second involves artificial inoculation. For grass endophytes, this approach involves inserting fungal mycelium from a pure culture, with a hypodermic needle, above the meristem of aseptically-grown seedlings. An alternative approach has been to inoculate calli from tissue culture with fungal mycelium and induce plantlet formation (Johnson *et al.*, 1986). The first method is more successful. A third method measures genetic variability of endophytes using isozyme or DNA markers and correlates this variation with host plant identity. The principal generalization which can be made so far is that seed-borne endophytic *Acremonium* spp. exhibit narrower host ranges than sexual endophytes that sporulate on their hosts.

In an early study, Latch and Christensen (1985) successfully transferred the endophyte of *Lolium perenne* and *E. typhina* from red fescue into tall fescue seedlings although they did not become seed-transmitted. Siegel *et al.* (1990) also reciprocally inoculated a variety of seed-borne endophytes into different grasses and were able to establish many new combinations. Interestingly, alkaloid production in these often differed from the original combinations, indicating that biosynthetic pathways are controlled to some extent by both partners. We have inoculated tall fescue seedlings with a variety of seed-borne endophytes from other grasses. Seven of 14 isolates initially infected the plants (checked after 2 months) but, 6 months later, only the endophyte originally from tall fescue still persisted (Leuchtmann and Clay, unpublished). Thus, short-term infections may not become permanent and seed-borne.

Ascospore-producing endophytes appear to be less specific to their hosts. Some hint of this can be gathered from host records. For example, *E. typhina* infects over 25 grass species in Britain (Sampson and Western, 1954). Similarly, *Balansia henningsiana* is reported to infect at least that many panicoid grasses (Diehl, 1950). However, one recognizable species may actually consist of many host-specific races that cannot be identified without doing cross inoculations. Nevertheless, reported instances of host shifts in field plots indicate cross-species infections can occur (Rykard *et al.*, 1985; Clay, 1990c). There are, however, species with narrow host ranges. *Balansia obtecta* is restricted to one host species while *B. cyperi* and *Atkinsonella hypoxylon* are restricted to single host genera (Clay, 1990b).

Inoculations of seedlings with pure fungal cultures have produced successful transfers among host species within a genus, among genera within a tribe and among tribes within grass subfamilies. Cross inoculations of seedlings from several populations of *Danthonia compressa* and *D. spicata* with isolates of *Atkinsonella hypoxylon* from those same populations gave no evidence of any specificity towards host population or species whatsoever (Leuchtmann and Clay, 1989a). Isolates from *D. compressa* actually exhibited a higher infection rate of *D. spicata* seedlings than isolates from *D. spicata*. Similarly, isolates of *Balansia cyperi* from *Cyperus virens* infected seedlings from its original host and seedlings from *C. pseudovegetus* equally well (Leuchtmann and Clay, 1988). We have also found that transfers of *E. typhina* among species of the closely related genera *Elymus* and *Hystrix* were very successful and independent of the original host (Leuchtmann and Clay, unpublished). Siegel *et al.* (1990) infected several different host species with isolates of *E. typhina* and several taxa of *Acremonium*.

Isozyme data on patterns of host-specific variation in isolates also suggest that cross-species transfers do occur in nature. Identical multilocus genotypes of both *A. hypoxylon* and *E. typhina* were isolated from different host species growing sympatrically (Leuchtmann and Clay, 1989b, 1990).

The close symbiotic relationship between seed-borne leaf endophytes and their host plants offers the possibility for tight specific coevolution and extreme host specificity (Thompson, 1989). These endophytes are transmitted entirely within maternal lineages without opportunity for exposure to other possible host species. Genes conferring broad host ranges would be of no value and may in fact impede specific adaptation to their host. The only possible exception to this that I can envisage would arise from interspecific hybridization. If F1 hybrids were predominantly backcrossed with the paternal species, the endophyte from the maternal host could find itself in a different genetic background. However, interspecific hybridization will usually occur only among closely related species to begin with so this may not present a problem for host-specific endophytes. In contrast, spore-producing endophytes maintain the capability of contagious transmission and appear to exhibit less host specificity. Thus, the seed-borne habit of many endophytes of grasses has been the primary factor favouring their specialization to single host species or genera.

Interactions of Mycophyllas and Mycorrhizas

Relatively little work has been conducted on the possible interaction between subterranean and aerial fungal symbionts of plants. The possible impact of fungal endophytes in plant leaves has only been appreciated in the past 10 years and most research has focused on direct plant–fungal interactions. Most grasses form associations with VA mycorrhizal fungi, so interactions with clavicipitaceous endophytes could be common in grasses. One would predict *a priori* that leaf endophytes could limit mycorrhizal growth by restricting their access to carbohydrates, whereas endophyte growth would be less limited by mycorrhizal infection because the former fungi obtain plant sugars before they are ever translocated from leaves.

Two preliminary studies suggest that endophyte infection in tall fescue does indeed inhibit colonization by mycorrhizal fungi. Chu-Chou *et al.* (1990) found that reproductive rates of two *Glomus* spp. were suppressed on mycophyllous compared with non-mycophllous tall fescue. In a separate study, tillers of endophyte-infected and uninfected tall fescue were rooted in sterile water and then transplanted into field soil in pots in the greenhouse. Counts of random mycorrhizal and non-mycorrhizal root segments using the grid-intercept method revealed that endophyte-infected plants had 40% lower colonization rates than endophyte-free plants ($P < 0.06$, Blaney and Clay, unpublished).

Further controlled experiments are necessary to explore the effect of the two fungal symbioses on plant growth. Simple experiments where the two types of symbiosis are controlled (four plant categories: non-mycorrhizal/non-mycophyllous, mycorrhizal/non-mycophyllous, non-mycorrhizal/mycophyllous, mycorrhizal/mycophyllous) would reveal the statistical effects of the two main effects and the interaction between them. A weakness of this design, and all glasshouse experiments, is that results may not reflect the complex biotic interactions occurring in the field. It is possible to plant endophyte-infected and uninfected grasses into a field site and their infection status does not change because horizontal, or contagious, transmission is not possible (Clay, 1990c). It is difficult to maintain non-mycorrhizal plants in realistic field conditions.

One unexpected interaction between mycorrhizal and endophytic fungi was reported by Barker (1987) from perennial ryegrass. Comparing non-mycorrhizal plants, mycophyllous plants were significantly more resistant to a major insect pest than non-mycophyllous plants, as has usually been reported. However, when the plants were mycorrhizal, the deterrent effects of endophyte-infection disappeared. These results are somewhat difficult to interpret since the plants would be expected to be mycorrhizal in field situations, yet field resistance to the same insect pest has been repeatedly documented.

Conclusions

The mycophyllas with well-demonstrated ecological effects on host plants and plant communities involve clavicipitaceous endophytes infecting grasses. Their limited host range, compared with mycorrhizal fungi, might appear to limit the significance of leaf endophytes in ecosystems. However, the grass family is one of the largest plant families, it covers most of the world's land surface and is economically more important than any other plant family. Therefore, the potential impact of the specific mycophylla between grasses and endophytes on ecosystems is very large. Actual impacts have been demonstrated only in a few cases involving grazing mammals and endophyte-infected grasses (Shaw, 1873; Bacon *et al.*, 1986). Much work has also been conducted on mycophyllas involving conifers and broadleaved trees (Petrini, 1986; Carroll, 1988). Again, the potential impact of endophyte symbiosis could also be large, especially with regard to insect outbreaks in monospecific communities. However, research on mycophyllas is very recent and relatively few scientists are involved. There are many unanswered questions and unexplored directions of research.

Mycophyllas have certain advantages over mycorrhizas for future research. The ability to conduct field experiments comparing the performance of infected and uninfected plants in realistic communities is difficult, if not impossible, for mycorrhizal associations because of the presence and horizontal spread of the fungi in the soil. In contrast, uninfected plants of grasses infected by seed-transmitted endophytes do not become contagiously infected in the field, allowing meaningful comparisons between infected and uninfected plants. Leaf endophytes also offer great promise in the area of plant biotechnology given their ease of culture, their high degree of host specificity and potential for seed transmission. Potential applications include pharmaceuticals (given the large numbers of biologically active compounds from grass endophytes), biological control against insects and other plant pest and as vectors for the introduction of foreign genes into grasses. Grasses have proved difficult to transform compared with many plant groups while filamentous fungi are relatively easy to transform. The major problem for utilizing mycophyllas in crop grasses is the potential toxicity to mammals of some of the fungal alkaloids.

In conclusion, consideration of the two types of plant–fungal mutualism emphasizes that individual plants represent complex ecosystems unto themselves. Researchers of mycorrhizas recognize this from the rhizosphere environment but leaves also represent habitats for many fungi and other microorganisms. When we isolate and focus on single interactions, the results may not be representative of the complex, real-world, system. How to resolve this problem is not clear. One approach is to investigate joint interactions, as outlined above for mycorrhizas and mycophyllas, and ask whether the interactions are more significant than the main effects in statistical analyses.

References

Allen, E.B. and Allen, M.F. (1990) The mediation of competition by mycorrhizae in successional and patchy environments. In: Grace, J.B. and Tilman, D. (eds), *Perspectives on Plant Competition*. Academic Press, NY, pp. 367–389.

Arachevaleta, M., Bacon, C.W., Hoveland, C.S. and Radcliffe, D.E. (1989) Effect of tall fescue endophyte on plant response to environmental stress. *Agronomy Journal* 81, 83–90.

Bacon, C.W. and Siegel, M.R. (1988) Endophyte parasitism of tall fescue. *Journal of Production Agriculture* 1, 45–55.

Bacon, C.W., Porter, J.K., Robbins, J.D. and Luttrell, E.S. (1977) *Epichloe typhina* from toxic tall fescue grasses. *Applied and Environmental Microbiology* 34, 576–581.

Bacon, C.W., Lyons, P.C., Porter, J.K. and Robbins, J.D. (1986) Ergot toxicity from endophyte-infected grasses: a review. *Agronomy Journal* 78, 106–116.

Barker, G.M. (1987) Mycorrhizal infection influences *Acremonium*-induced resistance to Argentine stem weevil in ryegrass. *Proceedings of the New Zealand Weed Pest Control Conference* 40, 199–203.

Bose, S.R. (1947) Hereditary (seed-borne) symbiosis in *Casuarina equisetifolia*. *Nature* 159, 512–514.

Boursnell, J.G. (1950) The symbiotic seed-borne fungus in the Cistaceae. I. Distribution and function of the fungus in the seedling and in the tissues of the mature plants. *Annals of Botany* 14, 217–243.

Bradshaw, A.D. (1959) Population differentiation in *Agrostis tenuis* Sibth. II. The incidence and significance of infection by *Epichloe typhina*. *New Phytologist* 58, 310–315.

Carroll, G.C. (1988) Fungal endophytes in stems and leaves: from latent pathogen to mutualistic symbiont. *Ecology* 69, 2–9.

Cheplick, G.P. and Clay, K. (1988) Acquired chemical defenses of grasses: the role of fungal endophytes. *Oikos* 52, 309–318.

Cheplick, G.P., Clay, K. and Wray, S. (1989) Interactions between fungal endophyte infection and nutrient limitation in the grasses *Lolium perenne* and *Festuca arundinacea*. *New Phytologist* 111, 89–97.

Chu-Chou, M., An, Z.Q., Hendrix, J.W., Zhai, Q. and Siegel, M.R. (1990) Effect of the *Acremonium* endophyte on endogonaceous mycorrhizal fungi reproducing on tall fescue. *Mycological Society of America Newsletter* 41, 10.

Clay, K. (1987a) The effect of fungi on the interaction between host plants and their herbivores. *Canadian Journal of Plant Pathology* 9, 380–388.

Clay, K. (1987b) Effects of fungal endophytes on the seed and seedling biology of *Lolium perenne* and *Festuca arundinacea*. *Oecologia* 73, 358–362.

Clay, K. (1988) Fungal endophytes of grasses: a defensive mutualism between plants and fungi. *Ecology* 69, 10–16.

Clay, K. (1990a) The impact of parasitic and mutualistic fungi on competitive interactions among plants. In: Grace, J.B. and Tilman, D. (eds), *Perspectives on Plant Competition*. Academic Press, NY, pp. 391–412.

Clay, K. (1990b) Fungal endophytes of grasses. *Annual Review of Ecology and Systematics* 21, 275–297.

Clay, K. (1990c) Comparative demography of three graminoids infected by systemic, clavicipitaceous fungi. *Ecology* 71, 558–570.

Clay, K. (1991a) Parasitic castration of plants by fungi. *Trends in Ecology and Evolution* 6, 162–166.

Clay, K. (1991b) Fungal endophytes, grasses, and herbivores. In: Jones, C., Krischik, V. and Barbosa, P. (eds), *Microorganisms, Plants and Herbivores*. John Wiley and Sons, NY. pp. 199–226.

Clay, K. and Leuchtmann, A. (1989) Infection of woodland grasses by fungal endophytes. *Mycologia* 81, 805–811.

Diehl, W.W. (1950) *Balansia and the Balansiae in America*. USDA, Washington, DC.

Harley, J.L. and Smith, S.E. (1983) *Mycorrhizal Symbiosis*. Academic Press, New York, NY.

Hawksworth, D.L. (1988) The variety of fungal–algal symbioses, their evolutionary significance, and the nature of lichens. *Botanical Journal of the Linnean Society* 96, 3–20.

Janzen, D.H. (1985) The natural history of mutualisms. In: Boucher, D.H. (ed.), *The Biology of Mutualism*, Oxford University Press, Oxford, pp. 40–99.

Johnson, M.C., Bush, L.P. and Siegel, M.R. (1986) Infection of tall fescue with *Acremonium coenophialum* by means of callus culture. *Plant Disease* 70, 380–382.

Jones, K. and Hendrix, J.W. (1987) Inhibition of root extension in tobacco by the mycorrhizal fungus *Glomus macrocarpum*, and its prevention by benomyl. *Soil Biology and Biochemistry* 19, 297–299.

Latch, G.C.M. and Christensen, M.J. (1985) Artificial infections of grasses with endophytes. *Annals of Applied Biology* 107, 17–24.

Latch, G.C.M, Christensen, M.J. and Samuels, G.J. (1984) Five endophytes of *Lolium* and *Festuca* in New Zealand. *Mycotaxon* 20, 535–550.

Leuchtmann, A. and Clay, K. (1988) Experimental infection of host grasses and sedges with *Atkinsonella hypoxylon* and *Balansia cyperi* (Balansiae, Clavicipitaceae). *Mycologia* 80, 291–297.

Leuchtmann, A. and Clay, K. (1989a) Experimental evidence for genetic variation in compatibility between the fungus *Atkinsonella hypoxylon* and its three host grasses. *Evolution* 43, 825–834.

Leuchtmann, A. and Clay, K. (1989b) Isozyme variation in the fungus *Atkinsonella hypoxylon* within and among populations of its host grasses. *Canadian Journal of Botany* 76, 2600–2607.

Leuchtmann, A. and Clay, K. (1990) Isozyme variation in the *Acremonium*/*Epichloe* fungal endophyte complex. *Phytopathology* 80, 1133–1139.

Lyons, P.C., Plattner, R.D. and Bacon, C.W. (1986) Occurrence of peptide and clavine ergot alkaloids in tall fescue grass. *Science* 232, 487–489.

Madej, C.W. and Clay, K. (1991) Avian seed preference and weight loss experiments: the effect of fungal endophyte-infected tall fescue seeds. *Oecologia* 88, 296–302.

Molina, R., Massicotte, H. and Trappe, J.M. (1992) Specificity phenomena in mycorrhizal symbioses: community ecological consequences and practical implications. In: Allen, M.F. (ed.), *Mycorrhizal Functioning: An Integrative Plant–Fungal Process*. Chapman and Hall, NY, pp. 357–423.

Petrini, O. (1986) Taxonomy of endophytic leaf fungi of aerial plant tissues. In: Fokkema, N.J. and van den Heuvel, J. (eds), *Microbiology of the Phyllosphere*. Cambridge University Press, London, pp. 175–187.

Purchase, I.F.H. (1974) *Mycotoxins*. Elsevier, Amsterdam.

Rayner, M.C. (1927) Mycorrhiza. *New Phytologist* Reprint 15. 246 pp.

Read, D.J. (1983) The biology of mycorrhiza in the Ericales. *Canadian Journal of Botany* 61, 985–1004.

Riesen, T. and Sieber, T. (1985) 'Endophytische Pilze von Winterweizen (*Triticum aestivum* L.).' Unpublished PhD dissertation, ETH, Zurich.

Rykard, D.M., Bacon, C.W. and Luttrell, E.S. (1985) Host relations of *Myriogenospora atramentosa* and *Balansia epichloe* (Clavicipitaceae). *Phytopathology* 75, 950–956.

Sampson, K. (1933) The systematic infection of grasses by *Epichloe typhina* (Pers.) Tul. *Transactions of the British Mycological Society* 18, 30–47.

Sampson, K. and Western, J.H. (1954) *Diseases of British Grasses and Forage Legumes*. Cambridge University Press, Cambridge.

Shaw, J. (1873) On the changes going on in the vegetation of South Africa. *Botanical Journal of the Linnean Society* 14, 202–208.

Siegel, M.R., Latch, G.C.M., Bush, L.P., Fannin, N.F., Rowan, D.D., Tapper, B.A., Bacon, C.W. and Johnson, M.C. (1990) Alkaloids and aphid response in grasses infected with fungal endophytes. *Journal of Chemical Ecology* 16, 3301–3315.

Smits, W.T.M. (1983) Dipterocarps and mycorrhiza, an ecological adaptation and a factor in forest regeneration. *Flora Malesiana Bulletin* 36, 3926–3937.

Thompson, J.N. (1982) *Interaction and Coevolution*. Wiley, NY.

Thompson, J.N. (1989) Concepts of coevolution. *Trends in Ecology and Evolution* 4, 179–183.

Thrower, L.B. and Lewis, D.H. (1973) Uptake of sugars by *Epichloe typhina* (Pers. ex. Fr.) Tul. in culture and from its host *Agrostis stolonifera* L. *New Phytologist* 72, 501–508.

Warcup, J.H. (1991) The *Rhizoctonia* endophytes of *Rhizanthella* (Orchidaceae). *Mycological Research* 95, 656–659.

White, J.F. and Morgan-Jones, G. (1987) Endophyte-host associations in forage grasses. IX. Concerning *Acremonium typhinum*, the anamorph of *Epichloe typhina*. *Mycotaxon* 29, 489–500.

Wilson, C.L. (1977) Management of beneficial plant diseases. In: Horsfall, J.G. and Cowling, E.B. (eds), *Plant Disease: an Advanced Treatise*. Academic Press, NY, pp. 347–362.

3 Why are some Plants more Mycorrhizal than Others? An Ecological Enquiry

A.H. Fitter and J.W. Merryweather

Department of Biology, University of York, York, YO1 5DD, UK

The vesicular-arbuscular mycorrhiza (VAM) is an ubiquitous symbiosis in the world's ecosystems, probably occurring in over two-thirds of vascular plant species (Trappe, 1987; Brundrett, 1991). Some ecosystems are dominated by plant species with other types of mycorrhiza, principally ericoid or ectomycorrhizal, and a few types of ecosystem may support predominantly non-mycorrhizal plants. Such variation can be related to the different roles of the various groups of mycorrhiza (Read, 1983), with VAM infection being the least specialized. The extent of VAM infection varies widely, however, both within and between ecosystems.

In this paper we wish to concentrate on variation in mycorrhizality at a finer scale: within ecosystems, within habitats, within species and even within individuals. In systems dominated by VAM plant species, it is nearly always possible to find not only some ectomycorrhizal or non-mycorrhizal species, but also a great range in the infection density of the roots of co-existing species (e.g. Gay *et al.*, 1982; Brundrett and Kendrick, 1988; Sanders and Fitter, 1992a). Some species are nearly always strongly infected, others often weakly so (Fig. 3.1). Even closely adjacent plants of well-infected species may exhibit considerable variation at any one time (Fig. 3.2) and every mycorrhizal worker is familiar with the variation in infection density that can be found between parts of the root system of an individual plant. This tendency for a range of infectedness to appear at all scales of observation suggests that the distribution of infection may be fractal.

Variation in infectedness poses a problem: if infection is beneficial, as is normally and reasonably assumed, why do some species, individual plants or roots have low levels of infection and some high? Is the variation in infection a reflection of the benefit that the plant can obtain from being mycorrhizal at that point in space or time? At one extreme, one might suggest that the intensity of infection was a highly controlled phenomenon, determined by the plant (and/or by the fungus) in response either to the fluxes of materials across the plant–fungus interface or to some other benefit or cost being exchanged. At the other, it would be possible to view much of the variation in infection as stochastic, the result of environmental pressures that determine the growth of the roots or the fungus and of the distri-

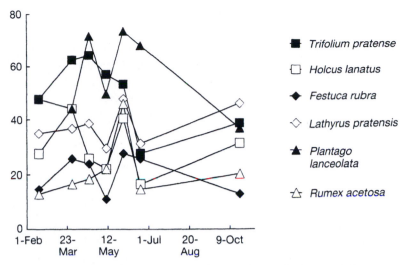

Fig. 3.1. Seasonal variation in VAM infection density in six co-existing grassland species at Wheldrake Ings, near York, UK (Sanders and Fitter, 1992a). Although there is much variation within a species, a consistent hierarchy is maintained.

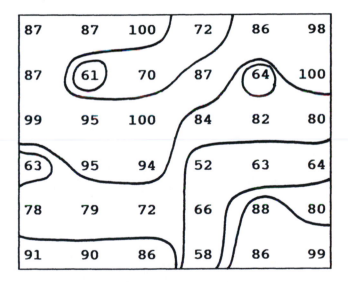

Fig. 3.2. The distribution of VAM infection density within a population of bluebell (*Hyacinthoides non-scripta*) over a 6 × 6 m area. Data from an unpublished study by J.W. Merryweather at Pretty Wood, Castle Howard, York.

bution of the fungus in space and time. The former view implies that there is a tight mutualism and that benefits to both sides should be readily demonstrable; the latter implies that benefits may occur more sporadically or be more diffuse, and would certainly be harder to detect.

The nature of the benefits to the partners is well established. The fungus is an obligate symbiont and can only receive carbon (and possibly other resources too) from the plant. The plant may receive a range of benefits, including the uptake of immobile ions (principally phosphate, but possibly also Zn^{2+}, Cu^{2+}, etc.), improved water relations (probably normally a consequence of improved P nutrition: Fitter, 1988) and protection from pathogens. Of these, it is normally held that the supply of phosphate is the most important. If P supply is accepted as the principal benefit, it is possible to carry out cost–benefit analyses to predict the behaviour of the symbiosis (Koide, 1990; Fitter, 1991). The main problem is deciding on a currency, since two distinct resources (C and P) are involved, and the conversion rate between them must be determined. We have simulated this by making assumptions about the relationships between (1) the rate of P uptake and leaf P concentration and (2) leaf P concentration and photosynthetic rate (Fitter, 1991). This makes it possible to measure benefit to the plant in carbon terms, as the extra photosynthate resulting from increased P inflow. Such an analysis ignores several important aspects of plant physiology, notably the tendency of photosynthetic rate to be determined by sink activity.

Simulations using this model imply that the frequency of infection points should be stabilized (Fig. 3.3) at an optimum which is not very variable over a wide range of soil P supply rates up to a critical value, above which plants should rapidly become non-mycorrhizal. The optimum values predicted by the model are similar to those quoted by several workers (Wilson, 1984; Amijee *et al.*, 1989) and

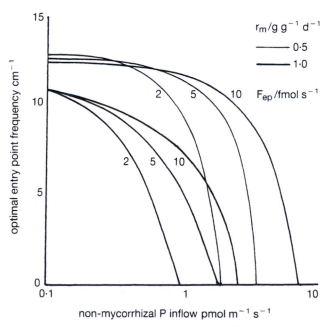

Fig. 3.3. Output from a simulation model which predicts optimal entry point density as a function of the P inflow that can be maintained by a non-mycorrhizal root system (i.e. by diffusion alone). From Fitter (1991). r_m = carbon cost to plant of fungal growth and respiration; F_{ep} = P influx per entry point.

support the suggestion that the plant may operate an internal control on the level of infection. It is well established that fertilization of plants with P tends to reduce infection (Thompson *et al.*, 1986), but less clear whether such responses are quantitative (the less P available, the more infection) or qualitative, as the model suggests.

This approach treats the fungus as a 'black box', ignoring the known diversity of fungal types that occurs in all VA mycorrhizal ecosystems. Some of these fungi may be more efficient symbionts, offering benefits greater than (or different from) others. Evidence supporting the idea of tight control of infection density comes from studies which show that different plant–fungus interactions may occur. For example, McGonigle and Fitter (1990) found that several coexisting grassland species were infected by both coarse and fine VA endophytes, but that *Holcus lanatus* was consistently more infected by fine endophyte (*'Glomus tenue'*) than were *Plantago lanceolata* or *Ranunculus acris*. Similarly Rosendahl *et al.* (1990) were able to recognize four distinct mycorrhizal types in the roots of several coexisting plants, while Sanders and Fitter (1992c) showed that different VAM fungal species sporulated when a number of coexisting grassland plants were grown

Table 3.1. Published maximum P inflows for plants grown under controlled and field conditions. In many cases (especially in field studies) these values represent peaks which were only briefly achieved; the average inflow is often much lower.

Conditions	Inflow pmol $m^{-1}s^{-1}$		Reference
	mycorrhizal	non-mycorrhizal	
Glasshouse/growth chamber			
Allium cepa	17.0	3.6	Sanders and Tinker, 1973
	11.0	3.0	Sanders *et al.*, 1977
	22.0	–	Smith *et al.*, 1986
Allium porrum	6.8	2.3	Warnock *et al.*, 1982
Trifolium subterraneum	35.0	–	Tester *et al.*, 1985
Pisum sativum	19.0	4.8	Fitter and Nichols, 1988
Sinapis alba	–	4.7	Fitter and Nichols, 1988
Trifolium repens	5.2	2.2	McGonigle and Fitter, 1988a
Agricultural			
Pisum sativum	7.8	2.9	Jakobsen, 1986
Fragaria × *ananassa*	43.5	–	Dunne and Fitter, 1989
Natural			
Holcus lanatus	1.4	0.4	McGonigle and Fitter, 1988b
	1.4	–	Sanders and Fitter, 1992b
Trifolium repens	1.8	–	McGonigle and Fitter, 1988a
Plantago lanceolata	8.3		Sanders and Fitter, 1992b
Rumex acetosa	5.8		
Trifolium pratense	5.0		
Lathyrus pratensis	4.3		
Festuca rubra	5.3		
Hyacinthoides non-scripta	1.8	–	J.W. Merryweather (unpub.)

either as field-collected plants in sterilized soil or from seed in soil taken from the field. Plant species are not, therefore, randomly infected by the available fungal species.

This evidence, therefore, points in the direction of the symbiosis being a closely regulated mutualism, but one that in turn requires the demonstration of well-defined benefits to the plant. The evidence for increased P uptake under cultural conditions is overwhelming, but this cannot be said for plants growing under natural field conditions. Such attempts as there have been to demonstrate such a benefit have often been unsuccessful (see Fitter, 1990). One reason for this may well be that the rates of growth of plants under natural conditions are often limited by factors other than P deficiency and are sufficiently low that P demand can be satisfied by the physicochemical processes of diffusion that the mycorrhizal symbiosis is so efficient at bypassing. We have measured P inflow in a number of species under natural conditions, and they are usually an order of magnitude less than those measured under agricultural or pot conditions (Table 3.1). It seems likely that these low P inflow values could be achieved without mycorrhizal mediation, although in practice much of the P uptake may still be via fungal hyphae if the plants are infected.

Where we have attempted to relate P inflow to infection we have usually been unsuccessful (McGonigle and Fitter, 1988a, b; Sanders and Fitter, 1992b). The only wild plant for which we have found a clear correlation between infected-ness and inflow has been bluebell *Hyacinthoides non-scripta* (Liliaceae), a bulbous plant with a very restricted system of coarse, unbranched roots, and therefore an ideal case for demonstrating mycorrhizal benefit (Merryweather and Fitter, unpublished data). Bluebell roots undergo a steady increase in infection density to a plateau in early spring, more or less at the time when the shoots first emerge above-ground allowing the plant to go into a positive carbon balance. During the subterranean phase of the life cycle (typically August–February in North Yorkshire), the plants lose both carbon and phosphorus and so experience nega-tive inflows; in the above-ground phase (February–July), when infection density is high, they have high P inflows. The two patterns are closely coincident (Fig. 3.4a), consequently giving a good correlation (Fig. 3.4b), but whether the relationship is causal is unclear.

It is not surprising that the nearest we have been able to come to demon-strating a link between infection density and P inflow is in a coarse-rooted species such as bluebell. The concept that root diameter is fundamental in determining mycorrhizal dependency is well established (Baylis, 1975), has good theoretical support based on the low root densities of coarse-rooted species and their con-sequent inability to obtain immobile ions such as phosphate (Nye and Tinker, 1977), and has received some experimental support (e.g. St John, 1980; Brundrett and Kendrick, 1988). Although Sanders (1991) was unable to demonstrate a link between P inflow and infection in a group of grassland species (Sanders and Fitter, 1992b), he did find that infection density was a function of root diameter (Fig. 3.5). Root morphology may therefore be part of the explanation for the range in infectedness found within ecosystems. In contrast to Sanders' result, however, when we explicitly tested the hypothesis by measuring root diameter (of terminal roots) and infection density in the ground flora in a group of deciduous wood-

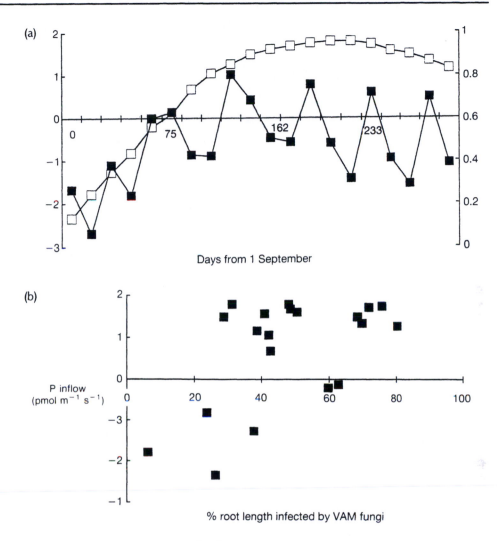

Fig. 3.4. (a) P inflow (pmol m^{-1} s^{-1}; ■, left axis) and infection density (percent root length infected, □, right axis) over one growing season in bluebell (*Hyacinthoides non-scripta*); the curve for P inflow represents a curve fitted to data for P content and root length. (b) The correlation between P inflow and infection density. VAM = vesicular-arbuscular mycorrhiza. Data from an unpublished study by J.W. Merryweather at Pretty Wood, Castle Howard, York.

lands in North Yorkshire, we could find no relationship (Fig. 3.6; G. Grimm, unpublished data). Indeed at the within-plant level, there was a negative relationship: the coarsest roots of individual plants were the least infected.

To attempt to detect a relationship with root diameter at a larger scale, we have analysed data from the Ecological Flora Project, a collation of ecological data from the entire British flora (Fitter, 1989). Plants with coarse roots do emerge from this as typically being mycorrhizal (Fig. 3.7), as predicted, but fine-rooted

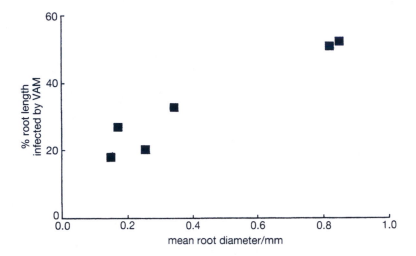

Fig. 3.5. The relationship between VAM infection density (mean over two seasons) and root diameter for six coexisting grassland species at Wheldrake Ings, near York, UK (Sanders, 1991).

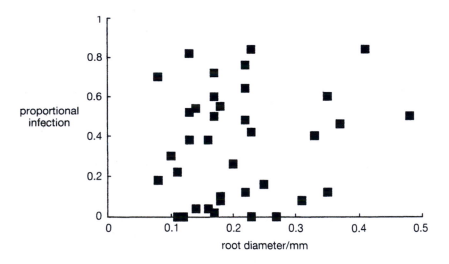

Fig. 3.6. The relationship between VAM infection density and root diameter (terminal roots only) for 21 species of the ground flora of four deciduous woods near Castle Howard, York. Unpublished data of G. Grimm.

plants show a wide range in infectedness, suggesting that other factors determine whether and to what extent they enter the symbiosis. It seems, therefore, that although there is certainly some relationship between infection and root morphology, it cannot explain much of the variation in infection seen at the finer scales of observation.

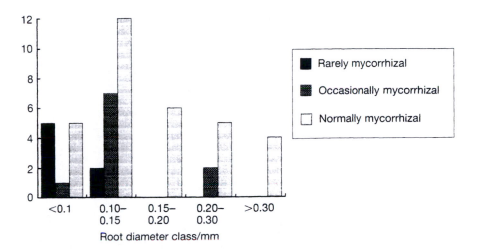

Fig. 3.7. Numbers of species in the British flora in each of three root diameter classes (terminal roots only) and three mycorrhizal categories. Data from the Ecological Flora database at York, where original sources are documented.

Explanations for variation in infection density based on P acquisition, therefore, leave much of the observed pattern unexplained: P inflows in the field are often too low to necessitate mycorrhizal intervention and the relationship with root morphology is only apparent at a large scale. It is probable that other benefits offered by VAM must be taken into account. This is certainly apparent from studies of the annual grass *Vulpia ciliata* in East Anglia. We have used the fungicide benomyl to reduce VAM infection in the roots of this species, which grows in open, sandy areas varying greatly in inoculum potential. Where infection density in control plants was extensive (*c.* 30%, reduced by about one-third by benomyl), we found a significant reduction in fecundity associated with benomyl application (Fig. 3.8a; Carey *et al.*, 1992). On a different site, where native infection was negligible, benomyl application significantly increased fecundity (Fig. 3.8b). The latter result implies that fungal pathogens may reduce seed production and, since these were presumably active at both sites, the true benefit offered by VAM at the first site may have been greater than that observed. However, there was no difference in P inflow between benomyl-treated and untreated plants at the first site, so that the mycorrhizal benefit must have been from some other source. At present the most likely candidate for that seems to be the possibility that VAM fungi may interact with other soil microbes and reduce their pathogenic impact on the roots. This is an avenue we are actively exploring.

There is evidence, therefore, for the view that the level of infection by VA mycorrhizas is an evolved character and that plants may regulate infection in their roots in relation to the benefits that they receive. However, other evidence points to an opposite conclusion: it is often difficult to demonstrate that a plant is benefiting from infection, possibly because benefits are diffuse or expressed irregularly. To resolve this conflict, we need more experiments on mycorrhizal function

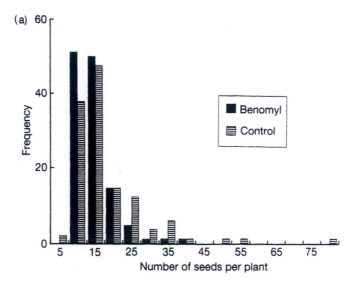

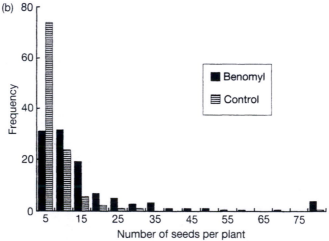

Fig. 3.8. Frequency distribution of fecundity (seed production per plant) for *Vulpia ciliata* ssp. *ambigua* at two sites in Norfolk, UK. Sprayed plants were treated with the fungicide benomyl. (a) Sandringham, 1988; (b) Mildenhall, 1989. Only the plants at Sandringham were infected by VAM fungi. From Carey *et al.* (1992).

under field conditions, aimed at explaining why most plants are mycorrhizal even under conditions where we cannot presently detect benefit from infection. Earlier, we suggested that the distribution of infection density in natural communities may have fractal properties, that is be expressed at all scales of measurement. This is not just a mildly curious observation; if true, it implies that there may be deterministic processes governing spatial patterns of infection that can be discovered, even though the pattern itself may be complex and apparently stochastic. It is those processes that mycorrhizal ecologists should seek to define.

Acknowledgements

Much of the work described here is that of co-workers in my laboratory. I am grateful to all of them, particularly where I have been able to refer to unpublished work, and especially to James Merryweather and Helen West, with whom I have discussed these ideas profitably. Some of the studies outlined here have been supported by the Natural Environment Research Council.

References

Amijee, F., Tinker, P.B. and Stribley, D.P. (1989) The development of endomycorrhizal root systems. VII. A detailed study of effects of soil phosphorus on colonization. *New Phytologist* 111, 435–446.

Baylis, G.T.S. (1975) The magnolioid mycorrhiza and mycotrophy in root systems derived from it. In: Sanders, F.E., Mosse, B. and Tinker, P.B. (eds), *Endomycorrhizas*. Academic Press, London, pp. 373–389.

Brundrett, M.C. (1991) Mycorrhizas in natural ecosystems. *Advances in Ecological Research* 21, 171–313.

Brundrett, M.C. and Kendrick, B. (1988) The mycorrhizal status, root anatomy and phenology of plants in a sugar maple forest. *Canadian Journal of Botany* 66, 1153–1173.

Carey, P.D., Fitter, A.H. and Watkinson, A.R. (1992) A field study using the fungicide benomyl to investigate the effect of mycorrhizal fungi on plant fitness. *Oecologia* (in press).

Dunne, M.J. and Fitter, A.H. (1989) The phosphorus budget of a field-grown strawberry (*Fragaria* × *ananassa* cv. Hapil) crop: evidence for a mycorrhizal contribution. *Annals of Applied Biology* 114, 185–193.

Fitter, A. H. (1988) Water relations of red clover *Trifolium pretense* L. as affected by VA mycorrhizal infection and phosphorus supply before and during drought. *Journal of Experimental Botany* 39, 595–603.

Fitter, A.H. (1989) An ecological flora. *Bulletin of the British Ecological Society* 20, 199–200.

Fitter, A.H. (1990) The role and ecological significance of vesicular-arbuscular mycorrhizas in temperate ecosystems. *Agriculture, Ecosystems and Environment* 29, 137–151.

Fitter, A.H. (1991) Costs and benefits of mycorrhizas: implications for functioning under natural conditions. *Experientia* 47, 350–355.

Fitter, A.H. and Nichols, R. (1988) The use of benomyl to control infection by vesicular-arbuscular mycorrhizal fungi. *New Phytologist* 109, 201–206.

Gay, P.E., Grubb, P.J. and Hudson, H.J. (1982) Seasonal changes in the concentrations of nitrogen, phosphorus and potassium, and in the density of mycorrhiza in biennial and matrix-forming perennial species of closed chalkland turf. *Journal of Ecology* 70, 571–593.

Jakobsen, I. (1986) Vesicular-arbuscular mycorrhiza in field-grown crops. III Mycorrhizal infection and rates of phosphorus inflow in pea plants. *New Phytologist* 104, 573–581.

Koide, R. (1990). Nutrient supply, nutrient demand and plant response to mycorrhizal infection. *New Phytologist* 117, 365–386.

McGonigle, T.P. and Fitter, A.H. (1988a) Growth and phosphorus inflows of *Trifolium repens* L. with a range of indigenous vesicular-arbuscular mycorrhizal (VAM) infection levels under field conditions. *New Phytologist* 108, 59–65.

McGonigle, T.P. and Fitter, A.H. (1988b) Ecological consequences of arthropod grazing on VA mycorrhizal fungi. *Proceedings of the Royal Society of Edinburgh* 94B, 25–32.

McGonigle, T.P. and Fitter, A.H. (1990) Ecological specificity of vesicular-arbuscular mycorrhizal associations. *Mycological Research* 94, 120–122.

Nye, P.H. and Tinker, P.B.H. (1977) *Solute Movement in the Soil-Plant System*. Blackwell Scientific, Oxford.

Read, D.J. (1983) The biology of mycorrhiza in the Ericales. *Canadian Journal of Botany* 61, 985–1004.

Rosendahl, S., Rosendahl, C.N. and Sochting, U. (1990). Distribution of VA mycorrhizal endophytes amongst plants of a Danish grassland community. *Agriculture, Ecosystems and Environment* 29, 329–336.

St John, T.V. (1980) Root size, root hairs and mycorrhizal infection: a re-examination of Baylis's hypothesis with tropical trees. *New Phytologist* 84, 483–487.

Sanders, F.E. and Tinker, P.B. (1973) Phosphate flow into mycorrhizal roots. *Pesticide Science* 4, 385–395.

Sanders, F.E., Tinker, P.B., Black, R.L. and Palmerley, S.M. (1977) The development of endomycorrhizal root systems. I. Spread of infection and growth promoting effects with four species of vesicular-arbuscular mycorrhizae. *New Phytologist* 78, 257–268.

Sanders, I.R. (1991) 'Seasonality, selectivity and specificity of vesicular-arbuscular mycorrhizas in grasslands.' Unpublished D Phil Thesis, University of York.

Sanders, I.R. and Fitter, A.H. (1992a) The ecology and functioning of vesicular-arbuscular mycorrhizas in co-existing grassland species. I. Seasonal patterns of mycorrhizal occurrence and morphology. *New Phytologist* 120, 519–524.

Sanders, I.R. and Fitter, A.H. (1992b) The ecology and functioning of vesicular-arbuscular mycorrhizas in co-existing grassland species II. Nutrient uptake and growth of vesicular-arbuscular mycorrhizal plants in a semi-natural grassland. *New Phytologist* 120, 525–533.

Sanders, I.R. and Fitter, A.H. (1992c) Evidence for differential responses between host/fungus combinations of vesicular-arbuscular mycorrhizas from a grassland. *Mycological Research* 96, 415–419.

Smith, S.E., Walker, N.A. and Tester, M. (1986) The apparent width of the rhizophere of *Trifolium subterraneum* L. for vesicular-arbuscular mycorrhizal infections: effects of time and other factors. *New Phytologist* 104, 547–558.

Tester, M., Smith, F.A. and Smith, S.E. (1985) Phosphate inflow into *Trifolium subterraneum* L.: effects of photon irradiance and mycorrhizal infection. *Soil Biology and Biochemistry* 17, 807–810.

Thompson, B.D., Robson, A.D. and Abbott, L.K. (1986) Effects of phosphorus on the formation of mycorrhizas by *Gigaspora calospora* and *Glomus fasciculatum* in relation to root carbohydrates. *New Phytologist* 103, 751–765.

Trappe, J.M. (1987) Phylogenetic and ecologic aspects of mycotrophy in the angiosperms from an evolutionary standpoint. In: Safir, G.R. (ed.), *Ecophysiology of VA Mycorrhizal Plants*. CRC Press, Baton Rouge, Florida, pp. 5–25.

Warnock, A.J., Fitter, A.H. and Usher, M.B. (1982) The influence of a springtail *Folsomia candida* (Insecta, Collembola) on the mycorrhizal association of leek *Allium porrum* and the vesicular-arbuscular mycorrhizal endophyte *Glomus fasciculatus*. *New Phytologist* 90, 285–292.

Wilson, J.M. (1984) Comparative development of infection by three vesicular-arbuscular mycorrhizal fungi. *New Phytologist* 97, 413–426.

4 What is the Role of VA Mycorrhizal Hyphae in Soil?

L.K. Abbott, A.D. Robson, D.A. Jasper and C. Gazey

Soil Science and Plant Nutrition, School of Agriculture, The University of Western Australia, Nedlands, 6009 WA, Australia

Hyphae of VA mycorrhizal fungi play key roles in the formation, functioning and perpetuation of mycorrhizas in natural and disturbed ecosystems. Hyphae in soil, originating from either an established hyphal network or from other propagules (spores, vesicles and root pieces) lead to the infection and subsequent colonization of roots. The distribution of hyphae and its relationship with sporulation will influence the location of propagules in relation to newly formed roots. The roles of the hyphae in phosphate uptake and soil stabilization are dependent on their distribution within the soil matrix in relation to the root surface. To understand these multiple roles of hyphae and their relative importance for mycorrhizas formed by different species of mycorrhizal fungi in agricultural and natural ecosystems is an immediate challenge.

Nicolson (1959) described hyphae of VA mycorrhizal fungi in soil and grouped them into two categories based on wall thickness and diameter. The aseptate thick-walled hyphae were considered to be the permanent mycelial network. In contrast, the majority of thin-walled hyphae appeared temporary. The thin-walled hyphae arose from thick-walled hyphae either as lateral branches or by repeated branching. It is likely that the latter develop into thick hyphae, but that the thin lateral branches do not (Nicolson, 1959). The hyphae of very few species of VA mycorrhizal fungi have been studied; most studies relate to the genus *Glomus*. Indeed, hyphae of only two species of *Acaulospora* and one species of *Scutellospora* have been studied quantitatively. Therefore it is not possible to make broad generalizations about the roles of hyphae of VA mycorrhizal fungi.

There is a wide range of methods that have been used to quantify the length and activity of VA mycorrhizal hyphae in soil (Sylvia, 1990); all methods are inherently tedious and imprecise. There is relatively little information available about the longevity of hyphae in soil (Schubert *et al.*, 1987; Sylvia, 1988; Hamel *et al.*, 1990).

The nature of the relationship between the growth of hyphae in soil and within a root is not well known. The quantity of hyphae associated with mycorrhizal roots

Table 4.1. The ratio of fungal biomass as estimated by chitin assay outside the root (E) to that within the root (I) for soyabean colonized by *Glomus fasciculatum* (derived from Bethlenfalvay *et al.*, 1982).

Weeks	E/I
3	10.25
5	5.25
7	2.49
10	1.33
13	1.28
16	0.27
19	0.21

(a)
Glomus sp. (WUM 10 (1))

(b)
Scutellospora calospora

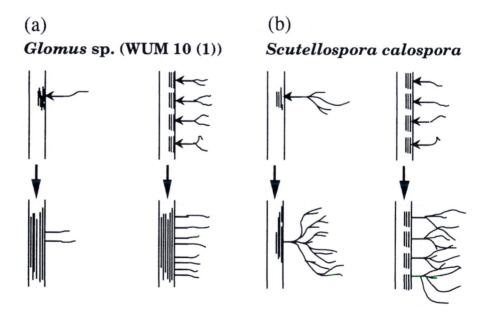

Fig. 4.1. Proposed patterns of development of hyphae within and outside roots for a *Glomus* sp. [WUMIO(1)] (**a**) and *Scutellospora calospora* (**b**) in relation to the density of infective hyphae at the root surface (→).

can vary widely (Sylvia, 1986). The amount of hyphae associated with roots can also vary with time (Table 4.1).

Abbott and Robson (1984) hypothesized that for *Glomus* sp. [WUM 10 (1) – previously referred to as *G. fasciculatum*], low initial levels of infective hyphae in the soil may lead to small amounts of hyphae in soil relative to in the root (Fig. 4.1a). In contrast, for high densities of infective hyphae, the exponential phase of colonization of roots may be extended in parallel with extensive development of hyphae in the soil. Such an association between the formation of hyphae in soil and within the root may not occur for *S. calospora* which consistently produces large quantities of hyphae in soil (Fig. 4.1b), irrespective of the density of hyphae within the root.

The differences in the formation of hyphae in soil by *Glomus* sp. [WUM 10 (1)] and *S. calospora* (Abbott and Robson, 1985; Jakobsen *et al.*, 1992; Fig. 4.2c) parallel differences in the spread of mycorrhizas for these fungi from a point of inoculation (Fig. 4.2a, b; Scheltema. *et al.*, 1985, 1987). The gradual spread of mycorrhizas by *S. calospora* contrasts markedly with the spread by *Glomus* sp. [WUM 10 (1)] which occurs on a distinct front. Mycorrhizal spread to roots away from a point of inoculation by *S. calospora* appears to be independent of the extent to which individual roots are colonized (see Fig. 4.1 above). However,

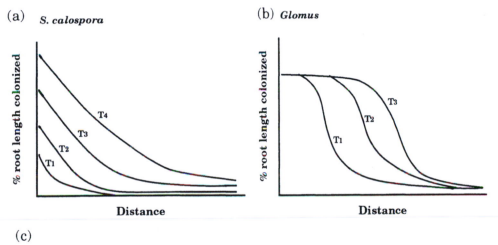

(a) *S. calospora*

(b) *Glomus*

(c)

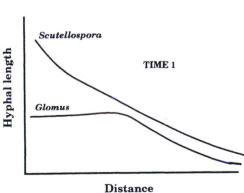

Fig. 4.2. Patterns of spread of mycorrhizas for *Scutellospora calospora* **(a)** (after Scheltema et al., 1987) and a *Glomus* sp. **(b)** (after Scheltema et al., 1985) in relation to the distribution of hyphae away from roots **(c)** (after Jakobsen et al., 1992); T_1, T_2, T_3, T_4 represent different harvest times.

Table 4.2. The spread of mycorrhizas below 6 cm from the soil surface and external hyphae production for two fungi in two soils (Abbott and Robson, unpublished data).

Soil	Fungus	% Root length colonized		cm hyphae per cm colonized root in the surface 6 cm of soil
		surface 6 cm of soil	below 6 cm from surface	
1.	S. calospora	67	33	1178
	Glomus sp.	75	4	39
2.	S. calospora	55	28	3992
	Glomus sp.	82	18	1726

for *Glomus* sp. [WUM 10 (1)], the growth of hyphae in soil and subsequent spread is dependent on rapid and extensive colonization within roots.

The infectivity of VA mycorrhizal fungi in soil can be severely reduced by disturbance of the soil (Jasper *et al.*, 1989). The extent to which this occurs depends on the species of fungus and its stage of development (Jasper *et al.*, Chapter 19 this volume).

Fungi differ in their relative response to soil characteristics (Sylvia, 1990). This may be due to direct effects on the fungi or indirect effects via the plant. The extent of spread of mycorrhizas to depth within the soil was related to the extent of mycorrhiza formation in surface soil for *Glomus* sp. [WUM 10 (1)] but not for *S. calospora* (Table 4.2). These observations on hyphae and mycorrhiza formation parallel those discussed earlier for these fungi (Figs 4.1 and 4.2).

There is a need to study further the longevity of hyphae in relation to the functioning of the symbiosis. Some studies indicate that the hyphae are relatively short-lived (Schubert *et al.*, 1987; Sylvia, 1988) but this is not always the case (Hamel *et al.*, 1990). Benefits in phosphate uptake may be available even for hyphae with relatively short life-spans, depending on the rate of depletion of phosphorus. Similarly for aggregation of soil (Tisdall and Oades, 1982), maintenance of living hyphae may not be crucial. Sporulation may be linked to phosphate uptake because hyphae that survive long enough to support spore formation have an opportunity to take up phosphate as well. The longevity of the thick-walled hyphae may be most important for providing opportunities for extended distribution of thin-walled hyphae that are likely to be specifically involved in phosphorus uptake and soil aggregate formation and stability.

Conclusion

Although few fungi have been studied, questions have been raised which, if answered, will provide further insight into the functioning of VA mycorrhizas. The interesting differences between fungi studied so far highlight the potential for explaining differences observed in the effectiveness of mycorrhizal fungi. Further knowledge of the effects of soil physical, chemical and biological properties on the formation and function of hyphae are needed in an integrated approach to clarify the role of VA mycorrhizal hyphae in diverse ecosystems.

Acknowledgements

We thank The Australian Wool Corporation, The Wheat Research Committee of Western Australia and AMIRA for their support of our mycorrhizal studies. Special thanks to Martine Scheltema for her valued contribution to the research.

References

Abbott, L.K. and Robson, A.D. (1984) The effect of root density, inoculum placement and infectivity of inoculum on the development of vesicular-arbuscular mycorrhizas. *New Phytologist* 97, 285–299.

Abbott, L.K. and Robson, A.D. (1985) Formation of external hyphae in soil by four species of vesicular-arbuscular mycorrhizal fungi. *New Phytologist* 99, 245–255.

Bethlenfalvay, G.J., Pacovsky, R.S., Brown, M.S. and Fuller, G. (1982) Mycotrophic growth and mutualistic development of host plant and fungal endophyte in an endomycorrhizal symbiosis. *Plant and Soil* 68, 43–54.

Hamel, C., Fyles, H. and Smith, D.L. (1990) Measurement of development of endomycorrhizal mycelium using three different vital stains. *New Phytologist* 115, 297–302.

Jakobsen, I., Abbott, L.K. and Robson, A.D. (1992) External hyphae of vesicular-arbuscular mycorrhizal fungi associated with *Trifolium subterraneum* L. 1. Spread of hyphae and phosphorus inflow into roots. *New Phytologist* 120, 371–380.

Jasper, D.A., Abbott, L.K. and Robson, A.D. (1989) Hyphae of a vesicular-arbuscular mycorrhizal fungus maintain infectivity in a dry soil, except when the soil is disturbed. *New Phytologist* 112, 101–107.

Nicolson, T.H. (1959) Mycorrhiza in the Gramineae 1. Vesicular-arbuscular endophytes, with special reference to the external phase. *Transactions of the British Mycological Society* 42, 421–438.

Scheltema, M.A., Abbott, L.K. and Robson, A.D. (1985) The spread of *Glomus fasciculatum* through roots of *Trifolium subterraneum* and *Lolium rigidium*. *New Phytologist* 100, 105–114.

Scheltema, M.A., Abbott, L.K. and Robson, A.D. (1987) The spread of mycorrhizal infection by *Gigaspora calospora* from a localized inoculum. *New Phytologist* 106, 724–734.

Schubert, A., Marzachi, C., Mazzitelli, M., Cravero, M.C. and Bonfante-Fasolo, P. (1987) Development of total and viable extraradial mycelium in the vesicular arbuscular mycorrhizal fungus *Glomus clarum* Nicol. and Schenck. *New Phytologist* 107, 183–190.

Sylvia, D.M. (1986) Spatial and temporal distribution of vesicular-arbuscular mycorrhizal fungi associated with *Uniola paniculata* in Florida foredunes. *Mycologia* 78, 728–734.

Sylvia, D.M. (1988) Activity of external hyphae of vesicular-arbuscular mycorrhizal fungi. *Soil Biology and Biochemistry* 20, 39–43.

Sylvia, D.M. (1990) Distribution, structure, and function of external hyphae of vesicular-arbuscular mycorrhizal fungi. In Box, J.E. and Hammond, L.C. (eds), *Rhizosphere Dynamics*. Westview Press, Boulder, Colorado. pp. 144–167.

Tisdall, J.M. and Oades, J.M. (1982) Organic matter and water-stable aggregates in soils. *Journal of Soil Science* 33, 141–163.

5 Contribution of Mycorrhizal Hyphae to Nutrient and Water Uptake of Plants

E. George[1], K. Häussler[1], S.K. Kothari[2], X.-L. Li[3] and H. Marschner[1]

[1]Institute of Plant Nutrition, University of Hohenheim, P.B. 700562, 7000 Stuttgart 70, Germany: [2]Central Institute of Medicinal and Aromatic Plants, Lucknow, India: [3]Department of Soil Science and Plant Nutrition, Beijing Agricultural University, China

Introduction

Effects of mycorrhizal fungi on plant growth can be both direct and indirect. Hyphae can, for example, take up nutrients such as P, translocate them to the host plant and thus directly increase host nutrient acquisition. On the other hand, infection of its root system by mycorrhizal fungi causes morphological and physiological changes in the host plant. For example, VA mycorrhizal plants may differ from non-mycorrhizal plants in their root length (Gnekow and Marschner, 1989). We have attempted to distinguish between direct and indirect effects in a series of experiments using a compartmented box system (Li et al., 1991a). By spatially separating the growing zones of roots and hyphae, the direct contributions of external hyphae to P, Cu and water uptake of the plant roots and to pH changes in the soil were measured.

Methodology

Plants were grown in boxes in which a central compartment containing the plant root system was separated from outer compartments on either side. The outer and inner compartments were in initial experiments separated by fine nylon mesh (mesh size 30 μm) to impede roots from growing into the outer compartment but to allow hyphae to pass (Schüepp et al., 1987; Kothari et al., 1990). By means of cotton wicks, water was supplied only to the central ('root') compartment. This ensured that nutrient movement by mass flow from the outer ('hyphal') to the inner 'root' compartment was minimized. In some experiments, the outer compartment was subdivided into a hyphal and a 'bulk soil' compartment using a membrane (0.45 μm pore size) not penetrable by hyphae (Li et al., 1991b). At the inner side of the membrane, a plane of hyphae developed, similar to the plane of roots at the inner side of the nylon mesh restricting the root compartment. For the experiments on water uptake, an additional modification provided a barrier to soil water

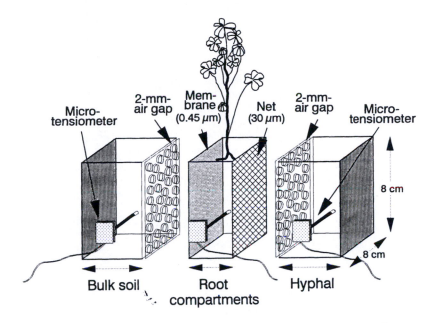

Fig. 5.1. Experimental pots for spatial separation of root and hyphae growing zones and for measuring hyphal water transport.

movement between hyphal and root compartments through the introduction of a 2-mm air gap (Fig. 5.1). Microtensiometers (Vetterlein *et al.*, unpublished) were placed centrally in each compartment, and soil water potentials continuously recorded with a datalogger.

Experiments were carried out using two partially sterilized soils, a Luvisol (loess subsoil, 3.8 mg kg^{-1} NaHCO$_3$-extractable P, pH[CaCl$_2$] 7.3) or a Cambisol (sandy loam, 3.5 mg kg^{-1} H$_2$O-extractable P, pH 6.2). Seeds of *Zea mays*, *Trifolium repens* or *Agropyron repens* were placed in the central compartment of the boxes along with VA mycorrhizal inoculum consisting of soil and roots infected with an isolate of *Glomus mosseae*. Non-mycorrhizal controls were inoculated either with sterilized roots together with a filtrate of the original inoculum, or with soil and roots from non-mycorrhizal plants.

Results and Discussion

In an experiment with non-mycorrhizal and VA mycorrhizal white clover on the Luvisol, the outer (hyphal) compartments were fertilized with 50 or 150 mg P [Ca(H$_2$PO$_4$)$_2$] per kg soil. In the non-mycorrhizal plants, P uptake did not respond to the P levels in the outer compartments. Mycorrhizal plants grew much larger and acquired much more P than non-mycorrhizal plants, particularly at the higher P level. The mean rate of P uptake by the hyphae was calculated as 0.3–0.4 pmol s^{-1} m^{-1} (Li *et al.*, 1991 a). At both P levels, hyphae extended up to the ends of the box, approximately 12 cm from the roots.

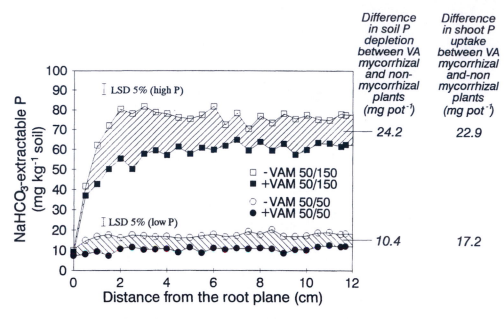

Fig. 5.2. Depletion profiles of extractable P in the outer (hyphal) compartments, difference in the soil P depletion in these compartments and difference in shoot P uptake between seven-week-old mycorrhizal (*Glomus mosseae*) and non-mycorrhizal white clover plants in a Luvisol as affected by low (50 mg P kg^{-1}) and high (150 mg P kg^{-1}) P fertilization in the outer compartments (Li *et al.*, 1991a).

Vesicular-arbuscular mycorrhizal hyphae in the outer compartment reduced levels of extractable soil P at all distances from the root plane (Fig. 5.2). In the outer compartment, the zone of P diffusion towards the plane of roots extended to approximately 1 cm. At the high level of P supply, the quantity of P extracted by the hyphae from the outer compartment corresponded well to the difference between the quantities of P in shoots of VA mycorrhizal and non-mycorrhizal plants. At the low P level, however, P uptake by VA mycorrhizal plants was greater than could be accounted for by measurement of soil depletion (Fig. 5.2). It is possible that VA mycorrhizal hyphae may possess special properties allowing more efficient uptake of soil P in addition to acting merely as diffusion sinks.

In another group of experiments in the Cambisol, again using white clover, hyphal compartments were separated from bulk soil compartments using membranes (0.45 μm pore size). Phosphate was depleted from the hyphal compartments across their whole width, while a distinct zone of P depletion developed from the plane of hyphae (membrane) into the bulk soil to which the hyphae had no access (Li *et al.*, 1991b). By analogy with the rhizosphere, the soil zone around a hypha which is influenced by its activity can be termed a 'hyphosphere'. In plants supplied with $(NH_4)_2SO_4$ as source of N, the soil pH was lowered not only in the rhizosphere but also in the hyphal compartment and in the 'hyphosphere' (Li *et al.*, 1991b). In soils where a pH decrease results in increased P availability, this may allow hyphae to take up additional P.

In addition to P, hyphae may also transport zinc (Zn) and copper (Cu). An experiment was carried out in which the hyphal compartments received three levels of P fertilization (Li *et al.*, 1991c). A substantial proportion of the Cu in the plants had been transported by the hyphae. The quantity of Cu transported was not directly proportional to the quantity of P being transported at the same time (Table 5.1). In consequence, according to the level of P supply the P:Cu transport ratio in the hyphae varied drastically from $37 \, mol \, mol^{-1}$ at low P supply to $912 \, mol \, mol^{-1}$ at high P supply. Much of the Cu delivered by the hyphae at low P supply remained in the roots while, with increased P supply an increasing proportion of the delivered Cu was translocated from the roots into the shoots (Table 5.1). This may indicate the presence of a mechanism in VA mycorrhizal roots regulating the release and transport of Cu from roots to shoots according to the shoot demand (Li *et al.*, 1991c).

The compartmented boxes can also be used to assess indirect mycorrhizal effects on the plant root system or on the rhizosphere. In a further experiment, maize plants were adequately fertilized with P in the root compartment. In addition to VA mycorrhizal and non-mycorrhizal plants, a sterile control was included. Because of the high P content of the root compartment, there were no differences in shoot dry weight between treatments, but manganese (Mn) concentrations in shoots and roots were distinctly lower in VA mycorrhizal plants and also in the sterile plants (Kothari *et al.*, 1991). Compared with the rhizosphere of non-mycorrhizal plants, the concentration of exchangeable Mn in the rhizosphere soil of VA mycorrhizal plants was lower as were the numbers of Mn reducing microorganisms, indicating a distinct shift in rhizosphere microorganism population.

To examine VA mycorrhizal effects on plant water relations, maize plants, grown in compartmented boxes, were well fertilized with P (Kothari *et al.*, 1990). There were no differences in shoot dry weight between VA mycorrhizal, non-mycorrhizal or control plants maintained under sterile conditions (sterile control). Over the 6-week growth period, mycorrhizal plants transpired more water than non-mycorrhizal or sterile control plants and had a higher rate of water uptake per unit root length (6.1, 13.4 and $6.9 \, nl \, s^{-1} \, m^{-1}$ in the non-mycorrhizal,

Table 5.1. Hyphal delivery and amount of P and Cu in roots and shoots of VA mycorrhizal white clover plants as affected by the supply of P [as Ca $(H_2PO_4)_2$] to the outer (hyphal) compartments. Figures within one column followed by the same letter are not significantly ($P < 0.05$) different by the Student–Newman–Keuls test (Li *et al.*, 1991c).

P supply (mg kg^{-1} soil)		Hyphal delivery (μmol pot^{-1})		Amount of P and Cu (μmol pot^{-1})			
				Roots		Shoots	
Roots	Hyphae	P	Cu	P	Cu	P	Cu
50	0	27a	0.73a	29a	0.97c	52a	0.23a
50	20	226b	0.57a	102b	0.61b	184b	0.43b
50	50	465c	0.51a	152c	0.45a	382c	0.52b

VA mycorrhizal and sterile plants, respectively). This effect may have resulted from a different shoot morphology of VA mycorrhizal plants, which had broader leaves leading to a higher leaf area per plant. Also, roots of VA mycorrhizal plants were much shorter, thicker and less branched than roots of non-mycorrhizal plants (Kothari *et al.*, 1990). This different root morphology may be a result of microbial activity or of effects of VA mycorrhizal infection on the root apical meristems.

Water transport through the hyphae is probably not the major cause of the greater rate of water uptake per unit root length of VA mycorrhizal plants. The difference in the mean rate of water uptake between VA mycorrhizal and non-mycorrhizal plants was approximately 7 nl m^{-1}s^{-1}, or 26 μl m^{-1}h^{-1}. With 60% of the roots colonized by the VA mycorrhizal fungus and with, on average, ten entry points per centimetre of infected root, this results in a calculated water inflow of 44 nl h^{-1} per entry point (Kothari *et al.*, 1990). With an hyphal inner diameter of 8 μm, this inflow requires a transport velocity of approximately 0.9 m h^{-1} in the hyphae, which appears very high. From further calculations (Kothari *et al.*, 1990) it was estimated that approximately 10% of the increased water inflow into mycorrhizal roots could be explained as due to direct water transport in hyphae. The higher specific water uptake rates of VA mycorrhizal roots may be due to higher transpirational demand and/or higher hydraulic conductivity of VA mycorrhizal roots.

To provide a more specific test of the water transport capability of hyphae, root and hyphal compartments were additionally separated by an air gap (Fig. 5.1). Hyphae were able to cross this air gap and grow into the outer compartments.

There was no difference in water loss from the outer compartments whether mycorrhizal hyphae were present or not. The roots rapidly absorbed water from the

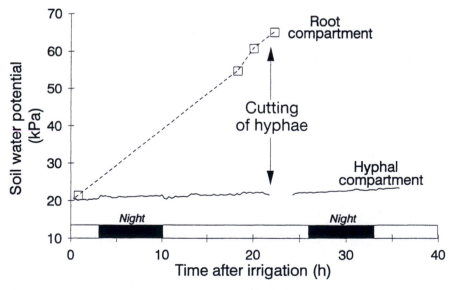

Fig. 5.3. Soil water potential in the root compartment and in the hyphal compartment of 10-week-old white clover plants grown in a Luvisol.

root compartment, but there was only a very small water loss from the hyphal compartment (Fig. 5.3), even when the soil in the root compartment had become very dry. Cutting the hyphal connections in the air gap and separation of the compartments did not reduce the small rate of water loss from the hyphal compartments (Fig. 5.3). This is a further indication that hyphae did not transport substantial amounts of water to the host plant.

Calculations suggest that only mycorrhizal fungi with much greater hyphal diameters than *Glomus mosseae* could maintain water transport rates of sufficient magnitude to contribute significantly to plant water uptake. In contrast to VA mycorrhizas, the rhizomorphs of many ectomycorrhizal fungi have a sufficiently large capacity to permit significant rates of water transport to host trees (Duddridge *et al.*, 1980). However, when *Thelephora terrestris* mycorrhizas on 2-year-old Norway spruce plants were tested in the compartmented boxes, no difference in water loss from hyphal compartments was observed (unpublished results).

Acknowledgements

We thank Dr F.E. Sanders for critical reading and comments.

References

Duddridge, J.A., Malibari, A. and Read, D.J. (1980) Structure and function of mycorrhizal rhizomorphs with special reference to their role in water transport. *Nature* 287, 834–836.

Gnekow, M.A. and Marschner, H. (1989) Influence of the fungicide pentachloronitrobenzene on VA-mycorrhizal and total root length and phosphorus uptake of oats (*Avena sativa*). *Plant and Soil* 114, 91–98.

Kothari, S.K., Marschner, H. and George, E. (1990) Effect of VA mycorrhiza and rhizosphere microorganisms on root and shoot morphology, growth and water relations of maize. *New Phytologist* 116, 303–311.

Kothari, S.K., Marschner, H. and Römheld, V. (1991) Effect of a vesicular-arbuscular mycorrhizal fungus and rhizosphere micro-organisms on manganese reduction in the rhizosphere and manganese concentrations in maize (*Zea mays* L.). *New Phytologist* 117, 649–655.

Li, X.-L., George, E. and Marschner, H. (1991a) Extension of the phosphorus depletion zone in VA-mycorrhizal white clover in a calcareous soil. *Plant and Soil* 136, 41–48.

Li, X.-L., George, E. and Marschner, H. (1991b) Phosphorus depletion and pH decrease at the root-soil and hyphae-soil interfaces of VA mycorrhizal white clover fertilized with ammonium. *New Phytologist* 119, 397–404.

Li, X.-L., Marschner, H. and George, E. (1991c) Acquisition of phosphorus and copper by VA-mycorrhizal hyphae and root-to-shoot transport in white clover. *Plant and Soil* 136, 49–57.

Schüepp, H., Miller, D. D. and Bodmer, M. (1987) A new technique for monitoring hyphal growth of vesicular-arbuscular mycorrhizal fungi through soil. *Transactions of the British Mycological Society* 89, 429–435.

6 Phosphorus Transport by External Hyphae of Vesicular-Arbuscular Mycorrhizas

I. Jakobsen

Plant Biology Section, Environmental Science and Technology Department, Risø National Laboratory, DK-4000 Roskilde, Denmark

Introduction

The external hyphae of vesicular-arbuscular (VA) mycorrhizas constitute an important pathway for phosphorus (P) transport through soil as they extend beyond the P depletion zones around the roots and gain access to P which is otherwise transported only by slow diffusion processes. The majority of growth responses to VA mycorrhizas are caused by the increased supply of P to the plant via the external hyphae. However, we do not know why some VA-mycorrhizal fungi differ in their ability to transport P and we know very little about interactions between hyphae and the various components of the soil ecosystem. If we wish eventually to manage the symbiosis in order to maximize the P supply from the mycobiont, an identification of the rate-limiting steps in the hyphal P transport will be requested.

Hyphal P transport may be considered as the result of three steps: hyphal uptake, translocation in hyphae and transfer across the symbiotic interface. Phosphorus uptake by the hyphae could be affected by the total length of viable hyphae, by the spread of hyphae into the soil, by release of any P solubilizing agent and by the uptake kinetics of the hyphae. Phosphorus is thought to be translocated in the hyphae in the form of polyphosphates and translocation rates could be affected by concentration gradients and cytoplasmic streaming. Transfer to the host may be influenced by the interface area and by interface enzymes in a broad sense including membrane transport systems. Each of these components may be affected by environmental factors, by the nutrient status and species of host plant, and by the species of mycobiont.

The metabolic events leading to P transport by VA mycorrhizal hyphae have recently been reviewed by Gianinazzi-Pearson and Gianinazzi (1989) and Smith and Smith (1990). The present paper considers mainly the overall hyphal P transport through soil to plant roots. Emphasis will be on methods for measuring hyphal P transport, on the relationship between P transport and spread of hyphae in soil by different VA mycorrhizal fungi and on a preliminary study of hyphal P transport under field conditions.

Measurement of Hyphal P Transport

Mycorrhizal effects on P supply to plants are often quantified as the difference in P content between mycorrhizal and non-mycorrhizal plants. This is a poor measure of the hyphal P contribution, as root morphology is normally altered by colonization (Hetrick, 1991) and the uptake characteristics of the roots themselves may be affected by colonization.

Phosphorus inflow, the P uptake per unit root length and time, is more suitable for measuring the P supplied by hyphae. The inflow to colonized parts of the roots due to hyphal uptake of P may be estimated as $I_M = 100 \times (I - I_C) \times (\% \text{ colonization})^{-1}$, where I_C is the inflow to roots of non-mycorrhizal plants and I the inflow to roots of colonized plants (Sanders *et al.*, 1977). Information on the causal relationships of potential differences in hyphal P transport by different fungi might be obtained if inflow values are compared with amounts of hyphae produced. Methods to quantify external hyphae are discussed by Sylvia (1992) and Hamel *et al.* (1990) compared three vital stains for assessing the length of metabolically active hyphae.

Phosphorus inflow to roots of onion (Sanders *et al.*, 1977) and growth responses in *Citrus* (Graham *et al.*, 1982) were related to the amount of hyphae produced by a number of associated VA-mycorrhizal fungi, but not in later studies by Abbott and Robson (1985) and Jakobsen *et al.* (1992a). The development of growth systems with separate hyphal and root compartments has allowed more detailed studies of hyphal P transport. Simple two-compartment systems of Petri-dish size were used for short-term studies of P transport by single fungi (Rhodes and Gerdemann, 1975; Cooper and Tinker, 1978, 1981). Schuepp *et al.* (1987) introduced a fine nylon mesh to separate a soil-filled hyphal compartment from the root compartment, and a hyphal compartment containing 6.5 kg of soil was used to study hyphal P transport in relation to hyphal rate of spread (Jakobsen *et al.*, 1992a). Hyphal P uptake from defined locations in the hyphal compartment can be measured indirectly as depletion of soil P (Li *et al.*, 1991). Hyphal P transport over defined distances can be measured directly by means of radiotracers (Rhodes and Gerdemann, 1975; Jakobsen *et al.*, 1992b).

A Study of Hyphal P Transport by Three VA mycorrhizal Fungi

Trifolium subterraneum L cv. Seaton Park was grown in association with *Scutellospora calospora* [WUM 12(2)], *Acaulospora laevis* [WUM 11(4)] or *Glomus* sp. [WUM 10(1)] in two-compartment systems in order to study the background for previously demonstrated differences in the ability of these fungi to improve P uptake and growth of the host (Jakobsen *et al.*, 1992a). A layer of ^{32}P-labelled soil with a high P-adsorption capacity was placed in the hyphal compartment at 0, 1.0, 2.5, 4.5 or 7.0 cm distance from the root compartment. Hyphae from roots with 80% of their length colonized by one of the three fungi were then allowed to grow into the hyphal compartments.

Plants were harvested after 37 days when both *Glomus* sp. and *A. laevis* had transported considerable amounts of ^{32}P to the host (Fig. 6.1); the highest

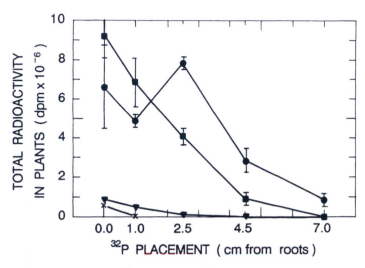

Fig. 6.1. Transport of ^{32}P into *Trifolium subterraneum* in symbiosis with *Acaulospora laevis* (●), *Glomus* sp. (■), *Scutellospora calospora* (▼) or without fungus (x).

radioactivity measured corresponds to about 7% of the ^{32}P supplied to each hyphal compartment. *Glomus* sp. transported most ^{32}P to the plant when the labelled soil layer was placed closest to the roots, whereas ^{32}P transport over distances greater than 1 cm was largest with *A. laevis*. Hyphae of *A. laevis* transported P to the plant from distances as large as 7 cm. The larger area under the curve for *A. laevis* than for *Glomus* sp. reflects the greater total transport of P from the hyphal compartment into the plant with *A. laevis*. The difference between these two fungi in P uptake from root-free soil is closely related to previously measured differences in their ability to spread into the soil (Jakobsen *et al.*, 1992a): rate of spread into a hyphal compartment was 3 mm d^{-1} for *A. laevis* but only 0.9 mm d^{-1} for *Glomus* sp. After 28 days of spread, *A. laevis* maintained a constant plateau of hyphal density up to 7 cm from the root compartment, while the hyphal density of *Glomus* sp. declined rapidly beyond 2 cm.

S. *calospora* transported much less ^{32}P to the host plant than did the two other fungi (Fig. 6.1). Its hyphal density fell off quickly with increasing distance from the root compartment, but this could not explain its low P transport. Consequently, a unit length of hypha of *S. calospora* appears to have a lower capacity for P transport than the two other fungi tested. This was further investigated by comparing radioactivity in hyphae washed from the 2.5 cm soil segments of the hyphal compartments closest to the roots. The radioactivity expressed on a hyphal dry weight basis was considerably higher with *S. calospora* than with the two other fungi, while the opposite relationship was found for radioactivity in roots and shoots (Table 6.1). The results indicate that the P transport of hyphae of *S. calospora* was not limited by hyphal uptake but rather by a restriction in hyphal translocation or in transfer of P to the plant. The latter possibility is in accordance with the observation that the frequency of arbuscule formation in colonized roots is lower with *S. calospora* than with *Glomus* sp. and *A. laevis* (J.N. Pearson,

Table 6.1. Specific radioactivity (d.p.m. mg^{-1} dry wt) in components of 37-day-old symbioses between *Trifolium subterraneum* and three VA mycorrhizal fungi.

Fungus	Shoots	Roots	Hyphae
S. calospora	20	6	21 150
A. laevis	1 593	919	5 200
Glomus sp.	798	481	6 900

personal communication). Consequently there is a need for a detailed investigation of the relationship between P transport and arbuscule number or area of active interface.

The same set of plants was used for a time-course study of transport of ^{32}P from the hyphal compartment to leaflets harvested every third day from initiation of the labelling period. The results agreed well with the measurements of total radioactivity in plants described above (see Jakobsen *et al.*, 1992b). In general, the combined use of a hyphal compartment and soil labelled with a radioisotope of P seems to be a useful tool for the selection of mycorrhizal fungi with well-defined characteristics for hyphal P transport. The two-compartment systems also allow for the isolated studies of interactions between external hyphae and physicochemical as well as biological soil components. There is a need to include as many of these components as possible in future work. Ultimately, the studies of hyphal P transport should be performed under field conditions.

Direct Measurements of Hyphal P Transport in the Field

Injection of ^{32}P into undisturbed field plots has been used to study P uptake by different mycorrhizas of birch (Dighton *et al.*, 1990) but in such experiments it is difficult to separate hyphal uptake from root uptake. In disturbed soils, e.g. the ploughed layer of arable land, direct measurements of hyphal P uptake can be obtained by means of the two-compartment concept without affecting the normal growth conditions too much. This was demonstrated in a preliminary experiment on a clay loam with a pH (0.01 M, CaCl$_2$) of 6.8 and 15 mg kg^{-1} of 0.5 M NaHCO$_3$-extractable P, where plots were either untreated or fumigated with 75 g dazomet m^{-2} (Jakobsen, unpublished). PVC tubes (length 28 cm, inside diameter 5.7 cm) with a perforated bottom inserted were buried to 20 cm soil depth in four replicate plots of each treatment (Fig. 6.2). Each tube had two wide windows on opposite sides from 5 to 15 cm soil depth. Bags of 37 μm nylon mesh were fitted inside each tube and, to avoid drying out of the soil surface at the windows, these were initially closed by inserting a 30 cm PVC tube (outside diameter 5.2 cm) into each mesh bag. Twenty-four pea plants were grown in three concentric circles around each PVC tube. Thirty days after sowing the roots of plants from untreated soil were heavily colonized (85%) while roots in fumigated soil remained non-mycorrhizal. The innermost PVC tube was removed and each mesh bag was filled with 700 g fumigated soil, uniformly labelled with 1850 kBq of carrier-free ortho-phosphate in aqueous solution.

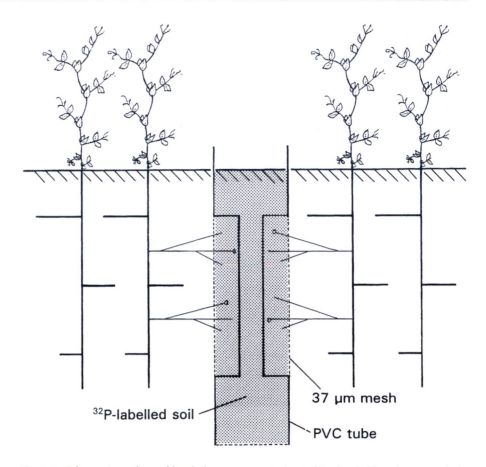

Fig. 6.2. Schematic outline of hyphal compartment buried in the field and surrounded by pea plants.

The time-course of transport of ^{32}P into the shoots was monitored on pairs of leaflets sampled from the youngest fully expanded leaves of plants from the two innermost plant rows at four times after the addition of labelled soil (Fig. 6.3). The specific radioactivity (c.p.m. g^{-1} leaf P) was similar in the two treatments at the first sampling, but higher in mycorrhizal than in non-mycorrhizal plants at later samplings. This difference was most likely due to hyphal transport of ^{32}P from the hyphal compartment. The ^{32}P in leaves of non-mycorrhizal plants could have occurred from uptake by root hairs penetrating the nylon mesh and from uptake of ^{32}P diffused from the hyphal compartment. These results indicate that the two-compartment system is readily applicable to the field situation although, in future work, it is important to confirm the hyphal growth into the hyphal compartment by direct measurements of hyphal length. Obviously, the size and shape of the hyphal compartment as well as the labelling method may be adapted to specific experimental requirements.

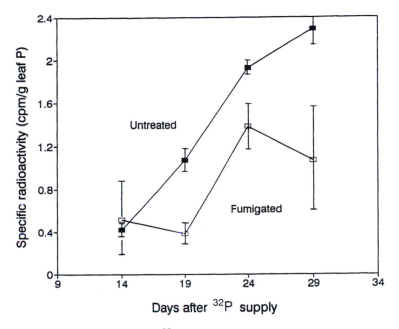

Fig. 6.3. The time course of ^{32}P transport from hyphal compartments into leaflets of peas grown in untreated or fumigated field plots.

Conclusions

The aim of this chapter has been to demonstrate that simple methods are now available which allow for the direct study of P transport by hyphae of VA mycorrhizal fungi, not only in pots, but also under field conditions. The two-compartment system combined with isotope labelling facilitate studies of growth and nutrient transport of external hyphae in relation to specific fungal charac-teristics and a wide range of interacting environmental factors. The methodology should also be applicable to other types of mycorrhizas.

References

Abbott, L.K. and Robson, A.D. (1985) Formation of external hyphae in soil by four species of vesicular-arbuscular mycorrhizal fungi. *New Phytologist* 99, 245–255.

Cooper, K.M. and Tinker, P.B. (1978) Translocation and transfer of nutrients in vesicular-arbuscular mycorrhizas. II. Uptake and translocation of phosphorus, zinc and sulphur. *New Phytologist* 81, 43–52.

Cooper, K.M. and Tinker, P.B. (1981) Translocation and transfer of nutrients in vesicular-arbuscular mycorrhizas. IV. Effect of environmental variables on movement of phos-phorus. *New Phytologist* 88, 327–339.

Dighton, J., Mason, P.A. and Poskitt, J.M. (1990) Field use of ^{32}P to measure phosphate uptake by birch mycorrhizas. *New Phytologist* 116, 655–661.

Gianinazzi-Pearson, V. and Gianinazzi, S. (1989) Phosphorus metabolism in mycorrhizas. In: Boddy, L., Marchant, R. and Read, D.J. (eds), *Nitrogen, Phosphorus and Sulphur Utilization by Fungi*. Cambridge University Press, Cambridge, pp. 227–241.

Graham, J.H., Linderman, R.G. and Menge, J.A. (1982) Development of external hyphae by different isolates of mycorrhizal *Glomus* ssp. in relation to root colonization and growth of troyer citrange. *New Phytologist* 91, 183–189.

Hamel, C., Fyles, H. and Smith, D.L. (1990) Measurement of development of endomycorrhizal mycelium using three different vital stains. *New Phytologist* 115, 297–302.

Hetrick, B.A.D. (1991) Mycorrhizas and root architecture. *Experientia* 47, 355–362.

Jakobsen, I., Abbott, L.K. and Robson, A.D. (1992a) External hyphae of vesicular-arbuscular mycorrhizal fungi associated with *Trifolium subterraneum*. 1. Spread of hyphae and phosphorus inflow into roots. *New Phytologist* 120, 371–380.

Jakobsen, I., Abbott, L.K. and Robson, A.D. (1992b) External hyphae of vesicular-arbuscular mycorrhizal fungi associated with *Trifolium subterraneum*. 2. Hyphal transport of ^{32}P over defined distances. *New Phytologist* 120, 509-516.

Li, X.-L., George, E. and Marschner, H. (1991) Extension of the phosphorus depletion zone in VA-mycorrhizal white clover in a calcareous soil. *Plant and Soil* 136, 41–48.

Rhodes, L.H and Gerdemann, J.W. (1975) Phosphate uptake zones of mycorrhizal and non-mycorrhizal onions. *New Phytologist* 75, 555–561.

Sanders, F.E., Tinker, P.B., Black, R.L.B. and Palmerly, S.M. (1977) The development of endomycorrhizal root systems. I. Spread of infection and growth-promoting effects with four species of vesicular-arbuscular endophyte. *New Phytologist* 78, 257–268.

Schuepp, H., Miller, D.D. and Bodmer, M. (1987) A new technique for monitoring hyphal growth of vesicular-arbuscular mycorrhizal fungi through soil. *Transactions of the British Mycological Society* 89, 429–435.

Smith, S.E. and Smith, F.A. (1990) Structure and function of the interfaces in biotrophic symbioses as they relate to nutrient transport. *New Phytologist* 114, 1–38.

Sylvia, D.M. (1992) Methods to quantify external hyphae of VAM fungi. *Methods in Microbiology* 24, 53–65.

7 Mycorrhizal Infection of Wild Oats: Parental Effects on Offspring Nutrient Dynamics, Growth and Reproduction

R. Koide* and X. Lu

Department of Biology, The Pennsylvania State University, University Park, PA 16802, USA

In many cases infection of roots by vesicular-arbuscular mycorrhizal fungi results in an increase in the uptake of phosphorus (P) from the soil. This can significantly alter the general nutrient, water and carbon economies of the host plant (Smith, 1980; Kucey and Paul, 1982; Koide, 1985; Allen and Allen, 1986). Much of the research on the mycorrhizal symbiosis has been concerned with effect of the fungus on plant growth or agronomic yield. Relatively little research has been performed on the role of the fungi in natural populations. In particular, although we have speculated much about the effect of infection on plant fitness, surprisingly few effects of infection on plant fitness have been documented.

Two important components of plant fitness include fecundity (seed production) and offspring vigour. We have previously shown that mycorrhizal infection can significantly increase seed production (Koide *et al.*, 1988; Bryla and Koide, 1990). The objective of this study was to determine whether mycorrhizal infection could influence the quality of the seeds produced and thus the offspring vigour. Variation in offspring vigour may determine the demographic patterns of mature vegetation (Harper, 1977; Cook, 1979). For example, individuals which germinate more rapidly or which grow faster may enjoy greater survivorship because they are larger (Solbrig, 1981; Weiner, 1988). They may also pre-empt a greater proportion of the available resources (Ross and Harper, 1972; Abul-Fatih and Bazzaz, 1979) thereby suppressing less competitive individuals and achieving greater fitness. In some cases, early seedling size may be correlated with eventual reproductive output (Edwards and Emara, 1969; Stanton, 1984; Roach, 1986).

Recently, Bolland and Paynter (1990) reported that for several legume species, increased seed P content was associated with increased vegetative dry matter and P contents of resultant plants. Austin (1966) demonstrated that seed yields were higher in those pea plants that were produced by seeds containing greater concentrations of P. Zhang *et al.* (1990) reported a good correlation between seed P

---

*Present address: Department of Horticulture, The Pennsylvania State University.

content and seedling root and shoot weight. Infection of plants by vesicular-arbuscular mycorrhizal fungi may enhance the uptake of P in many plant species and in many soil types (Smith, 1980). Mycorrhizal infection can thereby increase the P content of seeds (Jensen, 1983; Koide *et al.*, 1988, Bryla and Koide, 1990) and thus may have a substantial effect on offspring vigour.

Wild oats (*Avena fatua* L.) were used in this study. Mycorrhizal (M) infection of these oats by *Glomus intraradices* resulted in a significantly greater seed P content (for example in one experiment, M: 50 μg P seed^{-1}, non-mycorrhizal (NM): 32 μg P seed^{-1}). The difference in total P was largely due to a difference in the phytate P fraction. During germination of seeds in vermiculite, the P contained within the endosperm was mobilized into the developing seedling. As significantly more endosperm P was available in seeds produced by M parents compared with those produced by NM parents, significantly more P was allocated to the developing seedlings of M parents.

In a separate phosphorus-deficient soil experiment, associated with the superior phosphorus status of seedlings produced by M parents was a significantly enhanced rate of leaf expansion. Moreover, when grown in a phosphorus-deficient soil, offspring (all NM) produced by M parents had greater root surface and rhizosphere acid phosphatase, phytase and ATPase activities than those offspring (all NM) produced by NM plants. This was contrary to expectation since it is generally observed that phosphorus-deficient plants have higher activities of phosphatase and phytase.

When offspring plants (again, all NM) became reproductively mature the effects of mycorrhizal infection of the parent generation were still quite evident. For example, leaf area, shoot P concentration and P content, root N concentration and N content and root P content, and root:shoot ratio were all significantly greater for those offspring produced by M parents.

The reproductive characters of the offspring generation were also significantly affected by mycorrhizal infection of parents. The number of seeds per spikelet was greater for those offspring produced by M parents, but the weight of individuals seeds was significantly lower. The seeds produced by offspring of M parents also contained greater N concentrations, P concentrations and P contents than those produced by offspring of NM parents, suggesting that the effects of mycorrhizal infection of the first generation of plants might still be evident in mature third generation individuals.

The results of these experiments show that mycorrhizal infection of *Avena fatua* plants did have significant positive effects on offspring growth and reproduction that were associated with increases in seed P content. These effects were not consequences of different rates of seedling emergence as there were never significant effects of maternal treatment on emergence.

The P content of seeds is often correlated with the dry weight, P content and seed production of the resultant plants (Demirlicakmak and Kaufmann, 1963; Austin, 1966). Thus, differential maternal provisioning of seeds with P as seen here may have been involved in the effects seen in these studies. However, offspring from M parents produced seeds of higher nutrient content and lower weight than offspring from NM parents. This maternal effect on the quality of seeds produced by offspring is similar to the direct effect of mycorrhizal infection on the weight and P content of individual seeds produced by the infected generation. This is somewhat puzzling and is not easily explained by simple differential seed provisioning by the first generation.

These greenhouse and growth chamber experiments must be interpreted with some caution. Variability in resource availability from site to site in the field could diminish the importance of any variability in offspring vigour introduced by parental mycorrhizal infection. We suggest, however, that field work should be conducted to determine the importance of the parental effects observed in these experiments as it is possible that mycorrhizal infection influences plant population dynamics via effects on offspring vigour.

Acknowledgements

We thank the A.W. Mellon Foundation and the US National Science Foundation for financial support of this project. We also thank Mingguang Li, Carla Picardo, Robert Robichaux, Ian Sanders, Roger Schreiner and Margot Stanley for comments on earlier versions of this manuscript.

References

Abul-Fatih, H.A. and Bazzaz, F.A. (1979) The biology of *Ambrosia trifida* L. II. Germination, emergence, growth and survival. *New Phytologist* 83, 817–827.

Allen, E.B. and Allen, M.F. (1986) Water relations of xeric grasses in the field: Interactions of mycorrhizas and competition. *New Phytologist* 104, 559–571.

Austin, R.B. (1966) The influence of the phosphorus and nitrogen nutrition of pea plants on the growth of their progeny. *Plant and Soil* 24, 359–370.

Bolland, M.D.A. and Paynter, B.H. (1990) Increasing phosphorus concentration in seed of annual pasture legume species increases herbage and seed yields. *Plant and Soil* 125, 197–205.

Bryla, D.R. and Koide, R. (1990) Role of mycorrhizal infection in the growth and reproduction of wild vs. cultivated plants. II. Eight wild accessions and two cultivars of *Lycopersicon esculentum* Mill. *Oecologia* 84, 82–92.

Cook, R.E. (1979) Patterns of juvenile mortality and recruitment in plants. In: Solbrig, O.T., Jain, S., Johnson, G.B. and Raven, P.H. (eds), *Topics in Plant Population Biology*. Columbia University Press, New York, pp. 207–231.

Demirlicakmak, A. and Kaufmann, A.L. (1963) The influence of seed size and seed rate on yield and yield component of barley. *Canadian Journal of Plant Science* 43, 330–337.

Edwards, K.J.R. and Emara, Y.A. (1969) Variation in plant development within a population of *Lolium multiflorum*. *Heredity* 25, 179–184.

Harper, J.L. (1977) *Population Biology of Plants*. Academic Press, London.

Jensen, A. (1983) The effect of indigenous vesicular-arbuscular mycorrhizal fungi on nutrient uptake and growth of barley in two Danish soils. *Plant and Soil* 70, 155–163.

Koide, R. (1985) The effect of VA mycorrhizal infection and phosphorus status on sunflower hydraulic and stomatal properties. *Journal of Experimental Botany* 36, 1087–1098.

Koide, R., Li, M., Lewis, J. and Irby, C. (1988) Role of mycorrhizal infection in the growth and reproduction of wild vs. cultivated plants. I. Wild vs. cultivated oats. *Oecologia* 77, 537–543.

Kucey, R.M.N. and Paul, E.A. (1982) Carbon flow, photosynthesis, and N_2 fixation in mycorrhizal and nodulated Faba beans (*Vicia faba* L.). *Soil Biology and Biochemistry* 14, 407–412.

Roach, D.A. (1986) Variation in seed and seedling size in *Anthoxanthum odoratum*. *American Midland Naturalist* 117, 258–264.

Ross, M.A. and Harper, J.L. (1972) Occupation of biological space during seedling establishment. *Journal of Ecology* 60, 77–88.

Smith, S.E. (1980) Mycorrhizas of autotrophic higher plants. *Biological Reviews* 55, 475–510.

Solbrig, O.T. (1981) Studies on the population biology of the genus *Viola* II. The effect of plant size on fitness in *Viola sororia*. *Evolution* 35, 1080–1094.

Stanton, M.L. (1984) Seed variation in wild radish: effect of seed size on components of seedling and adult fitness. *Ecology* 65, 1105–1112.

Weiner, J. (1988) Variation in the performance of individuals in plant populations. In: Davy, A.J., Hutchings, M.J. and Watkinson, A.R. (eds), *Plant Population Ecology*. Blackwell Scientific Publications, Oxford.

Zhang, M., Nyborg, M. and McGill, W.B. (1990) Phosphorus concentration in barley (*Hordeum vulgare* L.) seed: influence on seedling growth and dry matter production. *Plant and Soil* 122, 79–83.

8 Mycorrhizas, Seed Size and Seedling Establishment in a Low Nutrient Environment

N. Allsopp and W.D. Stock

Department of Botany, University of Cape Town, Private Bag, Rondebosch 7700, South Africa

Introduction

Seed size is constrained by the conflicting demands that dispersal and establishment requirements make on the resources allocated to sexual reproduction (Fenner, 1985). Dispersal is enhanced by a decrease in seed size with parallel increases in seed number, while establishment is promoted by the production of energy and nutrient rich seeds which enable seedlings to achieve a critical size thereby enhancing resource acquisition and competitive ability. In natural environments seed size of a species represents a selective compromise and is a useful indicator of past environmental pressures on a species (Harper *et al.*, 1970; Haig, 1989).

Small seed size is often a characteristic of ephemeral plants while larger seeds are found among longer lived plants, especially those exposed to unfavourable environmental conditions for some of their early life (Salisbury, 1942; Baker, 1972). However there is a considerable range in seed sizes produced by perennial plants in any one environment which reflects different solutions to the problem of seedling establishment and dispersal in these environments.

In this chapter the relationship between the mycorrhizal condition and the range of seed size and nutrient content of perennial, sclerophyllous shrubs growing in the mediterranean climate region of the SW Cape, South Africa, are explored. The dominant vegetation type is a heathland known as fynbos. High species diversity and low nutrient, sandy, acid soils are characteristic of the fynbos. Seedling establishment occurs after fires during mild, rainy winters and spring, in an unshaded, bare soil environment.

Seed Size and Mycorrhizal Dependence

In contrast to other habitats where shade (Salisbury, 1942) or drought (Baker, 1972) are major factors threatening seedling survival, scarcity of soil nutrients may be the critical factor in this heathland environment. One solution to this problem is the

59

provisioning of large nutrient rich seeds. Another would be to rely on mycorrhizas to acquire scarce soil nutrients.

Sclerophyllous plants growing in the fynbos form a relatively homogeneous group with respect to life history and structural characteristics but display a range of response to mycorrhizas. The majority of perennial plant species form vesicular-arbuscular (VA) mycorrhizas (Lamont, 1983; Allsopp, unpublished data) and VA plants as a group may dominate the vegetation in the first 2–20 years following fire (Hoffman et al., 1987). However, plants forming ericoid mycorrhizas (Ericaceae, with approximately 600 species in the Cape heathlands) and plants forming no mycorrhizas are significant components of the vegetation. Ectomycorrhizas have not been found on any indigenous fynbos plants (Lamont, 1983; Allsopp, unpublished data). Non-mycorrhizal (NM) plants include members of the Proteaceae (sclerophyllous shrubs) and the Restionaceae (reed-like monocots), both of which may dominate communities, especially in mature vegetation (Kruger and Bigalke 1984; Hoffman et al., 1987). As the Restionaceae are monocotyledons and functionally dissimilar to the other groups they will be excluded from further consideration.

Members of the Proteaceae produce big seeds with a high nutrient content which enable large seedlings to establish independently of soil nutrients (especially phosphorus) (Stock et al., 1990). These plants produce an extensive root system and the NM Proteaceae produce specialized root systems known as cluster or proteoid roots which are very efficient at acquiring nutrients. Seedlings of other families may rely on mycorrhizas to acquire soil nutrients and it is hypothesized that mycorrhizal plants will produce smaller seeds than NM plants when the plants are perennials with similar structural and functional characteristics.

This hypothesis was tested by weighing and measuring the phosphorus (P) content of seeds of 46 perennial, sclerophyllous South African fynbos species. For all but the smallest seeds, seed mass and seed P content refer to the embryo mass and P content with the seed coat removed.

Seed size and P content are closely correlated ($r = 0.89$, $P < 0.0001$) (Fig. 8.1). Most of the largest seeded species are the NM Proteaceae followed by the VA Fabaceae (Fig. 8.1). Other VA species have smaller seeds while the Ericaceae, which are obligate ericoid mycorrhizal species, have the smallest seeds (Fig. 8.1). Thus for the dicotyledonous species there is a clear trend of increasing seed size and P content, with decreasing mycorrhizal dependency.

A consequence of smaller seed size of the mycorrhizal species is that they are probably dependent on becoming mycorrhizal in order to establish themselves. The corollary of having small seeds is that more seeds can be produced, which enhances dispersal to microsites suitable for establishment. It is not known if mycorrhizal inoculum is distributed evenly or patchily in fynbos environments, although VA infectivity is low (Berliner et al., 1989). If mycorrhizal inoculum is patchily distributed then wide seed dispersal should ensure that at least some seedlings will encounter suitable inoculum. However, evenly distributed inoculum may in itself have allowed for the selection of smaller seeds by mycorrhizal species. Seedling survival would be high despite poorly provisioned seeds, as long as seedlings have a chance of becoming mycorrhizal.

Large seeded Proteaceae are characterized by low seed set, and hence small

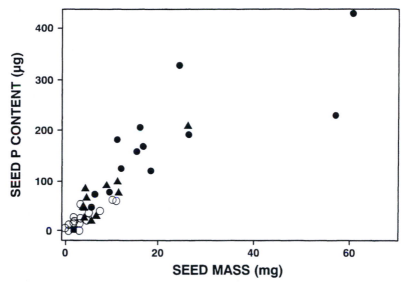

Fig. 8.1. The mass and phosporus content of seeds of perennial, sclerophyllous plants indigenous to the fynbos of the SW Cape, South Africa. Symbols: (●) Proteaceae, (▲) Fabaceae, (○) other VA forming families, (■) six species in the Ericaceae.

numbers of seed for dispersal (Esler *et al.*, 1989; Stock *et al.*, 1989), and in the absence of mycorrhizas are dependent on stored reserves for establishment. The relatively large seeds of the VA Fabaceae may be a consequence of their need to compete with other plants for establishment while waiting to become infected with both their microsymbionts.

Seed Size and Seedling Establishment of VA Species

Seedling establishment is a period when high dependency on mycorrhizas can be expected, especially by plants with smaller, poorly provisioned seeds (Koide, 1991). As fynbos VA plants exhibit a range of seed size, it is suggested that plants with smaller seeds with lower nutrient status (as measured by P content) will be more reliant on mycorrhizas for establishment. P was chosen as the nutrient of interest because of the close correlation between seedling P requirements and seed P content (Fenner and Lee, 1989). In addition, the low soil P content and its poor mobility will also make it among the most difficult resources for a seedling to acquire.

The prediction that among VA plants those with smaller seeds would exhibit greater mycorrhizal dependency was tested by growing 15 species with ($+$VA) or without ($-$VA) mycorrhizal inoculum on a low nutrient soil in pots in a greenhouse. After 40 weeks the plants were harvested, and plant mass and P content measured. VA mass response was measured by $(V_{+VA} - \bar{V}_{-VA})/\bar{V}_{-VA}$ where V is a variable such as mass and $\bar{V}$ is the mean (Bryla and Koide, 1990).

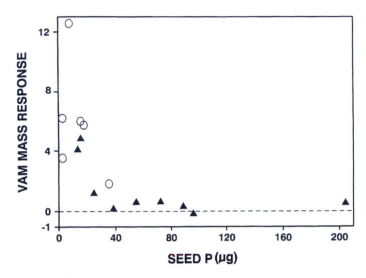

Fig. 8.2. Relative change in plant mass in response to vesicular-arbuscular mycorrhizal inoculation after 40 weeks' growth in a low nutrient soil in relation to the seed phosphorus content. Symbols: (▲) members of the Fabaceae, (○) members of other families.

If the VA mass response is plotted against seed P (Fig. 8.2), it can be seen that VA response decreases with increasing seed P. The difference in size of plants grown with and without VA inoculum decreases as seed P content increases. Thus larger seeds result in relatively larger seedlings than small seeds in the absence of mycorrhizas. Plants showing least response to VA are large seeded members of the Fabaceae while some of the smaller seeded Fabaceae are highly dependent on mycorrhizas for growth.

The prediction that smaller seeded plants are more dependent on VA inoculation holds. However, if the total P content of the $+VA$ and $-VA$ plants is compared, it is clear that while the $-VA$ plants of some species may be of similar size to the $+VA$ plants, the P content of $+VA$ plants is always significantly higher ($P < 0.05$) (Fig. 8.3). Very few of the $-VA$ plants were able to acquire P other than that provided by the seed (Fig. 8.3) and the nutrient use efficiency of the large seeded $-VA$ plants is high. These plants produce large seedlings independently of VA infection but their longer term survival may be jeopardized if they do not become mycorrhizal. Production of larger seeds by some VA species may be functionally analagous to the production of more numerous smaller seeds. Large seeds improve the chances of a seedling encountering VA temporally, while small seededness should provide seedlings with a spatial advantage in encountering VA during establishment.

As has already been shown, members of the Fabaceae produce larger seeds than other VA species. These plants are often the dominant species in the first few years following fire. This is probably due to their larger seed size which allows resultant seedlings to be large and highly competitive. However the consequences at the population level of supporting both the VA and nitrogen-fixing mutualisms

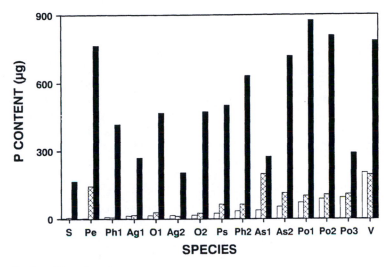

Fig. 8.3. The phosphorus content of seeds and 40-week-old seedlings grown with or without vesicular-arbuscular mycorrhizal (VA) inoculum on a low nutrient soil. Symbols: □ seed P, ▨ seedlings not inoculated with VA inoculum, ■ seedlings inoculated with VA inoculum. S *Staavia radiata* (Bruniaceae), Pe *Petalacte coronata* (Asteraceae), Ph1 *Phylica ericoides* (Rhamnaceae), Ag1 *Agathosma ovata* (Rutaceae), O1 *Otholobium hirtum* (Fabaceae), Ag2 *Agathosma gonaquensis* (Rutaceae), O2 *Otholobium fruticans* (Fabaceae), Ps *Psoralea pinnata* (Fabaceae), Ph2 *Phylica cephalantha* (Rhamnaceae), As1 *Aspalathus spinescens* (Fabaceae), As2 *Aspalathus linearis* (Fabaceae), Po1 *Podalyria sericea* (Fabaceae), Po2 *Podalyria cuneifolia* (Fabaceae), Po3 *Podalyria calyptrata* (Fabaceae), V *Virgilia oroboides* (Fabaceae).

in these low nutrient soils needs further investigation. An interesting feature of *Aspalathus* spp. is the formation of cluster roots in addition to vesicular-arbuscular mycorrhizas which may make them less dependent of the mycorrhizas for P uptake. Soil P taken up by − VA, *Aspalathus* spp. can be ascribed to the cluster roots.

Evidence has been presented to support the prediction that smaller seeded VA plants are more VA dependent as they require earlier infection for successful establishment. Even though larger seeded species produced relatively larger non-mycorrhizal seedling, all the species tested irrespective of seed size required VA infection for continued development in these low nutrient environments.

Conclusions

Among the sclerophyllous, perennial plants of a low nutrient mediterranean heath-land environment, seed size is related to the mycorrhizal habit. Conflict between seed size and dispersal is resolved according to the mycorrhizal dependency of the species. Obligate mycotrophic species have small seeds which ensure wide dispersal thus increasing the chance of encountering sites suitable for establishment. Less mycotrophic species invest in fewer, larger seeds with restricted dispersal which ensure establishment independently of the immediate environment in which the seed falls.

In different environments and among different functional guilds of plants, factors determining seedling establishment and seed size may vary. However, in view of the close link between mycorrhizas and the mineral nutrition of plants it is not unexpected that in a regularly disturbed, low nutrient environment the mycorrhizal status of the plants influences the size and provisioning of seeds.

References

Baker, H.J. (1972) Seed weight in relation to environmental conditions in California. *Ecology* 53, 997–1010.

Berliner, R., Mitchell, D.T. and Allsopp, N. (1989) The vesicular-arbuscular mycorrhizal infectivity of sandy soils in the south-western Cape, South Africa. *South African Journal of Botany* 55, 310–313.

Bryla, D.R. and Koide, R.T. (1990) Role of mycorrhizal infection on the growth and reproduction of wild vs. infected plants II. Eight wild accessions and two cultivars of *Lycopersicon esculentum* Mill. *Oecologia* 84, 82–92.

Esler, K.J., Cowling, R.M., Witkowski, E.T.F. and Mustart, P.J. (1989) Reproductive traits and accumulation of nitrogen and phosphorus during the development of fruit of *Protea compacta* R. Br. (calcifuge) and *Protea obtusifolia* Buck. ex Meisn. (calcicole). *New Phytologist* 112, 109–115.

Fenner, M. (1985) *Seed Ecology*. Chapman and Hall, London.

Fenner, M. and Lee, W.G. (1989) Growth of seedlings of pasture grasses and legumes deprived of single mineral nutrients. *Journal of Applied Ecology* 26, 223–232.

Haig, D. (1989) Seed size and adaptation. *Trends in Ecology and Evolution* 4, 145.

Harper, J.L., Lovell, P.H. and Moore, K.G. (1970) The shapes and sizes of seeds. *Annual Review of Ecology and Systematics* 1, 327–356.

Hoffman, M.T., Moll, E.J. and Boucher, C. (1987) Post-fire succession at Pella, a South African lowland fynbos site. *South African Journal of Botany* 53, 370–374.

Koide, R.T. (1991) Nutrient supply, nutrient demand and plant response to mycorrhizal infection. *New Phytologist* 117, 365–386.

Kruger, F.J. and Bigalke, R.C. (1984) Fire in fynbos. In: Booysen, P. de V. and Tainton, N.M. (eds), *Ecological Effects of Fire in South African Ecosystems*. Springer-Verlag, Berlin, pp. 67–114.

Lamont, B.B. (1983) Strategies for maximising nutrient uptake in two mediterranean ecosystems of low nutrient status. In: Kruger, F.J., Mitchell, D.T. and Jarvis, J.U.M. (eds), *Mediterranean-type Ecosystems: The Role of Nutrients*. Springer-Verlag, Berlin, pp. 246–273.

Salisbury, E.J. (1942) *The Reproductive Capacity of Plants*. Bell, London.

Stock W.D., Pate, J.S., Kuo, J. and Hansen, A.P. (1989) Resource control of seed set in *Banksia laricina* C. Gardner (Proteaceae). *Functional Ecology* 3, 453–460.

Stock, W.D., Pate, J.S. and Delfs, J. (1990) Influence of seed size and quality on seedling development under low nutrient conditions in five Australian and South African members of the Proteaceae. *Journal of Ecology* 78, 1005–1020.

9 The Nature of Fungal Species in Glomales (Zygomycetes)

J. Morton, M. Franke and G. Cloud

Division of Plant and Soil Sciences, 401 Brooks Hall, West Virginia University, Morgantown, WV 26506-6057 USA

Introduction

The discovery of over 150 species in Glomales, Zygomycetes (Morton and Benny, 1990) during the past 15 years represents just a small fraction of the diversity yet to be found world-wide. Difficulties have arisen in establishing criteria to distinguish isolates (defined here as parts of a fungal organism in a single rhizosphere sample) from species or one species from another. Mycorrhizologists have made the assumption that glomalean species are cohesive wholes, wherein properties of the isolate reflect the properties of all members of the species to which it belongs. This assumption has affected not only the design of experiments, but also explanations and generalizations concerning experimental results. A fundamental problem has been the lack of knowledge concerning what constitutes a species taxon.

Each organism found in nature can be recognized formally only if it is placed in the basal species category of the Linnaean classification system and assigned a Latin binomial name according to nomeclatoral conventions. This operation is based on measures of similarity and ordinarily does not require any knowledge of evolutionary pattern or process. Universality of Linnaean categories and a non-evolutionary basis for classification has fostered the erroneous impression that species of all organisms are the product of similar causal mechanisms and therefore are equivalent entities. But different groups of organisms possess different diversification mechanisms manifested as genetic, reproductive, morphological, developmental or ecological discontinuities. It is not surprising, therefore, that a wide range of species concepts have been formulated to account for prevalent mechanisms in each group (see Mishler and Donoghue, 1982 for a survey).

Morton (1990b) examined the 'species situation' for glomalean fungi and advocated a concept which included both phenetic and evolutionary considerations. However, his analysis did not resolve the nature of organismal variation. Moreover, he did not explicitly define grouping and ranking criteria for placement of organisms in a species taxon. In this chapter, we propose a species concept

65

which answers the following questions: What kinds of variation are important in recognizing a species taxon as a cohesive and discrete entity? What biological processes are responsible for origin and maintenance of that variation? How does this variation explain the role of species taxa in evolution of endomycorrhizal symbioses?

Nature of Variation

In the near-obligately asexual glomalean fungi, mutations in genetic information systems are the major source of new variation. Mutations need not, and often do not, correspond to phenotypic expression of that variation. Mutants simply are not expressed at one extreme or they produce multiplicative pleiotrophic effects at the other. Expressed variation usually is sorted by environmental selection pressures on growth and reproductive potentials. The phenotypic characters manifested from this sorting process are of two kinds. Characters *reversible* with changing environmental conditions show no evolutionary directionality. That is, they cannot be used to reconstruct historical events leading to origin of species and higher taxa because ancestral and advanced states are indistinguishable. Most of these characters are biochemical, behavioural, and physiological properties of an organism. Statistical analyses are required to sort out genetic causation from environmental effects because variation is visible as an unbroken continuum. Characters *irreversible* with changing conditions show evolutionary directionality and thus can be used to recognize discrete groupings in a phylogeny or genealogical hierarchy (formalized in Linnaean categories). These characters are considered to be *epigenetic* because they are integral components of developmental and differentiation pathways and are buffered against most genomic and environmental perturbations. As a result, they remain as discrete and stable units in space and time.

Organismal and Ecological Properties of Glomalean Fungi

Each fungal organism is an agglomeration of separate parts with diffuse organization and no discrete boundaries in space and time. It cannot be a single entity evolving as a cohesive whole. In glomalean fungi, an organism can be partitioned into two discrete modules based on: (1) nature, scope and stability of morphological diversity; (2) role in functional mycorrhizas; and (3) interactions with external environmental forces.

Interactor module

All somatic units of the fungal organism participate directly in a functional mycorrhiza and thus are interactors with the plant host, soil conditions, and other environmental forces. A filamentous hyphal network makes up a large portion of the module, which can grow indefinitely as long as carbon and nutrient resources are available in live roots. It consists of intraradical and extraradical components that differ in form and function. External hyphae differentiate into runner, absorptive, and infection components (Friese and Allen, 1991) and are capable of producing numerous spores and auxiliary cells (the latter only in Gigasporineae). Intraradical hyphae repeatedly form arbuscules and vesicles (the latter only in Glomineae). All suborganismic units, with the exception of spores, show little morphological diversification. The striking similarity between arbuscules in extant taxa and arbuscule-like structures in fossils from the Triassic (Stubblefield *et al.*, 1987) indicates homeostasis through geologic time. Arbuscules and other somatic units define only higher taxonomic groups (Morton and Benny, 1990) because historical and developmental constraints have conserved form and function as diversification occurred in the replicator module of fungal organisms (see below).

Despite constraints on morphological form and general function, considerable phenotypic plasticity in the interactor module still is possible in response to locally contingent conditions. Types of variation include differences in numerical abundance, size, branching architecture, longevity, and physiological interactions of each repeating unit of the interactor module (Smith and Gianinazzi-Pearson, 1988). Most of these characters are of little historical significance in defining taxonomic groups because they are labile with changing environmental conditions.

Organization of the interactor module prevents clear-cut distinctions between individuals and populations. Totipotency of filamentous hyphae provides a potential for mosaic genetic changes in different parts of the same organism. Somatic mutations can persist if they are conserved in replication, expressed in the phenotype, and favoured in selection processes. Some evidence that parts of an interactor module are genetically and physiologically heterogeneous was presented in a poster at this symposium (Rosendahl *et al.*, p. 400 this volume). The closest approximation of an *ecological* individual (as opposed to *genetic* individuals which are recognizable only with molecular probes) currently is an 'isolate', obtained from a defined sample volume as small as empirically practical.

Replicator module

Spores formed repeatedly from external (and sometimes internal) hyphae are not direct interactors in the symbiosis. Instead, a spore organizes the genomic information system of a fungal individual in a package that can survive environmental fluxes, disperse this information to a new location, and initiate formation of a new interactor module. Spores are the only components of glomalean fungi with enough diversifying morphological evolution to recognize taxonomic groups at the species level (Morton, 1988). Some characters such as shape, size, colour, hyphal features and wall thickness are reversible because they are expressed while spores are physiologically linked to the parent interactor module and subject to local selection pressures. These characters are realized as an overlapping continuum among genetically and geographically isolated fungal populations, even though they might appear discrete when only one or two populations of a putative species taxon are studied. Their value as historically informative characters is equivocal at this time.

Most stable diversification in glomalean species has occurred in differentiation of spore walls and subcellular components. We have measured the extent to which historical and developmental patterns constrain emergence of variation in subcellular spore characters of selected species isolates from all genera except *Sclerocystis*. Only a selected sampling of results is reported preliminarily here. We observed that phenotypically different elements (the various wall types defined by Walker (1983) and other workers) joined together only in certain combinations to form discrete complexes of independent origin (see Fig. 9.1 for examples in Gigasporaceae). Each complex appeared in an ordered and regular linear sequence that was causally linked: positional and other intrinsic properties of one stage constrained types of variation feasible in the next. Equally important, each complex was restricted to only certain positions in the differentiation sequence.

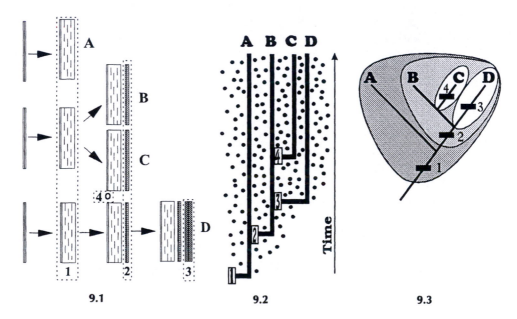

9.1   **9.2**   **9.3**

Fig. 9.1–3. Relationships in glomalean fungi between spore differentiation sequences in selected species, speciation patterns, and phylogeny reconstruction.
9.1. Sequence in differentiation of subcellular changes in spores of *Gigaspora margarita* Becker and Hall (A), *S. fulgida* Koske and Walker (B), *S. verrucosa* (Koske and Walker) Walker and Sanders (C), and a putative *S. calospora* (Nicol. and Gerd.) Walker and Sanders (D). See Morton (1988) for explanation of murographs. O = ornamentations. Each homologous character is bounded by a dashed rectangle and numbered 1–4. Adult spore wall organization does not match published descriptions; they are in the process of being redescribed.
9.2. An illustration of descendant species lineages splitting from ancestral ones. All spatially and genetically isolated interactor modules (large dots) are unified in a species lineage by the same spore differentiation pattern (from Fig. 9.1). They are not linked together in a phylogenetic pattern because most physiological and behavior characters are not stable enough to be hierarchically time-ordered.
9.3. Cladogram reconstructing pattern of common ancestry using numbers from Fig. 9.1 as shared derived characters. Different shaded regions encircle separate monophyletic groups.

Insertions early in differentiation, such as ornamentation patterns on structural wall elements or a unique transitory amorphous element discovered only in juvenile spores of *Scutellospora dipapillosa* did not alter subsequent differentiation sequences. Regulation of the sequence changed, however, in that timing of differentiation was compressed (heterochrony – see Gould (1977) for full treatment of this phenomenon) and spores with short or long sequences matured about the same time. The most surprising result of these studies was that many of the same phenomena observed in development of cellular, tissue and organ systems (Gould, 1977; Oster and Alberch, 1982) also occur at the subcellular level in spores of glomalean fungi. These characters are irreversible in different environments and thus can be used to reconstruct a phylogeny (Fig. 9.2). Despite asexual reproduction in disjunct interactor modules, epigenetic constraints on replicator variability clearly preserve historical continuity in ecologically different situations.

A Species Concept for Glomalean Fungi

Fossil evidence from geologic periods preceding the Triassic (Pirozynski and Dalpé, 1989), the arbuscule-like fossils from the Triassic (Stubblefield *et al.*, 1987), and the hypothesized cladistic pattern of character distribution in Glomaceae (Morton, 1990a) suggests that the *minimum* age for *all* species in Glomaceae is 250 million years. Intrinsic homeostatic properties of the fungal organisms and their ancient origin combine to impart long-lasting boundary conditions within which species evolved quite independent of other fungal groups (Morton, 1990b).

Species cohesiveness is assumed to be the result of dynamic interactive processes in most organisms, the most widely recognized one being gene flow (Mishler and Brandon, 1987). Near-obligate asexual reproductive strategy, however, precludes breeding relationships among interactor modules of glomalean fungi. Yet absence of gene flow does not hinder recognition of discrete species taxa. The key distinction here is the dichotomy in form and function between interactor and replicator modules. Only the latter is used to define species taxa, for very good reasons. Because spore morphological characters are developmentally constrained, particularly the subcellular elements formed during differentiation (Fig. 9.1), they are faithfully replicated in genetically, geographically and physiologically isolated interactor modules. All of these modules are recognized as members of the same species lineage (Fig. 9.2) because they produce the same spores as a result of common ancestry. Cladistic methods (Fig. 9.3) can be used to reconstruct patterns of shared ancestry based on these spore developmental sequences. They clearly are applicable to glomalean fungi (Morton, 1990a), despite problems associated with a faulty and somewhat scanty data base.

Species concepts are maximally informative if they specify historical groups that can be used to test hypotheses of genealogical relationships in other disciplines (e.g. biogeography, developmental biology, ecology, palaeontology, molecular biology). Past attempts to group glomalean organisms into a species taxon based on perceived similarities have proven equivocal because they were intuitive judgements rather than empirically-based deductions. Diagnostic characters were not analysed for their historical significance, an essential precondition if species taxa are to be biologically relevant entities rather than convenience groupings. Mishler and Brandon (1987) recognize two components in a complete species concept: (1) grouping criteria joining organisms from different locations in a historically unique taxon; and (2) ranking criteria to recognize the taxon as a species rather than place it in a higher or lower rank.

Grouping criteria

Characters used to group organisms into the same taxon must be homologous to be historically unique (derived from a single ancestral condition). Homology is validated only when a given character passes *both* tests of similarity and a test of phylogeny (Rieppel, 1988). Similarity based on correspondence in morphological form is inadequate for fungal characters because they are structurally and compositionally simple. Other tests are necessary, such as topographic correspondence (or similarity in adjacent structures) and shared developmental sequences (or the same beginnings and transformational changes of subcellular spore components – Fig. 9.1). Characters in different organisms which pass the test of similarity then are analysed cladistically. The most straightforward phylogenetic test is a minimum three-taxon comparison (to show that two of the three taxa are more closely related to each other than to the third). Relationships between these taxa can be ordered into monophyletic groups by outgroup analysis or the ontogenetic method (Morton, 1990a). Monophyletic groups are closed systems in that they include the ancestor in which a character originated and all descendants retaining that character (Mishler and Donoghue, 1982). In Fig. 9.2, for example, the two species C and D are separate monophyletic groups because each possesses a novel character and has no descendant lineages (new species have not evolved from them). At the next level of inclusiveness, B, C and D comprise a monophyletic group defined by appearance of character 2. Character 1 places all four taxa as a monophyletic group at the most general level of inclusiveness. From this analysis, all four characters are shown to be homologous, thus signifying historically unique changes delimiting real (rather than arbitrary) groups.

Ranking criteria

Monophyly is an essential criterion for grouping organisms according to unique historical events (homologies), but it cannot be used to rank these groups into the species category. If it were, then only terminal groups in a cladogram would be species taxa (monophyletic groups C and D in Fig. 9.2). Other monophyletic groups ((B, C, D) and (A, B, C, D)) would be placed in successively higher ranks because they include other groups (B and A, respectively). The problem with monophyly, in this case, is that taxa A and B go unrecognized as discrete entities. Cladistically, they are considered paraphyletic groups because they have no characters exclusively their own and they share novel characters (2 and 1, respectively) with descendant species. We consider them to be *time-extended* genealogical species (Griffiths, 1974) because they maintain only the ancestral characters and persist alongside diverging descendant species. This pattern of ancestor-descendant relationships (Fig. 9.2) is the result of the evolutionary autonomy between replicator and interactor modules. We hypothesize that character 2 emerged in spores of some totipotent subset of an organism belonging to species A. Because this new character was added on to an existing differentiation sequence (Fig. 9.1), no disruptions occurred in spore development or functions of the parent interactor module. The new species would proliferate if the variant survived, establish new individuals, and disperse into new habitats. These organizational features of a glomalean organism optimize ancestor-descendant coexistence following speciation events. Consequently, we predict that many fungal species taxa yet to be discovered will prove to be paraphyletic simply because extinction of all parts of a species is so improbable.

The most straightforward ranking criterion is to place the *smallest* diagnosable group that shares homologous *epigenetic* character(s) in a species taxon. The most predictable problem arising from this criterion is to rank organisms into a species which differ in morphological characters not adequately tested for their consistency and stability. For example, some isolates of *Glomus clarum* Nicolson & Schenck form only deep yellow spores, others are all white, and some are pigmented in a continuum between these extremes. *G. manihot* is grouped separately as a species mainly because all spores are yellow (Schenck *et al.*, 1984). These patterns change in descendant spores retrieved from successive propagative cycles (Morton, unpublished data), making colour unsuitable in grouping or ranking these particular individuals. The conclusion from this evaluation: isolates of *G. manihot* were erroneously grouped as distinct from those of *G. clarum* and all should be ranked together as a single species taxon (Morton, in preparation).

The interactor module does not possess the morphological diversity capable of distinguishing diagnosable groups as small as those determined by spore characters. Other characters, such as those obtained from molecular or physiological data sets, vary among isolates and have a high probability of undergoing transitions with changing local conditions (thus the unordered distribution of dots in Fig. 9.2). Conservative molecular characters may be found that group together two or more isolates if relatively invariant sequences of nuclear, ribosomal, or mitochondrial DNA can be found (e.g. Simon *et al.*, 1992). They are unlikely to be congruent with species concepts based on spore morphology, however, because of autonomy in function and mode of evolution of interactor and replicator modules.

In conclusion, we define species of glomalean fungi as: *the smallest assemblage of reproductively isolated individuals or populations diagnosed by epigenetic morphological or organizational properties of fungal spores that specify a unique genealogical origin and continuity of descent according to the criterion of monophyly.* Reconstruction of phylogeny based on this species concept is important to biologists interested in elucidating macroevolutionary phenomena (relationships between species and higher taxa). Each fungal isolate of a species has its own independent functional role, definable by ecological, molecular, or physiological properties of the interactor portion of the fungus in a mycorrhiza. Interactor modules of different species adapted to the same external environmental conditions are likely to be functionally similar, but be phylogenetically quite distant based on replicator characters. Ecologists, molecular biologists, biochemists and physiologists studying mycorrhizal interactions at the community level or below must focus on fungal isolates, even though these entities are not recognized formally in Linnaean systematics. They are the active units of evolutionary change and must receive formal recognition by adding supplemental information to a species name, usually as some type of coded system.

Acknowledgements

We thank Drs Rick Koske and Jane Gemma for constructive comments that aided in development of concepts. Funds to attend the conference were provided by National Science Foundation grant no. DIR-9015519.

References

Friese, C.F. and Allen, M.F. (1991) The spread of VA mycorrhizal fungal hyphae in the soil: inoculum types and external hyphal architecture. *Mycologia* 83, 409–418.

Gould, S.J. (1977) *Ontogeny and Phylogeny*. Belknap Press. Cambridge, MA.

Griffiths, G.C.D. (1974) On the foundations of biological systematics. *Acta Biotheoretica* 23, 85–131.

Mishler, B.D. and Brandon, R.N. (1987) Individuality, pluralism, and the phylogenetic species concept. *Biology and Philosophy* 2, 397–414.

Mishler, B.D. and Donoghue, M.J. (1982) Species concepts: a case for pluralism. *Systematic Zoology* 31, 491–503.

Morton, J.B. (1988) Taxonomy of VA mycorrhizal fungi: classification, nomenclature, and identification. *Mycotaxon* 32, 267–324.

Morton, J.B. (1990a) Evolutionary relationships among arbuscular mycorrhizal fungi in the Endogonaceae. *Mycologia* 82, 192–207.

Morton, J.B. (1990b) Species and clones of arbuscular mycorrhizal fungi (Glomales, Zygomycetes): their role in macro- and microevolutionary processes. *Mycotaxon* 37, 493–515.

Morton, J.B. and Benny, G.L. (1990) Revised classification of arbuscular mycorrhizal fungi (Zygomycetes): a new order, Glomales, two new suborders, Glomineae and Gigasporineae, and two new families, Acaulosporaceae and Gigasporaceae, with an emendation of Glomaceae. *Mycotaxon* 37, 471–491.

Oster, G. and Alberch, P. (1982) Evolution and bifurcation of developmental programs. *Evolution* 36, 444–459.

Pirozynski, K.A. and Dalpé, Y. (1989) Geological history of the Glomaceae with particular reference to mycorrhizal symbiosis. *Symbiosis* 7, 1–36.

Rieppel, O.C. (1988) *Fundamentals of Comparative Biology*. Birkhauser Verlag, Boston, Massachusetts, USA.

Schenck, N.C., Spain, J.L., Sieverding, E. and Howeler, R.H. (1984) Several new and unreported vesicular-arbuscular mycorrhizal fungi (Endogonaceae) from Colombia. *Mycologia* 76, 685–699.

Simon, L., Lalonde, M. and Bruns, T.D. (1992) Specific amplification of 18S fungal ribosomal genes from vesicular-arbuscular endomycorrhizal fungi colonizing roots. *Applied and Environmental Microbiology* 58, 291–295.

Smith, S.E. and Gianinazzi-Pearson, V. (1988) Physiological interactions between symbionts in vesicular-arbuscular mycorrhizal plants. *Annual Review of Plant Physiology and Molecular Biology* 39, 221–244.

Stubblefield, S.P., Taylor, T.N. and Trappe, J.M. (1987) Fossil mycorrhizae: a case for symbiosis. *Science* 237, 59–60.

Walker, C. (1983) Taxonomic concepts in the Endogonaceae: spore wall characteristics in species descriptions. *Mycotaxon* 18, 443–455.

Part Two

Ectomycorrhizas in Temperate and Boreal Forest Ecosystems

10 The Ecological Potential of the Ectomycorrhizal Mycelium

B. Söderström

Lund University, Department of Microbial Ecology, Helgonavägen 5, S-223 62 Lund, Sweden

Introduction

In order to understand the biology of the mycorrhizal symbiosis, a wide basic knowledge of the biology of both groups of organisms involved in the symbiosis is needed. Since the most spectacular results of the association involve the changed growth of the plant partner, most research interest has naturally been directed towards studies of effects on the host plant. However, during recent years there has been increasing interest in the biology of the fungal partner, in particular in the functioning of the external mycelium growing from the infected root into the surrounding soil. This has been manifested in a number of reviews (Read, 1985, 1989; Fogel, 1988). However, so far, rather little interest has been directed towards the interaction of the symbiosis with the surrounding soil environment. This contribution aims to elucidate some aspects of the relationship between the external ectomycorrhizal mycelium and the soil ecosystem.

Soil Carbon and Microbial Production

Primary production of boreal forests is generally considered to be mainly limited by low soil nutrient availability; the production rate can very often be increased considerably by adding one or more of the nutrients assumed to be lacking (Tamm, 1961). In Scandinavia, nitrogen application in particular has been proven to increase primary production (Tamm, 1961). This principle has been so well established that it has been considered to apply to biological production in the soil also. However, there are now indications that the saprophytic activity in many soils is limited not by inorganic nutrient availability but rather by energy deficiency. In a number of Swedish field sites Bååth *et al.* (1981) and Söderström *et al.* (1983) demonstrated decreased microbial activities after fertilization with nitrogen indicating that nitrogen is not a nutrient limiting growth of the saprophytic microorganisms in these soils. Bååth *et al.* (1978) found that whereas microbial biomass in microcosms did not increase in response to nitrogen additions a positive response was observed to glucose additions. Another example of soil carbon limitation is given by Norton and Firestone (1991). They estimated the metabolic status of the organisms by assessing the activity of the electron transport system using an artificial electron acceptor, INT. In the soil, 41% of the bacterial cells were active, while close to the young pine roots, which normally exude more carbon than do older roots, 57% were active. They concluded that 'potentially active microorganisms are present in relatively large quantities in both rhizosphere and non-rhizosphere soil but it is only in the C-rich environment adjacent to plant roots that these microorganisms have C substrate in excess of their requirement for maintenance'.

Most calculations on soil microbial energy demands have been based on microbial standing crop estimates. Recently, however, Bååth (1990) studied the growth rates of the bacteria in a sandy loam soil (4.4% organic matter) and in forest humus (73% organic matter) by successfully applying the [^{3}H]thymidine incorporation method to the soil system, a method ealier used mainly in aquatic ecosystems. He found that bacterial turnover times in these soils were between 50 and 100 h at 23°C. This is equivalent to a bacterial biomass turnover of 14–28 times per year at field temperature. The microorganisms in forest humus, which may have an organic matter content of 80–90%, thus appear to have a very low growth rate. This low growth rate is probably due to energy limitation since most organic material in a humus is highly recalcitrant and not easily available for microbial growth.

A further indication of soil carbon limitation is obtained by comparing microbial carbon demands with carbon inputs to the soil system. In Table 10.1 some estimates of soil microbial respiration in a Swedish pine forest are presented. The bacterial respiration estimate is based on a calculated bacterial biomass turnover (Bååth 1990) and the fungal respiration is calculated from measured active biomass (Söderström, 1979) and from a biomass respiration regression (Bååth and Söderström, 1988). These calculations, although approximate, are conservative. However, they give an indication of the possible relationship between microbial carbon demand and available carbon. As shown, 40–60% of the litter carbon input in this forest may be consumed by the soil microbial respiration alone.

Table 10.1. Estimated annual bacterial and fungal respiration rate and litter input at a 120-year-old pine forest in central Sweden (Jädraås). A carbon content of 40% and a growth efficiency of 60% of the microbial biomass is assumed.

Bacterial respiration*	25–50 gC m^{-2}
Annual fungal respiration†	50–75 gC m^{-2}
Estimated C litter input‡	200 gC m^{-2}

* Estimated from bacterial production using [^{3}H]thymidine incorporation in the soil and microscopical counts (Bååth, 1990).
† Estimated from live biomass estimates (Söderström, 1979) and biomass/respiration regression estimates (Bååth and Söderström, 1988).
‡ From Persson et al. (1980).

Available data thus strongly indicate that the saprophytic microorganisms in soil suffer from energy deficiency.

Carbon Export to Soil via Mycorrhizal Systems

The ectomycorrhizal fungi acquire most of their carbon from their plant partners (Harley and Smith, 1983). This carbon is used for sustaining the existing fungal biomass in the mycorrhizal root tips as well as the mycelial network in the soil, and also for producing new fungal biomass. In view of the calculations presented above, which indicate that there is a strong demand for carbon in the soil, the export of carbon to the soil via the mycorrhizal mycelial network, might be of considerable ecological significance. Few direct estimates of the carbon flow to the ectomycorrhizal mycelium are available in the literature. Söderström and Read (1987), using specially designed microcosms, found that the external mycelium of pine – *Suillus* mycorrhizas used up to 25% of the carbon assimilated. In a recent study by Norton *et al.* (1990) radioactively labelled carbon dioxide ($^{14}CO_2$) was fed to mycorrhizal ponderosa pine seedlings and around 30% of the radioactive carbon was found in the soil surrounding the roots, probably exported there by the mycelial network. No absolute measurements of assimilated carbon were made in this study, however. Similar experiments by Erland *et al.* (1991) demonstrated that about 5% of the originally added radioactive carbon was found in the surrounding soil during an 8-day incubation period. During the incubation an increasing proportion of the ^{14}C could not be accounted for and was probably respired. These laboratory studies thus indicate that 5–30% of the assimilated carbon may be used by the fungal partner for the production and sustenance of its external biomass.

The first field estimate of the amount of carbon needed by the fungal partner in ectomycorrhizal symbiosis was presented by Harley (1971) based on data given by Romell (1939) on fruitbody production in a Danish spruce forest. Harley estimated that '10% of the wood production was used by the fungus'. In the Pacific North West of America, Fogel and Hunt (1983) estimated the fungal demand to be 24% of the annual throughput and Vogt *et al.* (1982) calculated that 15% of the primary production was used by the fungi. All three estimates were calculated from biomass estimates in the field, either of fruitbodies, and/or mycorrhizal fine roots and soil fungal biomass. Similar data are also available from a Swedish pine forest from which a direct estimation of the assimilation rate is also available. In this particular forest, Linder and Axelsson (1982) estimated the annual carbon assimilation to be 5 800 kgC ha^{-1}. By use of soil fungal biomass data from Söderström (1979), fruitbody production data from Bååth and Söderström (1977), and fine root production data by Persson (1980) it is possible to estimate the total annual carbon demand of the fungi. If one assumes a 90% root infection rate, 40% of the dry weight of the infected fine roots is accounted for by fungal material, that the fungal production efficiency is 60% and that the carbon content of the fungal biomass is 40% of dry weight, the fungal carbon demand is calculated to be 830 kgC ha^{-1} yr^{-1}. According to such calculations 14% of the assimilated carbon is consumed by the fungal partner of the ectomycorrhizal systems.

If the calculations from the Swedish pine forest given above are correct, the mycorrhizal fungi may contribute up to 40% extra carbon to the soil system compared to the estimated litter input $(200 \, gC \, m^{-2} \, yr^{-1})$. This represents a significant proportion of the amount assimilated by the plants but it may also account for a significant proportion of the carbon potentially available for true saprotrophic life in the soil. The mycorrhizal carbon is distributed out into the soil via the external mycelial network. At present very little is known about how this carbon flow is regulated, whether the plant or the fungus is the most active partner in this regulation. Further, very little information is available on the fate of the carbon in the fungus. Obviously, it will be used for anabolic processes, but whether carbon compounds are leached from the fungus has not been studied.

One activity that has been studied in some detail during recent years is the production of proteolytic enzymes by mycorrhizal fungi. The ability of the fungi to attack proteinaceous material in the soil may allow the plant partner to have an indirect access to the organically bound nitrogen in the soil, although this has not yet been demonstrated to occur in the field. The nitrogen nutrition of the mycorrhizal symbiosis was recently reviewed by Read et al. (1989) and it seems clear that the constant supply of carbon will be advantageous to these fungi in any competitive situation in the soil. The proteolytic mycorrhizal fungi thus have the potential to play a significant role in the mineralizing processes in the soil.

The mycorrhizal mycelium is of course also a potential nutrient source for the soil saprophytes. However, very little is known about the longevity of the mycelium, or the extent to which it becomes available to the soil community as necromass. The factors which regulate mycorrhizal fungal biomass production are also still poorly understood. The large amounts of carbon exported to the soil via the mycorrhizal fungi and the poor knowledge of the fate of this carbon thus provide strong justification for continued research effort aimed at acquiring a better understanding of the dynamics of carbon flow to and through mycorrhizal mycelial systems.

The Mycosphere

The concept of the mycorhizosphere, that is the rhizosphere of a mycorrhizal root, is becoming well established (Fogel, 1988; Garbaye, 1991), and there are a number of studies demonstrating effects of the mycorhizosphere similar to those obtained in the rhizosphere. These include higher bacterial numbers, altered population structure, and even effects on particular defined species (Garbaye, 1991). However, there are fewer data on ecological or physiological effects. One example of an interesting interactive effect was demonstrated by Duponnois and Garbaye (1990) who showed direct growth stimulation or growth retardation of two mycorrhizal fungi by mycorhizosphere bacteria. These effects were only demonstrated in pure culture experiments and the authors were careful in the ecological interpretation of their data. However, the results clearly demonstrate the potential for interactions between fungi and bacteria.

Considering the amounts of carbon invested by plant hosts in their fungal partners, there is good reason to believe that a considerable proportion of the total soil fungal biomass is accounted for by mycorrhizal fungi. The concept of a mycorhizosphere could thus be widened into that of a 'mycosphere'. One example of such a mycosphere effect was demonstrated by Entry *et al.* (1991a) who studied the rate of needle decomposition in a Douglas fir forest in Oregon with respect to differences occurring in areas occupied by mycorrhizal mats produced by *Hysterangium setchellii*. The decomposition rate constant (k) for Douglas fir needle litter was 1.44 in the fungal mats compared to 1.3 in the surrounding soil and the authors speculated that this could be an effect of the much higher total microbial biomass (four times higher and with stronger seasonal fluctuations) in the mats; more carbon was available due to the mycorrhizal mycelium and thus a more active decomposing environment was created. Decomposition of pure cellulose was shown to be even more strongly affected by the ectomycorrhizal mats and the observed seasonal pattern was suggested to be due to changes in photosynthetic rate which in turn may have affected the amount of available carbon in the mats (Entry *et al.*, 1991b). The studies by Entry *et al.* (1991a, b) contradict the reports by Gadgil and Gadgil (1971, 1975) that mycorrhizal fungi may have a significant negative effect on litter decomposition rate. The latter results have proved difficult to reproduce and the hypothesis requires further testing in a range of different ecosystems. There seems to be little doubt that the ectomycorrhizal fungal mycelium can affect, as well as being affected by, the soil saprophytic organisms.

Concluding Remarks

The mycorrhizal symbiosis is extraordinarily widespread, it being present in almost all natural terrestrial ecosystems. To increase our understanding of the functional aspects of this symbiosis, a wider knowledge of the ecology of the mycorrhizal mycelium is needed. There seems to be no doubt that considerable amounts of carbon are exported to the soil ecosystem and that this energy input to the soil may have significant ecological effects. However, we need more information on the regulatory processes in the interface between the fungus and the plant partner. More detailed information is also needed about the interface between the mycorrhizal fungi and the soil ecosystem in order to evaluate possible effects of the interaction between the mycorrhizal mycelium and the true saprophytic microorganisms in the soil.

References

Bååth, E. (1990) Thymidine incorporation into soil bacteria. *Soil Biology and Biochemistry* 22, 803–810.

Bååth, E. and Söderström, B. (1977) An estimation of the annual production of basidio-mycete fruitbodies in a 120-year-old pine forest in Central Sweden. *Swedish Coniferous Forest Project. Uppsala. Internal Report* 67, 1–11.

Bååth, E. and Söderström, B. (1988) FDA-stained fungal mycelium and respiration rate in reinoculated sterile soil. *Soil Biology and Biochemistry* 20, 403–404.

Bååth, E., Lohm, U., Lundgren B., Rosswall, T., Söderström, B., Sohlenius, B. and Wiren, A. (1978) Effects of supply of nitrogen and carbon on development of soil organism populations and pine seedlings: a microcosm experiment. *Oikos* 31, 153–163.

Bååth, E., Lundgren B. and Söderström, B. (1981) Effects of nitrogen fertilization on the biomass and activity of fungi and bacteria in a podzolic soil. *Zentralblatt für Bakteriologie, Mikrobiologie und Hygiene*. I. Abteilung Originale C 2, 90–98.

Duponnois, R. and Garbaye, J. (1990) Some mechanisms involved in growth stimulation of ectomycorrhizal fungi by bacteria. *Canadian Journal of Botany* 68, 2148–2152.

Entry, J.A., Rose, C.L. and Cromack, K. Jr. (1991a) Litter decomposition and nutrient release in ectomycorrhizal mat soils of a Douglas fir ecosystem. *Soil Biology and Biochemistry* 23, 285–290.

Entry, J.A., Donnelly, P.K. and Cromack, K. Jr. (1991b) Influence of ectomycorrhizal mat soils on lignin and cellulose degradation. *Biology and Fertility of Soils* 11, 75–78.

Erland, S., Finlay, R. and Söderström, B. (1991) The influence of substate pH on carbon translocation in ectomycorrhizal and non-mycorrhizal pine seedlings. *New Phytologist* 119, 235–242.

Fogel, R. (1988) Interactions among soil biota in coniferous ecosystems. *Agriculture, Ecosystems and Environment* 24, 69–85.

Fogel, R. and Hunt, G. (1983) Contribution of mycorrhiza and soil fungi to nutrient cycling in a Douglas-fir ecosystem. *Canadian Journal of Forest Research* 13, 219–232.

Gadgil, R. and Gadgil, P. (1971) Mycorrhiza and litter decomposition. *Nature* 233, 133.

Gadgil, R. and Gadgil, P. (1975) Suppression of litter decomposition by mycorrhizal roots of *Pinus radiata*. *New Zealand Journal of Forest Science* 5, 33–41.

Garbaye, J. (1991) Biological interactions in the mycorhizosphere. *Experientia* 47, 370–375.

Harley, J.L. (1971) Fungi in ecosystems. *Journal of Ecology* 59, 653–668.

Harley, J.L. and Smith, S.E. (1983) *Mycorrhizal Symbiosis*. Academic Press, London and New York.

Linder, S. and Axelsson, B. (1982) Changes in carbon uptake and allocation patterns as a result of irrigation and fertilization in young *Pinus sylvestris* stand. In: Waring, R.H. (ed.), *Carbon Uptake and Allocation in Subalpine Ecosystems as a Key to Management*. Forest Research Laboratory, Oregon State University, pp. 38–44.

Norton, J.M. and Firestone, M.K. (1991) Metabolic status of bacteria and fungi in the rhizosphere of ponderosa pine seedlings. *Applied and Environmental Microbiology* 57, 1161–1167.

Norton, J.M., Smith, J.L. and Firestone, M.K. (1990) Carbon flow in the rhizosphere of Ponderosa pine seedlings. *Soil Biology and Biochemistry* 22, 449–455.

Persson, H. (1980) Death and replacement of fine roots in a mature Scots pine stand. In: Persson, T. (ed.), *Structure and Function of Northern Coniferous Forests – An Ecosystem Study*. Ecological Bulletin (Stockholm) 32, 251–260.

Persson, T., Bååth, E., Clarholm, M., Söderström, B. and Sohlenius, B. (1980) Trophic structure, biomass dynamics and carbon metabolism of soil organisms in a Scots pine forest. In: Persson, T. (ed.), *Structure and Function of Northern Coniferous Forests – An Ecosystem Study*. Ecological Bulletin (Stockholm) 32, 419–459.

Read, D.J. (1985) Structure and function of the vegetative mycelium of mycorrhizal roots. In: Jennings, D.H. and Rayner, A.D.M. (eds), *The Ecology and Physiology if the Fungal Mycelium*. Cambridge University Press, Cambridge, pp. 215–240.

Read, D.J. (1989) Ecological integration by mycorrhizal fungi. *Endocytobiosis* IV, 99–106.

Read, D.J., Leake, J.R. and Langdale, A.R. (1989) The nitrogen nutrition of mycorrhizal fungi and their host plants. In: Boddy, L., Marchant, R. and Read, D.J. (eds), *Nitrogen, Phosphorus and Sulphur Utilization by Fungi*. Cambridge University Press, Cambridge, pp. 181–204.

Romell, L.G. (1939) Barrskogens svampar och deras roll i skogens liv. *Svenska Skogs-vårdsföreningens Tidskrift* 37, 348–375.

Söderström, B. (1979) Seasonal fluctuations of active fungal biomass in the horizons of a podzolized pine forest soil in Central Sweden. *Soil Biology and Biochemistry* 11, 149–154.

Söderström, B. and Read, D.J. (1987) Respiratory activity of intact and excised ectomycorrhizal mycelial systems growing in unsterile soil. *Soil Biology and Biochemistry* 19, 231–236.

Söderström, B., Bååth, E. and Lundgren, B. (1983) Decrease in soil microbial activity and biomasses due to nitrogen amendments. *Canadian Journal of Microbiology* 29, 1500–1506.

Tamm, C.O. (1961) Nutrient uptake and growth after forest fertilization. In: *Proceedings of the VII International Congress of Soil Sciences*, Madison, USA, pp. 347–354.

Vogt, K.A., Grier, C.C., Meier, C.E. and Edmonds, R.L. (1982) Mycorrhizal role in net primary production and nutrient cycling in *Abies amabilis* ecosystems in western Washington. *Ecology* 63, 370–380.

11 Ectomycorrhizal Rhizomorphs: Organs of Contact*

R.J. Agerer

Institute for Systematic Botany, University of Munich, Menzingerstrasse 67, D-8000 München 19, Germany

Introduction

It has long been known that rhizomorphs of ectomycorrhizas and their mycelium are capable of uptake and transport of water and nutrients (e.g. Duddridge *et al.*, 1980; Finlay and Read, 1986; Finlay *et al.*, 1988; Kammerbauer *et al.*, 1989; Melin and Nilsson, 1950, 1953). As these structures provide contact between the ectomycorrhizal mantle and the soil, research on the intimate contact which exists at both ends of the rhizomorphs, that is, between the rhizomorph hyphae and soil particles at one end and between the rhizomorph hyphae and those of mantle at the other, is of importance in the study of nutrient uptake and transfer between the symbionts. Although rhizomorphs can be constructed in very different ways (Agerer, 1991a), they all have to facilitate; (1) efficient internal contact, (2) efficient contact with the soil, and (3) efficient contact with the mantle hyphae. Research into these areas has been initiated, and some preliminary results are presented.

Results and Discussion

Contact with soil

The type of contact of the hyphae with soil may be dependent upon the soil horizon in which the ectomycorrhizas grow. In the following, one example shows the contact of the hyphae with soil debris in the organic layer (*Sarcodon imbricatus*), the second in the mineral soil (*Hydnellum peckii*).

---

* Part XXXVIII of the series 'Studies on ectomycorrhiza'; part XXXVII: Brand, Ch. 18, in this volume.

Contact with soil debris

Ectomycorrhizas and rhizomorphs of *Sarcodon imbricatus* are often found in close contact with the excrement of earthworms (Agerer, 1991b). Anatomical studies revealed that those hyphae which are in close contact with these particles have considerably thickened walls. The distal ends of emanating hyphae can be thickened if in intimate contact with other debris (cf. Fig. 4 in Agerer, 1991b). The function of these cell-wall thickenings is not known. Pollen grains occurring in the organic layer have been found which were filled with repeatedly branching hyphae (Fig. 11.1d). This phenomenon has also been found with hyphae of *Thelephora terrestris* (Fig. 11.1c).

Contact with mineral soil

The ectomycorrhizas and rhizomorphs of *Hydnellum peckii* appear to occur preferably in mineral soil. Rhizomorphs and hyphal fans frequently envelope soil compartments that seem to be drier than the bulk soil (Agerer, in preparation). The hyphae in close contact with the mineral soil are heavily encrusted with soil grains, these are glued to gelatinous hyphal walls (Fig. 11.1b). The diameter of the hyphae can be increased up to four-fold. Sometimes mineral particles are completely enveloped by hyphae (Fig. 11.1a).

The two examples show that different methods of contact with the soil are possible and that this may be dependent upon the fungal species and the soil

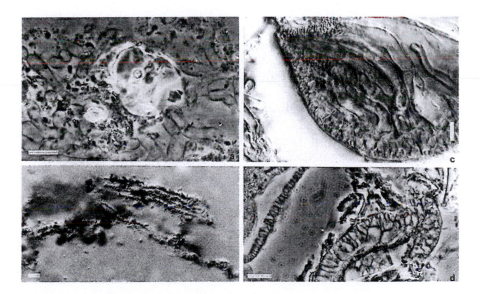

Fig. 11.1. Interaction of hyphae with soil debris. **a.** Mineral soil particle surrounded by hyphae (*Hydnellum peckii*, RA 11530). **b.** Mineral soil particles glueing to hyphae (*Hydnellum peckii*, RA 11530). **c.** Hyphae within a pollen grain (*Thelephora terrestris*, RA 10937). **d.** Hyphae within a pollen grain (*Sarcodon imbricatus*, RA 10916). Bars equal 10 μm; for further explanation see text.

horizon in which their ectomycorrhiza occur. Species-specific characteristics (e.g. wall-thickenings, or gelatinous hyphal walls) may influence how intimate the contact can be, probably also the kind of substances that can be exploited and the efficiency of the mycorrhizas in uptake and transfer of nutrients.

Contact with foreign fungal hyphae

Ectomycorrhizal hyphae apparently have the capability to envelop foreign hyphae with coils (see Fig. 5 in Agerer, 1991b). Sometimes hyphae of an ectomycorrhizal fungus can grow within rhizomorphs of another ectomycorrhizal fungus (Agerer, 1990, 1991d; Brand, Chapter 18 this volume). To date it is unknown what benefit the ectomycorrhizal fungi have from this intimate contact with other fungi.

Internal anatomical contact

The hyphae within rhizomorphs form close contact by several different mechanisms. These, can be seen separately or in combination. The hyphae can be glued together by gelatinous walls, they can be strongly intertwined, they can form backwardly growing ramifications, and anastomoses can bridge neighbouring hyphae (cf. various plates in Agerer, 1987–1991). Some observations on the two last topics are presented below.

Anastomoses

Anastomoses can be found as long hyphal bridges or as short regions of contact (Agerer, 1987–1991, 1991a). In ectomycorrhizal rhizomorphs of many species these anastomoses remain open but in others they can be secondarily closed by the formation of clamps or of simple septa (Agerer, 1987–1991, 1991a).

Such a secondary delimitation of two hyhae which have been bridged by anastomoses at first sight seems to be disadvantageous, but it is well known that the septa of higher fungi are characterized by a central pore, providing cyto-

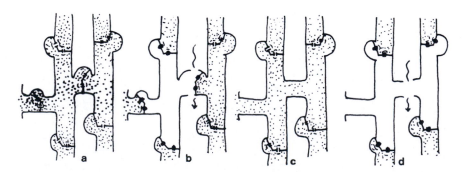

Fig. 11.2. Schematic drawings of hyphae connected by clamped (**a–b**) and open anastomoses (**c–d**). For explanation see text.

plasmatic continuity and pathways at least for solutions and even for organelles (Beckett *et al.*, 1974). Thus, through septate anastomoses transfer of substances can be maintained (Fig. 11.2a), though this is likely to be at a lower rate than through aseptate anastomoses. Septal pores can be plugged following the death or disruption of the cells (Beckett *et al.*, 1974). Consequently, disrupted anastomoses can be plugged preventing cytoplasmatic degeneration of the intact side of the anastomoses (Fig. 11.2b, d). After disruption of those anastomoses, which do not possess septa (Fig. 11.2c), both neighbouring cells would degenerate (Fig. 11.2d). Rhizomorphs with hyphae frequently interconnected by simple anastomoses probably lose a greater part of their longitudinal transport system if anastomoses are damaged (Fig. 11.2d).

Accordingly, anastomoses secondarily closed by septa, seem to have an ecological and functional advantage in comparison to open anastomoses if rhizomorphs and their hyphae are moved within the soil either during drying events or by the activity of soil animals. Interestingly, to date, closed anastomoses have been found more often in loosely woven than in compact rhizomorphs (cf. Agerer, 1987–1991). In the latter the hyphae are glued together or tightly intertwined.

Backwardly growing ramifications

Some ectomycorrhizal rhizomorphs are characterized by backwardly growing ramifications (Agerer, 1988, 1991a, b, c). A complex system can be formed by forwardly and backwardly branching hyphae, which can best be demonstrated diagrammatically (Agerer, 1991c). This mode of ramification could, in theory at least, have four different functions.

1. Facilitating rhizomorphal interconnections: when rhizmorphs which grow in the same main direction or towards one another meet, backwardly growing hyphae can establish a more intensive contact between the rhizomorphs. This would create a more mechanically resistant system (Fig. 11.3a–b) than if only forward growth of side branches occurred (Fig. 11.3c–d).
2. To achieve tight interconnections between hyphae within rhizomorphs: Mathematical models could be used to examine this hypothesis with respect to statics, viz. mechanical resistance of such a system during stretching events.
3. Formation of rhizomorphs only when necessary: Studies on the development of rhizomorphs have elucidated (Read *et al.*, 1985) that at the beginning of their formation hyphal fans are built, and later followed by the formation of thicker rhizomorphs. This is accompanied by the disappearance of hyphal fans. The first hyphae growing into the substrate, the leading hyphae, are later subsequently surrounded by frequently thinner backwardly as well as forwardly growing hyphae (Agerer, 1988, 1991a), thus, at first producing thin rhizomorphs within the broad hyphal front. Finlay and Read (1986) argue 'this intensive hyphal development may be associated with selective exploitation'. By the formation of backwardly growing hyphae rhizomorphs could be formed preferentially where selective exploitation would be advantageous, using the soil as signal. Such behaviour would provide economies of energy investment.

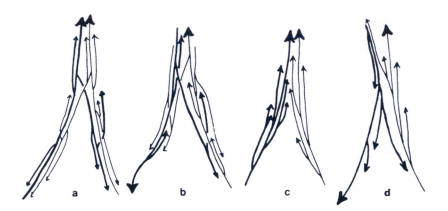

Fig. 11.3. Schematic drawings of hyphal growth (arrows) within meeting rhizomorphs (different thickness of lines indicates the hyphae of the two different rhizomorphs). **a–b** including backwardly growing branches. **c–d** backwardly growing branches lacking. **a, c** main direction of rhizomorphs identical (large arrowheads). **b, d** main direction of rhizomorphs opposed (large arrowheads). For further explanation see text.

4. Facilitating bidirectional transport: Buller (1933) has shown cytoplasmic streaming from older parts of hyphae towards the growing tips in both Ascomycetes and Basidiomycetes. The flow was always towards the hyphal tip without changes in direction and no bidirectional flow could be detected within the same hypha. If there is a bidirectional cytoplasmatic transport within rhizomorphs, which is assumed for the transport of soil derived solutes and carbohydrates (Finlay and Read, 1986), then the constant flow of cytoplasm towards hyphal tips may result both in proximal and distal transport within the same rhizomorph (Fig. 11.3a–b). Backwards growing hyphae could be appropriate to transport phosphate from the growing front to the mycorrhizas, while forward growing hyphae facilitate transport of carbohydrates from the mycorrhizas towards the growing hyphal front. The scenario becomes even more complex if Buller's (1933) observation, that a reversal flow can occur after the formation of new anastomoses and after cessation of hyphal growth, is correct. Research is required on the cytoplasmic streaming behaviour of individual hyphae within rhizomorphs to verify the outlined hypothesis.

Contact with mantle

Ectomycorrhizas that possess highly organized rhizomorphs with vessel-like central hyphae have not been seen to have such hyphae in the mantle (cf. Agerer, 1987). Therefore the question of an efficient transfer of solutes from these hyphae to those of the mantle is of interest. The transfer must be managed without damming-up of the transported substances or the advantage of the vessel-like hyphae will be reduced. Two possible answers to the problem are now known, in both cases exchange is apparently distributed over a large mantle area by repeatedly branched hyphae.

Contact with outer mantle layers

The vessel-like hyphae of the highly organized rhizomorphs of *Suillus bovinus* ramify delta-like at the transition zone to the mantle into several backwardly growing branches, which intertwine with the mantle hyphae (cf. Fig. 11 in Agerer, 1990). Thus, as in the smaller branches in a river delta, a large volume of material can be distributed by several small hyphal pathways over a large area of the ectomycorrhizal mantle.

Contact with inner mantle layers

The central hyphae in the rhizomorphs of *Sarcodon imbricatus*, are only slightly thicker than the marginal hyphae (Agerer, 1991b). They also repeatedly ramify proximally. The ultimate branches of these hyphae grow within the inner mantle layers and lie in close contact with the outer root surface (cf. Fig. 1e and Fig. 3c in Agerer, 1991b). In addition, some mantle hyphae apparently grasp the rhizomorphal hyphae by anastomoses which may enable them to take over the transported substances (cf. Fig. 3c in Agerer, 1991b), and to distribute the solutes over a large mantle area.

Conclusions

It is likely that there are several kinds of species-specific and probably ecologically and physiologically important contacts between rhizomorphs and soil. The internal organization of the rhizomorphs with respect to hyphal differentiation and with respect to their physical relation can reveal some hints for their function. Rhizomorphs may prevent damming-up of transported substances at the mantle by delta-like branching and intimate intertwining with the mantle hyphae. Additional investigations are urgently needed.

Acknowledgements

Critical comments on the manuscript by Dr Andrew Taylor are very much appreciated. The researches were financially supported by 'Deutsche Forschungsgemeinschaft'.

References

Agerer, R. (1987) *Colour Atlas of Ectomycorrhizae*. Einhorn-Verlag, Schwäbisch Gmünd.
Agerer, R. (1988) Studies on ectomycorrhizae XVII – The ontogeny of ectomycorrhizal rhizomorphs of *Paxillus involutus* and *Thelephora terrestris* (Basidiomycetes). *Nova Hedwigia* 47, 311–334.
Agerer, R. (1990) Studies on ectomycorrhizae XXIV – Ectomycorrhizae of *Chroogomphus*

helveticus and *C. rutilus* (Gomphidiaceae, Basidiomycetes) and their relationship to those of *Suillus* and *Rhizopogon*. *Nova Hedwigia* 50 (1-2), 1-63.

Agerer, R. (1991a) Characterization of ectomycorrhizae. *Methods in Microbiology* 23, 25-73. Academic Press, London.

Agerer, R. (1991b) Ectomycorrhizae of *Sarcodon imbricatus* on Norway spruce and their chlamydospores. *Mycorrhiza* 1, 21-30.

Agerer, R. (1991c) Comparison of the ontogeny of hyphal and rhizoid strands of *Pisolithus tinctorius* and *Polytrichum juniperinum*. *Cryptogamic Botany* 23, 85-92.

Agerer, R. (1991d) Studies on ectomycorrhize XXXIV – Mycorrhizae of *Gomphidius glutinosus* and of *G. roseus* with some remarks on Gomphidiaceae (Basidiomycetes). *Nova Hedwigia* 53 (1-2), 127-170.

Beckett, A., Heath, I.B. and McLaughlin, D.J. (1974) *An Atlas of Fungal Ultrastructure.* Longman, London.

Brand, F. (1991) Ektomykorrhizen an *Fagus sylvatica*. Charakterisierung und Identifizierung, ökologische Kennzeichnung und unsterile Kultivierung. *Libri Botanici* 2, 1-228.

Buller, A.H.R. (1933) *Researches on Fungi V.* Longmans, Green and Co. London.

Duddridge, J.A., Malibari, A. and Read, D.J. (1980) Structure and function of mycorrhizal rhizomorphs with special reference to their role in water transport. *Nature* 287, 834-836.

Finlay, R.D. and Read, D.J. (1986) The structure and function of vegetative mycelium of ectomycorrhizal plants II. The uptake and distribution of phosphorus by mycelial strands interconnecting host plants. *New Phytologist* 103, 157-165.

Finlay, R.D., Ek, H., Odham, G. and Söderström, B. (1988) Mycelial uptake, translocation and assimilation of nitrogen from 15N-labelled ammonium by *Pinus sylvestris* plants infected with four different ectomycorrhizal fungi. *New Phytologist* 110, 59-66.

Kammerbauer, H., Agerer, R. and Sandermann, H. Jr. (1989) Studies on ectomycorrhiza XXII – Mycorrhizal rhizomorphs of *Thelephora terrestris* and *Pisolithus tinctorius* in association with Norway spruce (*Picea abies*): formation *in vitro* and translocation of phosphate. *Trees* 3, 78-84.

Melin, E. and Nilsson, H. (1950) Transport of radioactive phosphorus to pine seedlings by means of mycorrhizal hyphae. *Physiologia Plantarum* 3, 88-92.

Melin, E. and Nilsson, H. (1953) Transfer of labelled nitrogen from glutamic acid to pine seedlings through the mycelium of *Boletus variegatus* (Sw.) Fr. *Nature* 171, 134.

Read, D.J., Francis, R. and Finlay, R.D. (1985) Mycorrhizal mycelia and nutrient cycling in plant communities. In: *Ecological Interactions in Soil*. Special Publications British Mycological Society 4, 193-217.

12 Uptake and Translocation of Nutrients by Ectomycorrhizal Fungal Mycelia

R.D. Finlay

Department of Microbial Ecology, University of Lund, Ecology Building, Helgonavägen 15, S-223 62 Lund, Sweden

Introduction

The importance of ectomycorrhizal fungal mycelia in translocating nutrients to their host plants forms a central axiom underlying many mycorrhizal studies, yet there have been few direct demonstrations of this process. Study of the mycelial phase is complicated by its delicate nature and inaccessibility, and quantitative estimates of nutrient fluxes are so far available for only a small number of elements. Different mechanisms and cellular pathways of translocation have been postulated but the relative importance of these under different conditions has yet to be determined. An extensive review of the subject is not possible here, instead some central problems are outlined and some new data are presented. Emphasis is placed on translocation of nutrients from fungus to host, and on nitrogen compounds in particular, but translocation of carbon compounds in the opposite direction will also be briefly considered.

Measurements of Mycelial Translocation

Much of the most precise data concerning translocation in the fungi has come from studies of mycorrhizal associations. Such systems offer ideal models in which to study translocation in that the plant provides a natural supply of the energy-rich carbon compounds necessary to support fungal growth and at the same time a strong, discrete sink for nutrients translocated in the reverse direction. Despite an extensive literature relating to the improved nutrient acquisition of mycorrhizal plants and a number of qualitative demonstrations of mycelial translocation there are few quantitative estimates of nutrient fluxes, only limited information about the form in which nutrients are translocated, and conflicting opinions about the relative importance of different mechanisms and cellular pathways.

    The first direct evidence of nutrient translocation through ectomycorrhizal mycelia was provided by the pioneering studies of Melin and Nilsson using stable and radioactive isotopes. In these experiments mycelial uptake and translocation of ^{32}P, ^{15}N, ^{45}Ca and ^{22}Na to pine seedlings of different species were demonstrated under sterile conditions using a range of mycorrhizal associations (see Melin et al., 1958 and references therein). The methods used were not suitable for quantitative measurement of fluxes or translocation velocity however, and no detailed information was given about the distances over which translocation occurred. Later experiments by Skinner and Bowen (1974) demonstrated mycelial translocation of ^{32}P over 12 cm in the field but only 12 mm in the laboratory. Further results of a more quantitative nature have since been obtained using laboratory observation chambers, developed by Read and his co-workers, in which different ectomycorrhizal associations are grown in thin layers of non-sterile peat or forest soil between sheets of transparent plastic. Using such systems, translocation of ^{14}C (Finlay and Read, 1986a), ^{32}P (Finlay and Read, 1986b; Finlay, 1989) and ^{15}N (Finlay et al., 1988, 1989) has been studied in a range of experiments using different ectomycorrhizal associations and shown to occur over distances of up to 40 cm, the distance being limited only by the physical size of the growth chamber. In recent experiments using similar systems (Finlay et al., unpublished) translocation of ^{32}P, ^{86}Rb (K) and ^{45}Ca was investigated in ectomycorrhizal systems of Fagus sylvatica infected with Paxillus involutus. Mycelial fluxes of these elements were calculated and are presented in Table 12.1, together with those calculated from earlier experiments using associations of Pinus and Larix species. The fluxes of P and K are similar to those measured for P in the earlier experiments but the values measured for Ca were almost two orders of magnitude lower.

Measurement of nutrient fluxes is complicated by a number of factors, the principal one being estimation of the cross-sectional area through which translocation occurs. Estimation of this is relatively easy in mycorrhizal species that show a high degree of mycelial differentiation, forming macroscopic structures such as strands, but more difficult in species that display diffuse mycelial gowth. In species forming mycelial strands with a high degree of internal differentiation there is the additional problem of estimating the relative cross-sectional areas of large, empty vessel hyphae and smaller, densely cytoplasmic hyphae and of making assumptions about the cellular pathway involved. The situation is complicated still further by

Table 12.1. Nutrient fluxes through mycelial strands estimated in different intact ectomycorrhizal associations using radioisotopes.

Plant	Fungus	Isotope	Flux
Pinus sylvestris Pinus contorta	Suillus bovinus*	^{32}P	$1.8–11.7 \times 10^{-10}$ mol P cm^{-2} s^{-1}
Larix eurolepis	Boletinus cavipes†	^{32}P	$13–153 \times 10^{-10}$ mol P cm^{-2} s^{-1}
Fagus sylvatica	Paxillus involutus‡	^{32}P	$36–47 \times 10^{-10}$ mol P cm^{-2} s^{-1}
Fagus sylvatica	Paxillus involutus‡	^{86}Rb	$16–28 \times 10^{-10}$ mol K cm^{-2} s^{-1}
Fagus sylvatica	Paxillus involutus‡	^{45}Ca	$0.3–0.5 \times 10^{-10}$ mol Ca cm^{-2} s^{-1}

Figures assume a cross-sectional area of 2.35×10^{-6} cm^2 per mycelial strand. * Finlay and Read (1986b); † calculated from data presented in Finlay (1989); ‡ Finlay et al. unpublished.

the possible existence of a lag phase during ^{32}P uptake (Finlay and Read, 1986b), which may lead to underestimation of flux values. Equilibration of the added label with unlabelled phosphorus already present in the translocation pathway may be important in this respect. Although the latter pool has been considered to be small (see Harley and Smith, 1983) discrimination between these two pools should be possible by labelling with ^{33}P before supplying ^{32}P, since the former isotope has a much lower emission energy than the latter. Continuous, non-destructive monitoring of isotope distribution using a β-scanner is also likely to permit more accurate investigation of the translocation dynamics. These methods are currently being developed in our laboratory. Finally, consideration must also be paid to the way in which nutrients are supplied, since in differentiated mycelia the primary absorptive function is probably fulfilled by the young mycelial margin while mature mycelial strands represent residual structures responsible mainly for long distance nutrient translocation from sites of absorption. Loading of nutrients into these two structures will thus be rather different, and supplying nutrients directly to the cut ends of single strands suffers the additional disadvantage that collapse of hyphal elements may occur. The estimates for ^{32}P presented above are similar to those obtained for vesicular-arbuscular mycorrhizal mycelium, in which a translocation mechanism based on cytoplasmic streaming has been suggested (Cooper and Tinker, 1981). Possible mechanisms of translocation in ectomycorrhizal mycelium are discussed below.

Mechanisms and Cellular Pathways of Translocation

The relative importance of different cellular pathways and mechanisms of translocation through ectomycorrhizal mycelia remains unclear. Possible mechanisms include diffusion, cytoplasmic streaming or cyclosis, and bulk flow of solution along turgor gradients, driven by pressure at the source or tension generated by evaporation at the sink. Perhaps the clearest evidence is that which exists for ^{14}C compounds translocated following the assimilation of $^{14}CO_2$ by host plants. Experiments by Söderström *et al.* (1988) suggested that carbohydrates might be translocated predominantly as mannitol and arabitol and stored as trehalose. Microautoradiographic analysis of the distribution of ^{14}C-labelled carbohydrates within mature mycelial strands of *Suillus bovinus* by Duddridge *et al.* (1988) revealed that label was mainly localized in the more densely cytoplasmic hyphae and within these predominantly associated with the cytoplasm rather than the vacuoles. Measurements by Finlay and Read (1986a) revealed an acropetal translocation velocity in excess of $20\,cm\,h^{-1}$ for ^{14}C-labelled compounds through the mycelium of *Suillus bovinus* and this is similar to rates of transport measured by Brownlee and Jennings (1982) in *Serpula lacrymans*. On the basis of the above observations it seems likely that, as in *Serpula*, movement of carbohydrates through ectomycorrhizal mycelia may take place by pressure driven bulk flow along a gradient of hydrostatic pressure generated by high internal solute concentrations. Active hyphal growth provides the carbon sink and assimilates move to the expanding mycelial front.

Translocation of tritiated water (^3H_2O) through the mycorrhizal mycelium of *Suillus bovinus* was demonstrated by Duddridge *et al.* 1980 who suggested that movement of water towards the plant probably takes place apoplastically through the central empty vessel hyphae. This is to be expected on structural grounds since the large diameters (6–20 μm) and absence of cross walls in mature vessel hyphae would confer high hydraulic conductivity. Vessel hyphae appear not to extend into the fungal sheath and it is possible that movement through the symplast may be more important in this region but it has been shown that there is no significant effect on hydraulic conductivity (Sands *et al.*, 1982). Recent experiments have also demonstrated the existence of an apoplastic transport pathway through the fungal sheath of ectomycorrhizal roots of *Pinus sylvestris* infected by *Suillus bovinus* using electron dense lanthanum nitrate and the dye sulphorhodamine G. So far direct visual evidence of bulk water flow through vessel hyphae is still lacking. Injection of dyes into vessel hyphae may provide useful direct evidence of the existence of a functional apoplastic pathway in mature mycelial strands. Mechanisms and pathways of carbohydrate and water movement so far postulated (Read, 1984) are highly analogous to transport through the vascular systems of plants. A turgor gradient is maintained through tension generated by evapotranspiration of the host plant and water moves apoplastically through large, empty vessel hyphae, comparable with the xylem vessels. Assimilates move in the opposite direction through densely cytoplasmic hyphae which surround the vessels and have thus been likened to phloem sieve tubes. The proposed mechanism is attractive in explaining bi-directional movement of water and carbon compounds through differentiated mycelial strands. However Jennings (1987) has pointed out that there is no *a priori* reason why solution flow should not occur in both directions within one hypha and argued that it is equally plausible that bulk flow may occur in the central hyphal volume with simultaneous cytoplasmic streaming, brought about by contractile elements, in the peripheral cytoplasm. Clearly further investigation of different possible translocation mechanisms is still required, particularly with respect to those species which show a low degee of mycelial differentiation.

It was originally proposed that dissolved nutrients such as phosphorus might move by bulk flow through the apoplastic pathway. No microautoradiographic evidence of this is available and Finlay and Read (1986b) suggested that ^{32}P movement through mycelial strands of *Suillus bovinus* occurred along a symplastic pathway. Fluxes of phosphorus measured in these experiments were similar to those measured in association with symplastic flow through VA mycorrhizal hyphae (Cooper and Tinker, 1981) and were unaffected by reducing transpiration of the host plants. There was some movement of label towards growing hyphal tips, in the opposite direction to that expected if transport were dominated by evapotranspirational effects, and it was concluded that movement of P might occur primarily by cytoplasmic streaming involving cyclosis. Inhibition of phosphorus movement by cytochalasin B in VA mycorrhizal fungi (Cooper and Tinker, 1981) is consistent with the operation of a contractile mechanism involving actin filaments but Harley and Smith (1983) have cautioned against extensive application of this substance to fungal cultures since it may interfere with phosphorus uptake as well as actin polymerization. Further experiments should clearly be performed using ectomycorrhizal systems in which cytoplasmic streaming inhibitors are applied at localized points in the translocation pathway, well-removed from the site of absorption. Changes in host transpiration influence both bulk flow and cytoplasmic streaming. Combined experiments involving the manipulation of transpiration rates and application of streaming inhibitors should therefore provide useful information about the relative importance of bulk flow and cytoplasmic streaming under different conditions.

Uptake, Assimilation and Translocation of Nitrogen

Mycelial uptake and assimilation of ^{15}N-labelled nitrogen has recently been shown in a range of ectomycorrhizal associations by Finlay *et al.* (1988, 1989). Evidence from these studies supports that from other studies using excised roots in identifying glutamine as the principal sink for assimilated nitrogen with alanine, arginine and aspartic acid/asparagine also acting as important sinks (see Martin and Botton, 1991). Identification of the earliest assimilation products is still necessary using time course experiments but evidence from the declining levels of enrichment and patterns of labelling suggest assimilation into glutamine at an early stage within the fungal mycelium and subsequent transport in amino acid form to the mycorrhizal root tips. Despite large differences in the size of the different amino acid pools, patterns of enrichment appear uniformly high indicating high aminotransferase activity. Incorporation of ammonium into carbon skeletons derived from trehalose and mannitol to form amino acids, which are released to the plant root, provides a route for the reverse translocation of carbon from fungus to host. Studies of systems with an intact mycelial phase have an important advantage that patterns of assimilation can be followed along the entire translocation pathway. Investigation of the spatial distribution of different assimilatory activities is also facilitated in associations showing 'metabolic zonation' (see Martin *et al.*, Chapter 41 this volume). Patterns of ^{15}N incorporation in intact ectomycorrhizal mycelium (Table 12.2) clearly indicate that the importance of arginine as a nitrogen sink differs according to the fungal species involved. Arginine is an important free amino acid

Table 12.2. Principal nitrogen pools and sinks for ^{15}N assimilation in free amino acids of some different ectomycorrhizal fungi.

	Rhizopogon roseolus	*Suillus bovinus*	*Paxillus involutus*	*Pisolithus tinctorius*	*Boletinus cavipes*
Amino acid					
Alanine	9.5	20.2 (25.0)	10.2	6.3	5.3 (7.5)
Serine	2.3	3.5 (2.0)	2.2	2.1	3.7 (3.5)
Asx	30.4	6.8 (9.1)	12.4	53.8	4.6 (6.3)
Glx	40.6	33.9 (48.8)	69.0	31.5	14.5 (25.5)
Arginine	n.d.	18.9 (9.2)	0.0	0.0	56.0 (35.2)
Ornithine	6.5	3.9 (5.0)	1.0	0.0	n.d. n.d.

Values are expressed as percentages of the total free amino acid nitrogen pool.
Figures in parentheses are percentages of the total free amino acid ^{15}N pool.
n.d. = not determined.

in *Cenococcum geophilum* (Martin, 1985), *Laccaria bicolor*, *Suillus bovinus* and *Boletinus cavipes*, but not in *Paxillus involutus* or *Pisolithus tinctorius*. Amino acids such as arginine neutralize the strong negative charge of polyphosphates and the influence of polyphosphate on the accumulation of basic amino acids in fungal vacuoles may be important in the regulation of P and N in the ectomycorrhizal sheath. Thus the association of arginine with polyphosphate in these fungi may lead to a joined hyphal translocation of N and P (Martin and Botton, 1991). Further experiments are required to confirm this.

Concluding Remarks

Different translocation mechanisms and cellular pathways have so far been proposed for translocation of carbohydrates, water and mineral nutrients in ectomycorrhizal mycelia. The relative importance of these under different conditions remains to be determined however. Further accurate estimation of nutrient fluxes is necessary to determine the relative efficiency of translocation of different substances in different directions. This is especially important with regard to eventual genetic manipulation and selection of 'efficient' mycobionts with which to increase plant yield. Detailed flux measurements are also necessary to determine differences in translocation efficiency between fungal species showing different levels of mycelial differentiation.

References

Brownlee, C. and Jennings, D.H. (1982) Long distance translocation in *Serpula lacrymans*: velocity estimates and continuous monitoring of induced perturbations. *Transactions of the British Mycological Society* 79, 143–148.

Cooper. K.M. and Tinker, P.B. (1981) Translocation and transfer of nutrients in vesicular arbuscular mycorrhizas. IV. Effect of environmental variables on movement of phosphorus. *New Phytologist* 88, 327–339.

Duddridge, J.A., Malibari, A. and Read, D.J. (1980) Structure and function of mycorrhizal rhizomorphs with special reference to their role in water transport. *Nature* 287, 834–836.

Duddridge, J.A., Finlay, R.D., Read, D.J. and Söderström, B. (1988) The structure and function of the vegetative mycelium of ectomycorrhizal plants. III. Ultrastructural and autoradiographic analysis of inter-plant carbon distribution through intact mycelial systems. *New Phytologist* 108, 183–188.

Finlay, R.D. (1989) Functional aspects of phosphorus uptake and carbon translocation in incompatible ectomycorrhizal associations between *Pinus sylvestris* and *Suillus grevillei* and *Boletinus cavipes*. *New Phytologist* 112, 185–192.

Finlay, R.D. and Read, D.J. (1986a) The structure and function of the vegetative mycelium of ectomycorrhizal plants. I. Translocation of ^{14}C-labelled carbon between plants interconnected by a common mycelium. *New Phytologist* 103, 143–156.

Finlay, R.D. and Read, D.J. (1986b) The structure and function of the vegetative mycelium of ectomycorrhizal plants. II. The uptake and distribution of phosphorus by mycelial strands interconnecting plants. *New Phytologist* 103, 157–165.

Finlay, R.D., Ek, H., Odham, G. and Söderström, B. (1988) Mycelial uptake, translocation and assimilation of nitrogen from ^{15}N-labelled ammonium by *Pinus sylvestris* plants infected with four different ectomycorrhizal fungi. *New Phytologist* 110, 59–66.

Finlay, R.D., Ek, H., Odham, G. and Söderström, B. (1989) Uptake, translocation and assimilation of nitrogen from ^{15}N-labelled ammonium and nitrate sources by intact ectomycorrhizal systems of *Fagus sylvatica* infected with *Paxillus involutus*. *New Phytologist* 113, 47–55.

Harley, J.L. and Smith, S.E. (1983) *Mycorrhizal Symbiosis*. Academic Press, London.

Jennings, D. H. (1987) Translocation of solutes in fungi. *Biological Reviews* 62, 215–243.

Martin, F. (1985) ^{15}N-NMR studies of nitrogen assimilation and amino acid biosynthesis in the ectomycorrhizal fungus *Cenococcum graniforme*. *FEBS Letters* 182, 350–354.

Martin, F. and Botton, B. (1991) Nitrogen metabolism of ectomycorrhizal fungi and ectomycorrhiza. *Advances in Plant Pathology*, Volume 8. Academic Press, London.

Melin, E., Nilsson, H. and Hacskaylo, E. (1958) Translocation of cations to seedlings of *Pinus virginiana* through mycorrhizal mycelium. *Botanical Gazette* 119, 243–246.

Read, D.J. (1984) The structure and function of the vegetative mycelium of mycorrhizal roots. In: Jennings, D.H. and Rayner, A.D.M. (eds), *The Ecology and Physiology of the Fungal Mycelium*. Symposium of the British Mycological Society. Cambridge University Press, Cambridge, pp. 215–240.

Sands, R., Fiscus, E.L. and Reid, C.P.P. (1982) Hydraulic properties of pine and bean roots with varying degrees of suberization, vascular differentiation and mycorrhizal infection. *Australian Journal of Plant Physiology* 9, 559–569.

Skinner, M.F. and Bowen, G.D. (1974) The uptake and translocation of phosphate by mycelial strands of pine mycorrhizas. *Soil Biology and Biochemistry* 6, 53–56.

Söderström, B., Finlay, R.D. and Read, D.J. (1988) The structure and function of the vegetative mycelium of ectomycorrhizal plants. IV. Qualitative analysis of carbohydrate contents of mycelium interconnecting host plants. *New Phytologist* 109, 163–166.

13 Mycorrhizal Mat Communities in Forest Soils

R.P. Griffiths and B.A. Caldwell

Department of Forest Science, Peavy Hall, Oregon State University, Corvallis, Oregon, 973310–3804, USA

Ectomycorrhizal Mats as Model Systems

The role of ectomycorrhizal (EM) fungi in forest soil function and tree productivity has become an increasingly important topic for research in light of possible direct large-scale degradation of EM populations in the forests of Europe (Arnolds, 1991). Although there is a rapidly growing body of information being assembled from laboratory studies using pure culture syntheses, there is relatively little known about how these fungi function in the field. EM species of *Gautieria* and *Hysterangium* that form distinctive hyphal or rhizomorph mats have been observed in forests ranging from the subtropical (*Eucalyptus* in Australia) to boreal forests in Alaska (Castellano, 1988; Griffiths *et al.*, 1991b; Griffiths, unpublished observations). The actual quantitative impact of these mat communities on the forest floor remains largely unknown, although Cromack *et al.* (1979) reported up to 27% of temperate coniferous forest mineral soils could be colonized by a single species, *H. setchellii*.

These mats present a novel solution to the problem of how to measure the impact of an EM fungus on its immediate surroundings by greatly magnifying the influence of a single fungal species. The mat communities studied are generally dominated by a single EM species which can have a biomass equivalent to up to half the mass of the soils with which they are associated (Ingham *et al.*, 1991). This level of impact permits one to study the influence of these fungi on associated soil by comparing the biology and chemistry of soils colonized by these fungi with soils that are not so colonized. This approach has now been used to document large differences between mat and non-mat soils over seasonal cycles and in different fungi located in different areas of the Pacific Northwest (Cromack *et al.*, 1988; Griffiths *et al.*, 1990, 1991a).

Accelerataed Mineral Weathering

An early study of fungal mats produced by presumptive EM fungi in Finland (Hintikka and Naykki, 1967) demonstrated that the mats have the capability of removing plant nutrients from soils and significantly altering soil chemistry. Cromack *et al.* (1979) found elevated oxalate concentrations and evidence for advanced clay weathering in *H. setchellii* mats in the Pacific northwest. The quantitative impact of these mat communities on the forest floor remains largely unknown, although more recently, Entry *et al.* (1987) and Rose *et al.* (1991) have reported elevated DTPA extractable cations associated with mycorrhizal mat communities. Elsewhere in this volume, we report elevated concentrations of a number of cations as well as oxalate and dissolved organic carbon (DOC) in the soil solution of these mats (Griffiths *et al.*, p. 380, this volume). These data all strongly suggest that organic acids are responsible for accelerated weathering of the mineral soils by these mat communities.

Processing of Detrital Nutrient Resources

In distinguishing 'early' and 'late' successional stage EM fungi, Dighton and Mason (1985) proposed that 'late' stage fungi could better utilize organic nutrient pools. Within the EM mat communities, we have found significantly elevated levels of several major soil enzymes which are thought to be responsible for processing complex detrital carbon, nitrogen and phosphorus (Table 13.1).

Table 13.1. Soil and litter enzyme activities in forest ectomycorrhizal mat communities.

Site/community	Cellulase*	Peroxidase†	Phosphatase‡	Proteinase§
Coast Range, Oregon (Douglas fir)				
Non-mat soil	1.34a	2.68a	22.2a	219a
Gautieria monticola	1.54a	87.6b	48.5b	161a
Hysterangium setchellii	3.25b	75.2b	58.3b	414b
Cascade Mountains, Oregon (Douglas fir)				
Non-mat soil	0.66a	0.78a	17.1a	148a
G. monticola	5.82b	98.5b	41.7b	193a
H. coriaceum	4.81b	57.4b	37.0b	376b
H. crassirhachis	4.92¶	58.6¶	35.8¶	289¶
Mendicino CA (Eucalyptus)				
Non-mat litter	2.20a	not detected	46a	391a
H. gardneri	4.94b	not detected	165b	831b

Enzyme activities: *μmol glucose g^{-1} h^{-1}; †change in A_{460} g^{-1} min^{-1};
‡μmol p-nitrophenol g^{-1} h^{-1}, §nmol tyrosine eq. g^{-1} h^{-1}.
a,b,c: for given site, different letters indicate significantly different ($P \leq 0.05$) enzyme activities.
¶single mat sampled.

Most striking of the mat enzyme activities were the elevated levels of peroxidase. The ability of mat-forming EM fungi to decolorize aromatic dyes *in vitro* (Table 13.2) provides presumptive evidence for the ability to degrade lignin (Glenn and Gold, 1983). While this may be important in processing litter, it has also been noted that lignolytic peroxidases can also degrade humic acid (Blondeau, 1989), suggesting a possible mechanism whereby these fungi could gain access to pools of nitrogen and phosphorus contained in recalcitrant soil organic matter.

Of the many papers published on phosphatase production by mycorrhizal fungi, virtually all have used phosphomonoester substrates. However, most organic phosphorus in living tissues, and thus fresh detritus, occurs as phosphodiesters. The abundance of phosphomonoesters, relative to phosphodiesters, in soil organic matter suggests that the phosphodiester pool is more important during litter decomposition. We have found that several species of mat-forming EM fungi are capable of hydrolysing RNA, a major phosphodiester (Table 13.2), providing strong circumstantial evidence supporting Went and Stark's direct nutrient cycling hypothesis (Dighton, 1991).

While the *in vitro* utilization of protein by EM fungi has been demonstrated (e.g. Abuzinadah and Read, 1986; Hutchinson, 1990), detrital proteins may frequently exist as complexes with reactive polyphenols that are resistant to enzyme action (Read, 1991). We have found several mat-forming EM fungi that are capable of growing on an insoluble tannic acid-protein complex. There are also reports of growth of EM fungi on insoluble chestnut tannin–protein complexes.

At this time, we do not know if the elevated mat enzyme activities are due to ectomycorrhiza directly, or enhanced saprophytic populations within the mat community. The ability to culture these fungi for comparison of *in vitro* and *in situ*

Table 13.2. Selected physiological characteristics of pure cultures 'early' successional stage and mat-forming ectomycorrhizal fungi.

EM fungus	Ribonuclease	Reaction on tannin–protein complex[1]	Polyaromatic dye decoloration[2]	
			RBB	R-478
'Early' stage				
Laccaria laccata	–	–	–	–
Hebeloma crustuliniforme	–	–	–	–
Thelephora terrestris	–	–	–	–
Mat-forming				
Chondrogaster sp.	+	+	+	+
Gautieria monticola	+	+	+	+
Hysterangium coriaceum	not tested	+	not tested	
H. gardneri	+	(+)	+	+

Key: 1 reactions:– no visible reaction, (+) Bavendamm-like darkening, + clearing of precipitate.
2 – presumptive assay for lignolytic activity (Glenn and Gold, 1983): RBB, Remazol Brilliant Blue; R-478 (Sigma Chemical Co.).

enzymes make the mat communities novel model systems in which to establish the direct role for these specialized associations in forest nutrient cycling.

Establishment of a Distinct Soil Microhabitat

If these mat-forming mycorrhizal fungi are capable of degrading organic polymers, it is likely that they have some mechanism to transport organic nitrogen and phosphorus released during the degradative process to the host tree. Since the EM fungus has other sources of organic carbon during most of the year, it is likely that only those organic compounds which contain plant nutrients are removed from the soils. If this were the case, we would expect to find higher C:N ratios in mat soils than in non-mat soils: an observation we have made repeatedly in our studies (Griffiths *et al.*, 1990, 1991a). The ratio of chloroform fumigation CO_2 to mineralizable N in *H. setchellii* and *G. monticola* mats over several seasons are consistently higher than in mineral soils not colonized by these fungi (Griffiths *et al.*, 1990, 1991a).

There are at least three possible explanations for the large differences in these ratios: (1) there is a proportionately greater concentration of bacteria than fungi

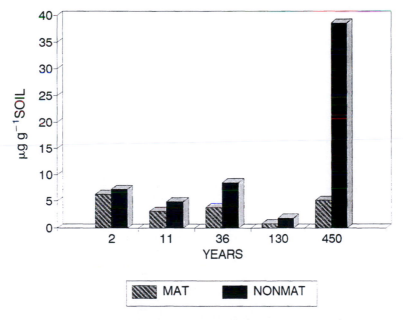

Fig. 13.1. Labile nitrogen in mat and non-mat soils collected in Douglas fir stands that had been disturbed at different times in the past. The years given are the years since the site was burned (130 and 450 years) or harvested and burned (2, 11, and 36 years). The stand that had been harvested and burned 2 years previously still contained approximately 10% of the original old-growth trees. The values given are means of five observations for each treatment. The units are μg nitrogen per g dry weight soil.

in non-mat soils; (2) the fungal mat community is selectively removing organic nitrogen from these forest soils; and/or (3) the fungal community is releasing organic compounds with high C:N ratios into these soils (i.e. organic acids). Direct counts of bacteria and fungi (excluding rhizomorph material) shows no consistent enrichment of bacteria over fungi in mat soils (Ingham et al., 1991), thus alternative (1) appears unlikely. If the organic N is selectively removed, one would expect a reduction in organic N. If the fungal community is releasing compounds of high C:N ratio one would expect to find elevated organic carbon concentrations in mat soils but no difference in nitrogen concentrations. In a recent study of labile carbon and nitrogen in G. monticola mats in a chronosequence of Douglas fir forest soils, we found the pattern one would expect if both of these activities were taking place (Figs 13.1 and 13.2).

If mat-forming EM fungi produce exoenzymes which break down organic polymers, they must be in competition with heterotrophic soil bacteria and saprophytic fungi for the break-down products. The question is, have the mat-forming fungi produced conditions in the soil that favour their growth relative to that of competing microorganisms? We have repeatedly observed lower pH values in mat soils and have observed that the mat soils are typically drier and much more hydrophobic than non-mat soils (Griffiths et al., 1990, 1991a). These are conditions that, in general, should favour fungi over bacteria.

We have also observed other chemical differences between mat and non-mat soils which may give EM fungi the competitive edge over other soil microorganisms for organic resources. Quantities of water-soluble phenolics were significantly

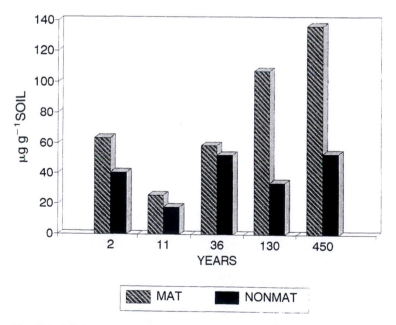

Fig. 13.2. Labile carbon in mat and non-mat soils collected in Douglas fir stands as in Fig. 13.1. The units are μg carbon per g dry weight soil.

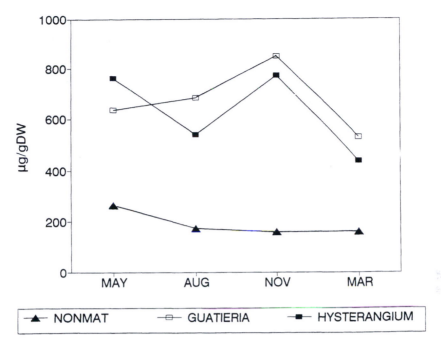

Fig. 13.3. Water-soluble phenolic compounds found in soils collected in *Gautieria monticola* and *Hysterangium setchellii* mats and associated non-mat soils. Soil samples were collected in different mats at each sampling time. The given values are the mean of five soils taken from each soil type. The units are μg gallic acid equivalents per g dry weight soil.

elevated in mat soils during all four seasons (Figure 13.3). In addition, amounts of siderophore-like compounds were also elevated in two of four seasonal samplings in mat soils (Griffiths, unpublished data). Since production of siderophores by one organism can limit Fe^{3+} availability to others this may be a mechanism by which these organisms control competition within mat communities. The high concentrations of oxalate observed in EM mat soil solution may have a similar function although the main function of this compound is most likely to be in mineral weathering (see Griffiths *et al.*, p. 380 this volume). It is known that oxalate can chelate Fe^{3+} in addition to other micronutrients which may have an effect similar to that of siderophores.

In addition to these features, differences in cation chemistry and quantities of DOC have been observed between mat and non-mat soils (Entry *et al.*, 1987; Caldwell *et al.*, 1991) as well as differences in respiration rates (Griffiths *et al.*, 1990, 1991a), and enzyme activities (Griffiths *et al.*, 1987). Microbial biomass levels, as determined by both direct and indirect means, are also higher in mat soils (Ingham *et al.*, 1991). Taken together, these observations suggest that EM mat communities may act as unique soil habitats in forest soils and may play a role in maintaining higher species diversity within forest ecosystems. Significant differences in populations of protozoa, nematodes and microarthropods

(Cromack *et al.*, 1988) between mat and non-mat soils suggest that this may be the case.

In summary, recent work on EM mat communities in the Pacific Northwest of the USA has demonstrated that: (1) these mats are sites of accelerated mineral soil weathering; (2) the fungi involved have the potential to provide access to detrital nitrogen and phosphorus resources; (3) they selectively remove organic nitrogen from the soils, and (4) form specialized habitats in forest soils which could increase overall biological diversity.

Acknowledgements

This work was supported by US Department of Agriculture 86-FYT-9-0224 and National Science Foundation BSR87-17849 and BSR91-06784.

References

Abuzinadah, R.A. and Read, D.J. (1986) The role of proteins in the nitrogen nutrition of ectomycorrhizal plants. I. Utilization of peptides and proteins by ectomycorrhizal fungi. *New Phytologist* 103, 481–493.

Arnolds, E. (1991) Decline of ectomycorrhizal fungi in Europe. *Agriculture, Ecosystems and Environment* 35, 209–244.

Blondeau, R. (1989) Biodegradation of natural and synthetic humic acids by the white rot fungus *Phanerochaete chrysosporium*. *Applied and Environmental Microbiology* 55, 1282–1285.

Castellano, M.A. (1988) The taxonomy of the genus *Hysterangium* (Basidiomycotina, Hysterangiaceae) with notes on its ecology. PhD thesis, Oregon State University, Corvallis, OR. 238 pp.

Cromack, K., Sollins, P., Graustein, W.C., Speidel, K., Todd, A.W., Spycher, G., Li, C.Y. and Todd, R.L. (1979) Calcium oxalate accumulation and soil weathering in mats of the hypogeous fungus *Hysterangium crassum*. *Soil Biology and Biochemistry* 11, 463.

Cromack, K., Jr., Fichter, B.L., Moldenke, A.M., Entry, J.A. and Ingham, E.R. (1988) Interactions between soil animals and ectomycorrhizal fungal mats. *Agriculture, Ecosystems and Environment* 24, 161–168.

Dighton, J. (1991) Acquisition of nutrients from organic resources by mycorrhizal autotrophic plants. *Experientia* 47, 362–369.

Dighton, J. and Mason, P.A. (1985) Mycorrhizal dynamics during forest tree development. In: Moore, D., Casselton, L.A., Wood, D.A. and Frankland, J.C. (eds), *Developmental Biology of Higher Fungi*. Cambridge University Press, Cambridge, pp. 117–139.

Entry, J.A., Rose C.L., Cromack, K., Jr., Griffiths, R.P. and Caldwell, B.A. (1987) The influence of ectomycorrhizal mats on chemistry of a coniferous forest soil. In: Sylvia, D.M., Hung, L.L. and Graham, J.H. (eds), *Mycorrhizae in the Next Decade: Practical Applications and Research Priorities*. North American Conference on Mycorrhizae. Gainesville, Florida, USA.

Glenn, J.K. and Gold, M.H. (1983) Decolorization of several polymeric dyes by the lignin-degrading basidiomycete *Phanerochaete chrysosporium*. *Applied and Environmental Microbiology* 45, 1741–1747.

Griffiths, R.P., Caldwell, B.A., Cromack, K., Jr., Castellano, M.A. and Morita, R.Y. (1987) A study of the chemical and microbial variables in forest soils colonized with *Hysterangium setchelli* rhizomorphs. In: Silva, D.M., Hung, L.L. and Graham, J.H. (eds), *Mycorrhizae in the Next Decade*. University of Florida, Gainesville, Florida, USA, p. 196.

Griffiths, R.P., Caldwell, B.A., Cromack, K., Jr. and Morita, R.Y. (1990) Douglas-fir forest soils colonized by ectomycorrhizal mats: I. Seasonal variation in nitrogen chemistry and nitrogen cycle transformation rates. *Canadian Journal of Forest Research* 20, 211–218.

Griffiths, R.P., Caldwell, B.A., Ingham, E.R., Castellano, M.A. and Cromack, K., Jr. (1991a) Comparison of microbial activity in ectomycorrhizal mat communities in Oregon and California. *Biology and Fertility of Soils* 11, 196–202.

Griffiths, R.P., Castellano, M.A. and Caldwell, B.A. (1991b) Ectomycorrhizal mats formed by *Gautieria monticola* and *Hysterangium setchellii* and their association with Douglas-fir seedlings, a case study. *Plant and Soil* 134, 255–259.

Hintikka, V. and Naykki, O. (1967) Notes on the effects of the fungus *Hydnellum ferrugineum* (Fr.) Karst. on forest soil and vegetation. *Communicationes Instituti Forestalis Fenniae* 62, 1–23.

Hutchinson, L.J. (1990) Studies on the systematics of ectomycorrhizal fungi in axenic culture. II. The enzymatic degradation of selected carbon and nitrogen compounds. *Canadian Journal of Botany* 68, 1522–30.

Ingham, E.R., Griffiths, R.P., Cromack, K., Jr. and Entry, J.A. (1991) Comparison of direct versus fumigation-flush microbial biomass estimates from ectomycorrhizal mat and non-mat soils. *Soil Biology and Biochemistry* 23, 465–71.

Read, D.J. (1991) Mycorrhizas in ecosystems. *Experientia* 47, 376–90.

Rose, C.L., Entry, J.A. and Cromack, K., Jr. (1991) Nutrient concentrations in *Hysterangium setchellii* fungal mats in western Oregon coniferous soil. *Soil Biology and Biochemistry* (in press).

14 Ecological Role of Specificity Phenomena in Ectomycorrhizal Plant Communities: Potentials for Interplant Linkages and Guild Development

R. Molina[1], H.B. Massicotte[2], and J.M. Trappe[2]

[1]*USDA Forest Service, Pacific Northwest Research Station, Forestry Sciences Laboratory, 3200 Jefferson Way, Corvallis, OR 97331, USA; [2]Department of Forest Science, Oregon State University, Corvallis, OR 97331, USA*

Introduction

Conventional wisdom generalizes that most ectomycorrhizal (EM) fungi are broad-host-ranging and express little host specificity. However, EM fungi vary from narrow to broad in host range. Similarly, EM hosts may associate with relatively few to very high numbers of fungus species; several plant taxa show strong co-evolution with particular fungal taxa. Consequently, not all host–fungus combinations are compatible. The overlap between hosts for shared compatibility with common fungi will determine their linkage potential (i.e. interplant fungal connection). Determining this linkage potential is critical to evaluating the mycorrhizal mediation of plant interactions, particularly in EM dominated ecosystems composed of diverse hosts.

Read *et al.* (1985) and Newman (1988) have reviewed the ecological implications of the fact that plants are connected by common mycorrhizal fungi; interplant movement of carbohydrates and minerals via shared hyphae and potential dependence of understory seedlings on adult trees have received primary attention. Implicit in these scenarios are the availability and function of compatible fungi, a likely prospect if the connected plants are of the same species. But if the plants are of different species, their ability to be connected depends on their overlap in compatibility with available fungi. In natural EM forests, the overlap in compatibility can affect the successional dynamics of the vegetation, particularly following disturbance (Amaranthus and Perry, 1989) and may influence the migration of mutualists during global climate change (Perry *et al.*, 1990). Similarity indices of compatible mycorrhizal fungi can also be useful to describe and quantify guild relationships among plants based on mycorrhizal interactions. Readers are referred to Molina *et al.* (1992) for a detailed discussion of specificity phenomena in mycorrhizas and their ecological implications.

Specificity Phenomena in Mycorrhizal Associations

In plant pathology, specificity typically implies a highly restricted association between species of host and species or strain of a pathogen. It is not surprising, therefore, that reviews on mycorrhiza often generalize that mycorrhizal associations do not express strong specificity. Such a generalization is misleading, as pointed out by Molina *et al.* (1992). Six phenomena that encompass the spectrum of specificity in mycorrhizal associations can be recognized:

1. *Dependency vs. independency* – defines whether plants form or do not form mycorrhiza;
2. *Facultative vs. obligate symbionts* – defines the ability or inability of symbionts to complete life cycles in the absence of mycorrhiza formation;
3. *Fidelity to a class of mycorrhiza* – recognizes that most symbionts typically form one class of mycorrhiza (e.g. vesicular-arbuscular vs. ectomycorrhiza), but that some can form two or more classes;
4. *Host range of mycorrhizal fungi* – describes a spectrum that varies from a narrow range, typically restricted to a host genus, to an intermediate range, often restricted to a host family, to a broad range, typically extending across diverse host families and orders;
5. *Host receptivity* – defines the numbers and diversity of mycorrhizal fungi accepted by a particular host, also ranging from narrow (low number) to broad (high number);
6. *Ecological specificity* – the influence of biotic and abiotic factors on ability of plants to form functional mycorrhizas with particular fungi in natural soils. The remainder of this chapter will concern itself only with the latter three phenomena.

Host Range and Receptivity

The mycological literature richly documents the degrees of specificity in EM associations based on host–sporocarp associations as summarized in Table 14.1. In genera such as *Amanita* and *Laccaria*, most species tend towards a broad host range. In others such as *Suillus*, *Leccinum*, *Naucoria*, and the Gomphidiaceae the tendency is towards the intermediate to narrow host ranges. Many genera show relatively even distribution of species from narrow to broad host-ranging. Most hypogeous basidiomycetes show stronger host restriction than epigeous species, thereby emphasizing the close co-evolution of these specialized fungi with their hosts.

Narrow host range is typically expressed at the host genus level for individual fungus species. From an ecological standpoint, this can be the equivalent of specificity at the species level when ranges of host species within a genus do not overlap. For example, *Suillus lakei* may form EM with all species of *Pseudotsuga*, but none of the *Pseudotsuga* spp. occur together in the same ecosystem. True host restriction at the species level has been documented for dipterocarps in Malaysia (Smits, 1983 and Chapter 38 this volume).

Table 14.1. Estimated percentages of species of selected Basidiomycota with narrow, intermediate, or broad host range by fruiting habit, family and genus (taken from Molina et al., 1992). Narrow = restricted to a host genus; intermediate = restricted to angiosperms, gymnosperms, or a single host family; broad = associates with diverse hosts.

Habit, Family	Genus	Narrow %	Intermediate %	Broad %
Epigeous habit				
Amanitaceae	Amanita	5	35	60
Boletaceae	Suillus	90	10	
	Boletus	20	35	45
	Leccinum	80	20	
	Small genera	30	30	40
Cortinariaceae	Cortinarius	20	55	25
	Dermocybe	30	45	25
	Hebeloma	25	45	30
	Inocybe	20	45	35
	Naucoria	65	35	
Gomphidiaceae	Gomphidius	65	35	
	Chroogomphus	60	40	
Hygrophoraceae	Hygrophorus	35	50	15
Russulaceae	Lactarius	40	35	25
	Russula	25	45	30
Sclerodermataceae	Scleroderma	25	45	30
Tricholomataceae	Laccaria		20	80
	Tricholoma	20	45	35
Hypogeous habit				
Boletaceae	Alpova	40	60	
	Rhizopogon	50	50	
Cortinariaceae	Hymenogaster	50	50	
	Thaxterogaster	70	30	
Gomphidiaceae	Brauniellula	100		
	Gomphogaster	100		
Russulaceae	Gymnomyces	50	50	
	Macowanites	50	50	
	Martellia	60	40	
Sclerodermataceae	Scleroderma	25	75	
Tricholomataceae	Hydnangium	100		

Host receptivity likewise varies from narrow to broad, but compiling these data is more difficult for hosts than for fungi. Some genera such as *Alnus* seem to have relatively few fungal associates (Molina, 1981) while others like *Pseudotsuga* may have 2000 or more reported fungal associates (Trappe, 1977). Still others such as arbutoid hosts in the genera *Arbutus* and *Arctostaphylos* can be widely receptive to EM fungi, even to those fungi that otherwise typically express restricted host ranges (Molina and Trappe, 1982a).

Differences in host range and receptivity are revealed in mycofloristic studies comparing EM fungus communities in different forest types. For example, Bills *et al.* (1986) examined the fungi in stands of red spruce (*Picea rubens*) and adjacent stands of mixed hardwoods (*Fagus*, *Quercus*, and *Betula*). Of 54 fungus species collected, 19 were only in spruce, 27 only in mixed hardwoods, and eight in both forest types. Lee and Kim (1987) similarly documented a range of narrow to broad host ranges among EM fungi associated with pure stands of eight EM tree genera in Korea. Sporocarp associations may not always accurately reflect the ability of fungi to form EM in the field with other hosts (Trappe, 1962; Molina and Trappe, 1982b) so the potential for mycorrhizal formation with a variety of hosts under field conditions needs examination.

Ecological Specificity

Harley and Smith (1983) coined the term 'ecological specificity' to emphasize that actual mycorrhizal development and therefore host–fungus associations in nature may differ strongly from those produced experimentally, particularly in pure culture syntheses. We agree and expand the concept of ecological specificity to include the influence of biotic and environmental interactions on the development and function of mycorrhizas in natural soils. Interactions between plants (e.g. companion or neighbouring plant influences), between fungi (competitive to facilitative), or between plants and fungi (e.g. stimulation of spore germination) exemplify components of the biotic complex that determines whether certain plants and fungi enter into mycorrhizal symbioses. Ectomycorrhizal fungi vary in physiological response to environmental conditions, so environment will influence specificity under field conditions. For example, Danielson and Visser (1989) used the term 'soil specificity' to indicate the importance of soil factors in influencing mycorrhizal development by certain fungi. The ecological conditions that modulate the development of particular EM host–fungus associations must be understood to develop models of functional guilds in EM ecosystems.

Massicotte and Molina (unpublished data) have documented companion plant influences on ectomycorrhizal specificity and shared mycobionts between diverse EM hosts. In one study, they inoculated species of *Pseudotsuga*, *Pinus*, *Tsuga*, *Picea*, and *Abies* with spores of several *Rhizopogon* species; hosts were grown and inoculated in both monocultures and dual-host culture combinations. The genus *Rhizopogon* contains four sections: *Villosuli*, *Fulviglebae*, *Amylopogon*, and *Rhizopogon* (Smith and Zeller, 1966). Molina (1980) previously found in pure culture syntheses that species in sections *Villosuli* and *Fulviglebae* were strongly specific to *Pseudotsuga* whereas many species in sections *Rhizopogon* and *Amylopogon* showed a preference for *Pinus*. Results from the spore inoculation study supported conclusions from the pure culture syntheses but also provided additional insights about specificity. Several *Rhizopogon* species formed EM only on one host species, namely *Pseudotsuga* or *Pinus*; when inoculated into dual host cultures, these fungi failed to form EM with companion plants even though hyphae grew vigorously from EM of the specific host and made abundant contact with companion plants. However, some *Rhizopogon* species that were restricted to *Pinus* in monoculture were able to spread in dual culture on to the companion tree species. Thus, in this study, we see clear examples of narrow, restricted host specificity, but also that companion plants can influence the mycorrhizal potential of EM fungi. We conclude that the mycelia of an EM fungus actively growing in the soil and drawing energy from a photosynthetic host can have a different colonizing potential and thus host range than spores of the fungus residing in the soil.

In a second study, Massicotte and Molina (unpublished data) grew *Pseudotsuga menziesii*, *Abies grandis*, *Pinus ponderosa*, and *Lithocarpus densiflora* in pots of soils collected respectively from a recent forest clearcut, a 25-year-old *Pseudotsuga menziesii* plantation, and an adjacent undisturbed 90–160-year-old stand of Pinaceae mixed with EM hardwoods in southwestern Oregon. This region is rich in EM host plants, many of which occur in well defined successional sequences following disturbance (Amaranthus, Chapter 27, this volume). Seedlings were again grown in both mono – and polycultures (all four together). While nearly half of the fungi, represented by EM types, proved compatible with all four host species, incompatibility was also expressed in a variety of combinations. Of the 18 EM types recovered from all soils and hosts, eight formed with all host species (broad host range), four formed with only one host (specific to either *Pseudotsuga* or *Pinus*), and six others formed with two to three hosts (intermediate host range). Each of these three categories of host range occurred fairly evenly in soils from the three forest sites. However, the number of EM types initiated varied if monoculture or polycultures were considered separately. We observed a trend towards fewer mycorrhizal types recovered in host polycultures than recovered in monoculture assays. Thus, interactions between different plant species may influence the development and predominance of particular EM types.

Exploring the EM Guild Concept with Specificity Phenomena

The above results demonstrate that indeed many EM hosts can have a wide overlap in EM fungus associates and thus a strong mycorrhizal linkage potential. Perry *et al.* (1989a) describe these ecologically linked EM hosts and fungi of southwestern Oregon as functional guilds: 'Associations for mutual aid and the promotion of common interests', in this case as defined by common below-ground mutualists. Such a holistic guild concept provides a framework for exploring the contributions of root mutualisms to ecosystem resiliency but does not illuminate the individual contributions of EM fungus species that have vastly different ecologies in space and time. Also, the emphasis placed on overlap for shared mutualists may unduly emphasize the ecological importance of broad-host-ranging fungi. We have documented that restricted host range is a widespread phenomenon among EM fungi. In the Pacific Northwest, several *Rhizopogon* species specific to *Pseudotsuga menziesii* are common associates throughout much of the life history of the tree and have been experimentally demonstrated to provide drought tolerance to seedlings, an important ecological function in a region with seasonal drought.

To develop a functional model of ectomycorrhizal guilds, we must integrate our understanding of specificity phenomena, particularly ecological specificities, with our understanding of plant community dynamics. For example, do broad host range fungi play different functional roles than host specific fungi during community succession? Perry *et al.* (1989b) found that inoculation of *Pseudotsuga/Pinus* seedling polycultures with host specific fungi (*Rhizopogon* species) reduced interspecies competition in comparison to when the companion plants were mycorrhizal with broad host range fungi alone (*Thelephora*, *Laccaria* and *Hebeloma* species). We hypothesize that host specific fungi provide a biological mechanism to partition soil resources, thereby providing an exclusive avenue for different plant species to obtain nutrients and enhance their fitness. The total EM guild is benefited by reduced competition for shared mutualists (Molina *et al.*, 1992).

Guild structure and function in EM ecosystems is complex and multilayered in plant–fungus interactions through time; a simple vision of interconnected plants is unrealistic. Unravelling functional aspects of guilds such as competitive interactions, facilitations, or other outcomes in relation to specificity phenomena poses a great challenge for mycorrhizal ecologists. Given the current emphasis on biodiversity and ecosystem management, research efforts in these areas will provide valuable tools for resource managers to conserve the biological complexity of the below-ground ecosystem.

References

Amaranthus, M.P. and Perry, D.A. (1989) Interaction effects of vegetation type and Pacific madrone soil inocula on survival, growth, and mycorrhiza formation of Douglas-fir. *Canadian Journal of Forest Research* 19, 550–556.

Bills, G.F., Holtzmann, G.I. and Miller, O.K. Jr. (1986) Comparison of ectomycorrhizal Basidiomycete communities in red spruce versus northern hardwood forests of West Virginia. *Canadian Journal of Botany* 64, 760–768.

Danielson, R.M. and Visser, S. (1989) Host response to inoculation and behaviour of introduced and indigenous ectomycorrhizal fungi of jack pine grown on oil-sands tailings. *Canadian Journal of Forest Research* 19, 1412–1421.

Harley, J.L. and Smith, S.E. (1983) *Mycorrhizal Symbiosis*. Academic Press, London. 483 pp.

Lee, K.J. and Kim, Y.S. (1987) Host specificity and distribution of putative ectomycorrhizal fungi in pure stands of twelve tree species in Korea. *Korean Journal of Mycology* 15, 48–69.

Molina, R. (1980) Patterns of ectomycorrhizal host-fungus specificity in the Pacific Northwest. Unpublished PhD thesis, Oregon State University, Corvallis, OR. 279 pp.

Molina, R. (1981) Ectomycorrhizal specificity in the genus *Alnus*. *Canadian Journal of Botany* 59, 325–334.

Molina, R. and Trappe, J.M. (1982a) Lack of mycorrhizal specificity by the ericaceous hosts *Arbutus menziesii* and *Arctostaphylos uva-ursi*. *New Phytologist* 90, 495–509.

Molina, R. and Trappe, J.M. (1982b) Patterns of ectomycorrhizal host specificity and potential among Pacific Northwest conifers and fungi. *Forest Science* 28, 423–458.

Molina, R., Massicotte, H.B. and Trappe, J.M. (1992) Specificity phenomena in mycorrhizal symbioses: Community-ecological consequences and practical implications. In: Allen, M.F. (ed.), *Mycorrhizal Functioning: An Integrative Plant–Fungal Process*. Chapman and Hall, New York, pp. 357–423.

Newman, E.I. (1988) Mycorrhizal links between plants: Functioning and ecological significance. *Advances in Ecological Research* 18, 243–270.

Perry, D.A., Amaranthus, M.P., Borchers, J.G., Borchers, S.L. and Brainerd, R.E. (1989a) Bootstrapping in ecosystems. *BioScience* 39, 230–237.

Perry, D.A., Margolis, H., Choquette, C., Molina, R. and Trappe, J.M. (1989b) Ectomycorrhizal mediation of competition between coniferous tree species. *New Phytologist* 112, 501–511.

Perry, D.A., Borchers, J.G., Borchers, S.L. and Amaranthus, M.P. (1990) Species migrations and ecosystem stability during climate change; the belowground connection. *Conservation Biology* 4, 266–274.

Read, D.J., Francis, R. and Finlay, R.D. (1985) Mycorrhizal mycelia and nutrient cycling in plant communities. In: Fitter, A.H., Atkinson, D., Read, D.J. and Usher, M.B. (eds), *Ecological Interactions in Soil*. Blackwell Scientific Publications, Oxford, pp. 193–217.

Smith, A.H. and Zeller, S.M. (1966) A preliminary account of the North American species of *Rhizopogon. Memoirs of the New York Botanical Gardens* 14, 1–178.

Smits, W.Th.M. (1983) VIII. Dipterocarps and mycorrhiza, an ecological adaptation and a factor in forest regeneration. *Flora Malesiana Bulletin* 36, 3926–3937.

Trappe, J.M. (1962) Fungus associates of ectotrophic mycorrhizae. *Botanical Review* 28, 538–606.

Trappe, J.M. (1977) Selection of fungi for ectomycorrhizal inoculation in nurseries. *Annual Review of Phytopathology* 22, 203–222.

15) Effects of Liming on Pine Ectomycorrhiza

S. Erland

Lund University, Department of Microbial Ecology, Helgonavägen 5, S-223 62 Lund, Sweden

Introduction

Research on the effects of pollutants on different parts of the forest ecosystem and measures against these effects has been, and is currently being, carried out by several research groups in Sweden. The work presented here is a small part of a project financed and directed by the Swedish National Environmental Protection Agency. The aim has been to find out more about the effects of forest soil liming on ectomycorrhizal fungi.

Ideally applied research considering the effects of a specific treatment should lead to results which can be used, on a general basis, to determine whether or not to recommend the treatment on a large scale. It is, however, very important to stress that results which show how mycorrhizal efficiency would be affected by a certain lime application cannot be expected at present since basic knowledge of mycorrhizal function is still poor and the available techniques for studying the symbiosis are limiting. It is not easy to identify the species of a mycobiont involved in a given mycorrhizal association. If possible at all, identification involves laborious microscopic analysis (Agerer, 1987). As a further complication mycorrhizal fungi sometimes have a patchy pattern of distribution within the forests (Danielson and Visser, 1989; Dahlberg and Stenström, 1990) and large numbers of samples are required for acceptable statistical significance. At best a field study of mycorrhizal roots will show whether there has been a change in mycorrhizal types or of total infection due to a treatment.

Different fungi may, however, vary in their ability to provide the host plant with mineral nutrients and in their ability to improve stress tolerance (Harley and Smith, 1983). Their carbon demands are also likely to be different. These properties probably also change with different environmental conditions for the same fungus. To examine such qualitative effects further, a laboratory approach which allows greater control of environmental factors is needed. It is, however, desirable to keep the laboratory conditions as close to those of the natural system as possible. Too many conclusions about mycorrhizal symbiosis have been based

113

on data from pure culture of fungi, excised root tips or mycorrhizal plants grow-ing with the extramatrical mycelium in unnatural substrates (Harley and Smith, 1983).

The effects of liming on ectomycorrhiza were studied in a range of field and laboratory experiments. The host plant was *Pinus sylvestris* and fungi were iso-lated from root tips and used for experiments, even when the species could not be identified. Efforts were made to provide the extramatrical mycelium with as natural a substrate as possible.

The results presented should be seen as providing information on how some different ectomycorrhizal fungi, occurring in symbiosis with pine, react to liming. They also provide some guidance as to which methods can best be used to inves-tigate these matters further.

Results and Discussion

A summary of the different experiments and the most important results is pre-sented in Fig. 15.1.

Most of the fungi studied tolerated a broad pH range (3.8–7) both when grow-ing in pure culture on agar and on sterilized peat (II), when infecting from an inoculum introduced into unsterile humus (III) as well as when infecting from pro-pagules in the humus, both in the laboratory (I) and under field conditions (IV). The one exception found in this work was that of *Piloderma croceum* which did not grow at pH values higher than around 6. However, a pH increase to around 5 did not seem to affect mycorrhizal infection negatively on an overall quantitative basis. The highest percentage infection of the root systems in the laboratory bioassay (I) was obtained at this pH.

According to the data from the experiments where mycorrhizal plants were grown in split root chambers with different substrate pH values (II), even *P. croceum* could grow at pH 7 when peat at pH 3.8 was present in the same system. Thus, all fungi studied seemed able to tolerate growth in substrates of variable pH from pH 3.8 to 7, probably through compensatory effects of the host plants. It was obvious from the pH measurements of humus samples from the treated field plots (IV) that pH in the field areas was very variable. Twelve and 18 months respectively after ash and lime had been carefully spread by hand, ashed areas had humus pH values varying between pH 4 and 7.5 with a mean of 6.4, while in limed areas, where the mean pH was 5.2, samples with pH values from 3.6 to 7 were found.

Only one fungus, *Paxillus involutus*, reacted differently to lime and ash amendments (III). Both its ability to grow saprophytically and to infect plant roots were markedly decreased in humus amended with ash compared with lime. This may have been a result of the high zinc contents in the ash since it has been shown that *P. involutus* is sensitive to zinc (Denny and Wilkins, 1987). Careful analysis to ensure that such contamination is low is thus needed before ash can be con-sidered for application on large scale.

The same mycorrhizal types were found in the field bioassay (IV) as in the laboratory bioassay (I) with the addition of a 'white' type. There were differences

Pinus sylvestris L. forest in Torrmyra, south Sweden. Stand age ~45 yrs. Humus pH (H_2O) 3.8

Field treatments; wood-ash: 3.0 and 7.5 tonnes ha^{-1}, dolomite-lime: 2.0 and 5.0 tonnes ha^{-1}

I. (Erland and Söderström, 1990)

Laboratory bioassay with *Pinus sylvestris* seedlings grown in humus in a pH range from 4 to 7.5 for 4 months.

* Five different mycorrhizal types were found with different infection optima over the pH interval: *Piloderma croceum.*, *Cenococcum* sp. + three unidentified types designated pink, brown and coralloid.

* All types were found over the whole pH interval except *Piloderma croceum* which disappeared at pH > 6.

* The total mycorrhizal infection level increased from 70% at pH 4 to nearly 100% at pH 5.2 and then decreased to 40% at pH 7.5.

IV. (Erland and Söderström, 1991b)

Field bioassay with 1 year *Pinus sylvestris* seedlings grown in field plots for 4 months.

* Six mycorrhizal types were found, the same as in the laboratory assay + one white type.

* The pink type was twice as common in lime as in ash and control *Piloderma croceum* was more abundant in lime than in ash.

* The pH in the humus layer was very variable after the treatment.

I. (Erland and Söderström, 1990)

Pure culture isolation of fungi from mycorrhizal roots and synthesis of mycorrhiza with pure culture isolates.

II. (Erland, Söderström and Andersson, 1990)

Mycelial growthrates in pure culture and in symbiosis with *Pinus sylvestris* were tested for *Piloderma croceum*, pink and white long isolates.

* All isolates grew in pure culture in sterile peat.

* Radial growth rates of all mycelia were 2–5 times higher in symbiosis.

* *Piloderma croceum* did not grow at pH over 5.5 on sterile peat in pure culture but grew well at pH 7.3 in symbiosis.

V. (Erland, Finlay and Söderström, 1991)

Effects of substrate pH on translocation and partition of ^{14}C-labelled plant assimilates to the extramatrical mycelium of the pink fungus.

* No signifcant differences in plant uptake of labelled ^{14}C or translocation to the mycelium were found whether the peat pH was 3.8 or 5.2.

* Little label (0.3%) was present in the peat around non-mycorrhizal plants whereas peat colonized by mycelium contained a maximum of 5% of the ^{14}C originally supplied to the plant.

III. (Erland and Söderström, 1991a)

Growth and host plant colonization by five inoculated ectomycorrhizal fungi in lime and ash treated, unsterile humus.

* *Paxillus involutus* grew saprophytically in pH 3.6 untreated humus and pH 4.1, 4.8 limed humus.

* *S. bovinus* and 'white long' mycelia degraded on humus. No mycorrhizal infections.

* *Piloderma croceum* infected to pH 6 and the pink fungus in the whole pH interval regardless of treatment. No saprophytic growth.

VI. (Erland and Finlay, 1992)

Colonization of new host plants from inoculated plants. Effects of incubation time and temp.

* Colonization patterns after 6 months at 12°C resembled those after 2 months at 20°C.

* *Piloderma croceum* infections were still increasing after 6 months at 20°C, pink infection was the same as after 2 months and *Paxillus involutus* decreased after 2 months.

Fig. 15.1. Summary of the results from a range of field and laboratory experiments examining the effects of liming on pine ectomycorrhiza.

in the relative abundance of different mycorrhizal fungi on the plants which had
grown at corresponding pH values in the field areas for 4 months, and the ones
which had grown in the laboratory for the same amount of time (IV). These differences were mainly explained as the result of alterations in the growth rates
and competitive abilities of fungi, when growing from already colonized roots
compared with growth from fungal propagules in the humus. The positive effects
on gowth rates of mycorrhizal fungi growing in symbiosis compared with growth
in pure culture were clearly demonstrated (II). Some of these differences could,
however, be due to a difference in temperature during the incubation period.
The laboratory temperatures were 18°C day and 15°C night whereas the field
temperatures were probably lower during at least part of the experiment.

Effects of incubation time and temperature on colonization of root tips by
Piloderma croceum, *Paxillus involutus* and the pink fungus were examined (VI).
Colonization of the root systems which had grown at 12°C for 6 months was
similar to that on plants grown at 20°C for 2 months. The proportion of root tips
colonized by each fungus varied substantially at each harvest, *Paxillus involutus*
reached a maximum early in the experiment, the pink isolate being intermediate,
while *Piloderma croceum* infection was still increasing after 6 months in both
systems. These patterns corresponded with the relative radial growth rates of the
different fungi.

Since there are many ectomycorrhizal fungi which grow more slowly than
the ones we tested in experiment VI, and it took 6 months to detect infection of
P. croceum on a plant 2.5 cm away from a plant inoculated with the same fungus
at 12°C, it may well be that one growth season is not long enough to detect slow
growing fungi on seedlings, even though the fungi are present on older plants, and
capable of infecting the seedlings in the field. In experiment IV there was at least
one fungus, designated 'white long', which was found on the roots of mature trees,
but which never occurred on the bioassay seedlings.

The importance of incubation time was also illustrated by Dahlberg and
Stenström (1991). They examined the mycorrhizal infection on *Pinus sylvestris*
seedlings planted in clearcut and forest plots in central Sweden; 2, 4 and 17 months
after outplanting. Seedlings planted in June had seven different mycorrhizal types
in August, eight in October and 19 different types in September the following
year.

Studies of mycorrhizal infection of seedlings planted in soil from different
types of forests after treatments with artificial acid irrigation, sometimes in combination with liming, have been widely performed but with contadictory results.
Such contradictions may be explained by differences between the soils used in
the experiments. Probably the main effects on the symbiosis are not due to the pH
value as such (within reasonable limits) but merely a result of effects on nutrient
availability, availability of heavy metals or other growth limiting substances in the
soil.

The field bioassay (IV) demonstrated a large increase in the abundance of a
pink mycorrhizal fungus after liming to a mean pH of 5.2. Further studies were
performed to investigate the influence of substrate pH on carbon allocation to this
fungus (V), but significant differences between pH 3.8 and 5.2 could not be
shown.

A limitation in this work, and in mycorrhizal work in general, is the lack of knowledge considering the ecological relevance of the particular fungal isolates studied in the laboratory experiments. Knowledge of the below ground population dynamics of these fungi is poor, mainly because of the difficulties in identifying the mycobionts in mycorrhizal root tips. It is, however, well known that different species vary in their ability to provide nutrients to host plants and increase stress tolerance of the host plants (Harley and Smith, 1983). Sometimes the variation between different strains of the same species can be even larger (Malajczuk *et al.*, 1990; Sen, 1990).

Conclusions

Quantitative effects on ectomycorrhiza after liming nutrient poor pine forest humus to a pH around 5 mainly seemed to affect the relative distribution of mycorrhizal types, without resulting in a total decrease in infection. Data from the laboratory assay (I) even showed an increase in total infection from 70% at pH 4 to nearly 100% at pH 5.

There are contradictory results from different experiments where mycorrhizal infection has been studied in relation to application of acid precipitation and lime, depending on soil type. These results strongly suggest that effects of the treatments on nutrient availability together with effects on availability of heavy metals (e.g. Al), affect the mycorrhizal symbiosis more than the actual substrate pH. Therefore results obtained from one forest soil type cannot be directly applied to other soils.

The qualitative effects on the function of the symbiosis resulting from the above mentioned soil changes are still very poorly understood. Further research is needed and consideration must be paid to providing natural growth conditions for the extramatrical mycelium.

Further effort is needed to study the ecological importance of the fungal isolates which are used in laboratory studies. Somatic incompatability, isozyme methods and molecular biological methods such as RFLP and PCR may provide useful tools in identifying the fungi and allow studies of population dynamics in the field.

Acknowledgements

This work was financially supported by the Swedish National Environmental Protection Agency and the Swedish Natural Science Research Council. I thank my co-workers, Professor B. Söderström, Dr R.D. Finlay, Ms S. Andersson and Ms A.-M. Sonnerfeldt.

References

Agerer, R. (1987) *Colour Atlas of Ectomycorrhizae*. Einhorn-Verlag, Munich.

Dahlberg, A. and Stenström, E. (1991) Dynamic changes in nursery and indigenous mycorrhiza of *Pinus sylvestris* seedlings planted-out in forests and clearcuts. *Plant and Soil* 136, 73–86.

Danielson, R.M. and Visser, S. (1989) Effects of forest soil acidification on ectomycorrhizal and vesicular-arbuscular mycorrhizal development. *New Phytologist* 112, 41–47.

Denny, H.J. and Wilkins, D.A. (1987) Zinc tolerance in *Betula* spp. III. Variation in response to zinc among ectomycorrhizal associates. *New Phytologist* 106, 535–544.

Erland, S. (1990) Effects of liming on pine ectomycorrhiza. Doctoral Thesis, Dept. of Microbial Ecology, University of Lund, Sweden.

Erland, S. and Finlay, R.D. (1992) Effects of temperature and incubation time on the ability of three ectomycorrhizal fungi to colonize *Pinus sylvestris* roots. *Mycological Research* 96, 270–272.

Erland, S. and Söderström, B. (1990) Effects of liming on ectomycorrhizal fungi infecting *Pinus sylvestris* L. I. Mycorrhizal infection in limed humus in the laboratory and isolation of fungi from mycorrhizal roots. *New Phytologist* 115, 675–682.

Erland, S. and Söderström, B. (1991a) Effects of liming on ectomycorrhizal fungi infecting *Pinus sylvestris* L. III. Saprophytic growth and host plant infection at different pH values by some ectomycorrhizal fungi in unsterile humus. *New Phytologist* 117, 405–411.

Erland, S. and Söderström, B. (1991b) Effects of lime and ash treatments on ectomycorrhizal infection of *Pinus sylvestris* L. seedlings planted in a pine forest. *Scandinavian Journal of Forest Research* 6, 519–525.

Erland, S., Söderström, B. and Andersson, S. (1990) Effects of liming on ectomycorrhizal fungi infecting *Pinus sylvestris* L. II. Growth rates in pure culture at different pH values compared to growth rates in symbiosis with the host plant. *New Phytologist* 115, 683–688.

Erland, S., Finlay, R.D. and Söderström, B. (1991) The influence of substrate pH on carbon translocation in ectomycorrhizal and non mycorrhizal pine seedlings. *New Phytologist* 119, 235–242.

Harley, J.L. and Smith, S.E. (1983) *Mycorrhizal Symbiosis*. Academic Press, London.

Malajczuk, N., Lapeyrie, F. and Garbaye, J. (1990) Infectivity of pine and eucalypt isolates of *Pisolithus tinctorius* on roots of *Eucalyptus urophylla in vitro*. 1. Mycorrhiza formation in model systems. *New Phytologist* 114, 627–631.

Sen, R. (1990) Isozymic identification of individual ectomycorrhizas synthesized between Scots pine (*Pinus sylvestris* L.) and isolates of two species of *Suillus*. *New Phytologist* 114, 617–626.

16 Variations in Field Response of Forest Trees to Nursery Ectomycorrhizal Inoculation in Europe

F. Le Tacon[1], I.F. Alvarez[2], D. Bouchard[1], B. Henrion[1], R.M. Jackson[3], S. Luff[3], J.I. Parlade[2], J. Pera[2], E. Stenström[4], N. Villeneuve[5] and C. Walker[6]

[1]*Laboratoire de microbiologie forestière, INRA, Champenoux, 54280 Seichamps, France: [2]Department of Plant Pathology, (IRTA), Centre de Cabrils, 08348 Cabrils, Barcelona, Spain: [3]School of Biological Sciences, University of Surrey, Guildford, Surrey, UK: [4]Department of Forest Mycology and Pathology, Swedish University of Agricultural Sciences, Uppsala, Sweden: [5]Centre de recherche en biologie forestière, Université Laval, Quebec, Canada: [6]Forestry Commission, Northern Research Station, Roslin, Midlothian, UK*

Introduction

In Europe, as in many temperate countries, it was long thought that there was little need for ectomycorrhizal inoculation. However, cases in which ectomycorrhizal inoculation is beneficial after outplanting are now being found frequently. Unfortunately the variation in field response of forest trees to nursery inoculation with ectomycorrhizal fungi is very high (Le Tacon *et al.*, 1987, 1988).

In this chapter we describe field experiments, which have been conducted during the last decade in western Europe, and attempt to explain why the results can be positive or negative. The most important factor influencing success or failure is the vigour of the natural ectomycorrhizal population.

Ectomycorrhizal Status of European Reforestation Sites

Ectomycorrhizal fungi can be found on about 90% of the trees in temperate forests. Each tree species can be associated with many fungal species. Stands of mixed host species are normally richer in ectomycorrhizal fungal species than are pure stands. Thus for example, in a mixed natural stand of oak, beech and birch the fungi common to all three hosts are found together with those specific to each one. Many of the species common to the three native trees are also able to form mycorrhizas with exotic conifers, such as Douglas fir. However the native fungal strains have been selected after a long co-evolution with their habitual hosts, and while they are perfectly adapted to these and to the prevailing ecological conditions, including the other microorganisms of the rhizosphere, they may not be the most suitable for exotic forest species.

It is difficult to obtain good evidence concerning the survival of ectomycorrhizal fungi after the removal of their original hosts. However, the deforestation that has occurred in Europe has resulted in the disappearance of the majority of the ectomycorrhizal fungi from cultivated soils. Nevertheless sites devoid of original host plants are not completely devoid of ectomycorrhizal inoculum because air-borne spores can be carried over very long distances. Consequently in all European reforestation sites there is a natural inoculum, but it may be one of limited value to introduced exotic tree species.

Description and Results of Eight Field Experiments

The results of eight field experiments each carried out under distinctive conditions of past land-use practice are described. In each case presentation of the major results of the experimental inoculation is preceded by a brief site description. Previous land use involved either unmanaged natural vegetation (Expt 1), sites which had been previously cultivated for agricultural use (Expts 2, 3, 4), sites which had been planted with a previous coniferous crop (Expts 5, 6) or sites which had previously supported a natural or semi-natural forest (Expts 6, 7 and 8).

Except for the first and the sixth experiment the control seedlings are naturally infected during the nursery phase, mostly by the ubiquitous fungus *Thelephora terrestris*.

Experiment 1. (Previously uncolonized natural site)

> Location: North of Scotland UK – Shin. Annual precipitation: 1700 mm. Elevation: 170 m. Soil: Deep peat site drained to 25–40 cm. Vegetation: *Scirpus* sp. *Juncus* sp. *Sphagnum* sp. *Calluna vulgaris*. Fertilization: Unground rockphosphate – 100 kg ha^{-1} (13.3 kgP ha^{-1}) – 500 kg ha^{-1} (66.3 kgP ha^{-1}). Host species: Sitka spruce. Inoculum species: *Thelephora terrestris* FC 005. *T. terrestris* TT3, *Laccaria proxima* FC 003. (Walker, 1991)

There were ten treatments, but we report results of only four. The results are expressed as total height or as differences from the control in total height.

At the lower level of phosphorus (P), growth of Sitka spruce with one *Thelephora* strain did not differ from that of the control, while with another strain of *Thelephora* and *L. laccata* growth was depressed (Fig. 16.1). At the higher level of P fertilization the three inoculated fungi produced increases of growth in spruce relative to that in controls (Fig. 16.2). Six years after outplanting there were *L. proxima* fruit bodies in all the *Laccaria* plots. In *Thelephora* inoculated plots the

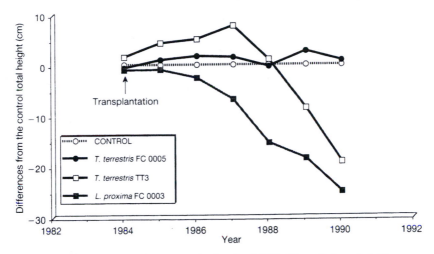

Fig. 16.1. Experiment 1: Shin 105 p 84 Scotland (UK). Growth (tree height expressed as differences from the control) of Sitka Spruce (*Picea sitchensis*) containerized seedlings mycorrhizal with four different fungal strains, six years after outplanting in a drained and fertilized (100 kg unground rock phosphate ha^{-1} or 13.3 kg P ha^{-1}) deep peat.

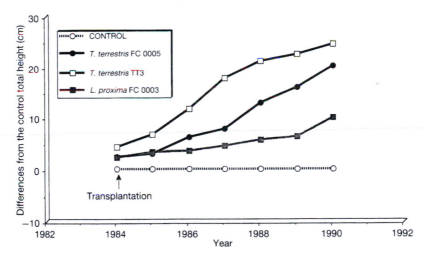

Fig. 16.2. Experiment 1: Shin 105 p 84 Scotland (UK). Growth (tree height expressed as differences from the control) of Sitka Spruce (*Picea sitchensis*) containerized seedlings mycorrhizal with four different fungal strains, six years after outplanting in a drained and fertilized (500 kg unground rock phosphate ha^{-1} or 66.3 kg P ha^{-1}) deep peat.

mycorrhizas were exclusively of *Thelephora*. In the control plots the short roots were non-mycorrhizal or mycorrhizal with *T. terrestris*.

It is evident that in such poor sites without sufficient added P the inoculation can have a depressive effect. The effect of the inoculation produces positive responses at higher levels of P application. However, the effects of inoculation are even then not very striking, because site fertility is inadequate even with 500 kg of unground phosphate. Since it is likely that these introduced fungal strains were not particularly efficient, despite the low pressure of competition from naturally-occurring mycorrhizal fungi, it is remarkable that responses to inoculation occurred.

Experiment 2. (Previously cultivated site 1)

Location: Central France – Royeres. Elevation: 750 m. Annual precipitation: 1700 mm. Soil: Brown podzolic soil over granite improved by prolonged cultivation. Vegetation: *Calluna* heathland. Fertilization: None or 100 kgP ha^{-1}. Host Species: Douglas fir. Inoculum Species: *Hebeloma cylindrosporum*. (Le Tacon *et al.*, 1988)

On this *Calluna* heathland site 2-year-old bare-root seedlings of Douglas fir which were mycorrhizal either with the naturally occurring fungus *Thelephora terrestris* or artificially inoculated with *Hebeloma cylindrosporum*. *H. cylindrosporum* had a positive effect during the nursery phase, but after outplanting the growth of the seedlings infected with *H. cylindrosporum* was very poor compared with that of the control (Figs 16.3 and 16.4). *H. cylindrosporum* appeared not to be adapted to the site conditions and did not survive after transplantation. The seedlings were therefore without mycorrhizas soon after transplanting and remained so for several years. As a consequence, 8 years after outplanting, these plants were 30% smaller than the controls without fertilization.

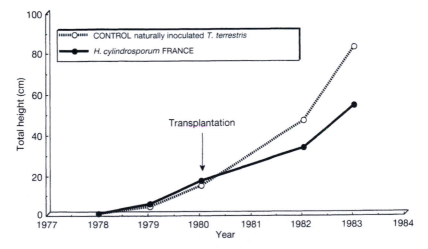

Fig. 16.3. Experiment 2: Royeres (France). Growth (tree height) of Douglas fir (*Pseudotsuga menziesii* (Mir)) bare-root seedlings (Peyrat Le Chateau nursery) mycorrhizal with two different fungal strains, two years after outplanting in a non-fertilized *Calluna* heathland.

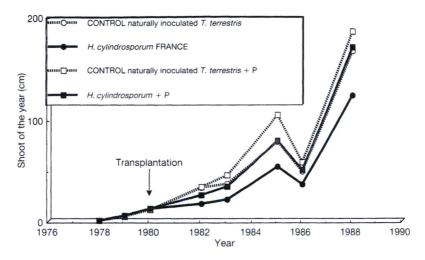

Fig. 16.4. Experiment 2: Royeres (France). Growth (shoot of the year) of Douglas fir (*Pseudotsuga menziesii* (Mir)) bare-roots seedlings (Peyrat Le Chateau nursery) mycorrhizal with two different fungal strains, five years after outplanting in a non-fertilized or fertilized (100 kg ha^{-1}) *Calluna* heathland.

Experiment 3. (Previously cultivated site 2)

> Location: Central France – Longechaud. Elevation: 580 m. Annual precipitation: 1700 mm. Soil: Brown podzolic soil over granite, improved by prolonged cultivation. Vegetation: Grassland (*Agrostis* sp. *Poa* sp. etc). Fertilization: None. Host species: Douglas fir. Inoculum species: *Laccaria laccata* S 238 A USA. (Le Tacon *et al.*, 1988)

On this site 2-year-old bare-root seedlings of Douglas fir were either naturally infected with *T. terrestris* and an unidentified *Suillus* sp., or inoculated with *Laccaria laccata* S 238, a strain from the west coast of USA and slightly contaminated by the same *Suillus* as in the control. The strain S 238 is now known to be a *Laccaria bicolor* isolate (Armstrong *et al.*, 1989).

At the end of the nursery phase, the *Laccaria* inoculated seedlings had greater height than the control. While this difference was still visible 6 years after outplanting, the inoculation with *Laccaria* did not result in a further growth increase after transplantation (Fig. 16.5). This is probably attributable to the fact that *Laccaria* mycorrhizas disappeared immediately after outplanting, while those of *Thelephora* and *Suillus* sp. survived. *Laccaria* sporophores were never seen at the site. The disappearance of *Laccaria* may have been attributable to the effects of the grass cover. In contrast to the responses seen in the previous experiment the absence of inoculant mycorrhizas had no depressive effect on Douglas fir growth, possibly because the inoculated seedlings, like the control, were slightly mycorrhizal with *Suillus* sp. before planting.

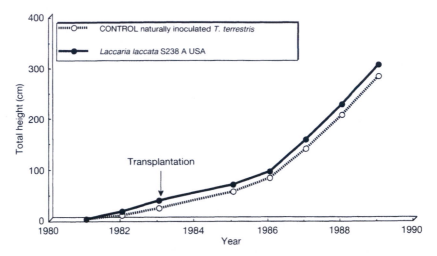

Fig. 16.5. Experiment 3: Longechaud (France). Growth (tree height) of Douglas fir
(*Pseudotsuga menziesii* (Mir)) bare-roots seedlings (Peyrat Le Chateau nursery)
mycorrhizal with two different fungal strains, five years after outplanting in a
previously cultivated soil (grassland).

Experiment 4. (Previously cultivated site 3)

> Location: Central France – Ballandeix. Elevation: 650 m. Annual precipita-
> tion: 1300 mm. Soil: Brown podzolic soil over granite, improved by prolonged
> cultivation. Vegetation: Bracken (*Pteridium aquilinum*). Fertilization: None.
> Host species: Douglas fir. Inoculum species: *Laccaria laccata* S 283 A, USA,
> *Laccaria bicolor* 81306 France. (Le Tacon *et al.*, 1988)

Two-year-old bare-root Douglas fir seedlings were used, the controls being
naturally mycorrhizal with *Thelophora terrestris*, the remaining seedlings being
inoculated either with *Laccaria laccata* S 238, from USA, or with *Laccaria bicolor*,
a French strain isolated from a plantation of Douglas fir. All the treatments were
slightly contaminated at the end of the second year of the nursery phase by the
same unidentified *Suillus* sp., nevertheless the differences between the three
treatments were always significant. The advantages conferred on seedlings by
inoculation with the two *Laccaria* strains were striking (Fig. 16.6). If the results are
represented on the basis of differences in annual shoot growth increment the
response to the two strains of *Laccaria* is seen to be distinct (Fig. 16.7). Immediately
after outplanting, during the period of transplantation shock, the French *Laccaria*
strain had a depressive effect, but 7 years after outplanting the two strains of
Laccaria were equally effective, and produced much better growth than the controls. These two inoculant fungi survived very well at this site. Three years after
outplanting 60% of the trees previously inoculated with *L. laccata* S 238 produced
sporophores of that fungus, as did 42% of those inoculated with the French strain
of *L. bicolor*. In the control there were no *Laccaria laccata* fruit bodies and only

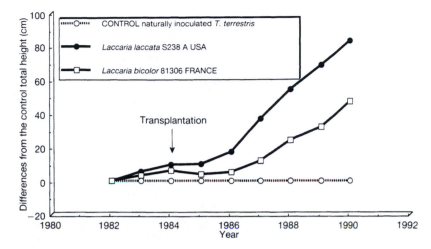

Fig. 16.6. Experiment 4: Ballandeix (France). Growth (tree height expressed as differences from the control) of Douglas fir (*Pseudotsuga menziesii* (Mir)) bare-roots seedlings (Peyrat Le Chateau nursery) mycorrhizal with three different fungal strains, six years after outplanting in a previously cultivated soil.

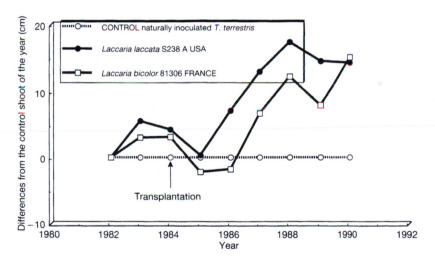

Fig. 16.7. Experiment 4: Ballandeix (France). Growth (shoot of the year expressed as differences from the control) of Douglas fir (*Pseudotsuga menziesii* (Mir)) bare-roots seedlings (Peyrat Le Chateau nursery) mycorrhizal with three different fungal strains, six years after outplanting in a previously cultivated soil.

0.5% of the trees produced *L. bicolor* sporophores. Six years after transplantation the inoculation by *L. laccata* S 238 led to a doubling of the total volume of wood, whereas the French strain of *L. bicolor* has increased it by about 50% (Fig. 16.8).

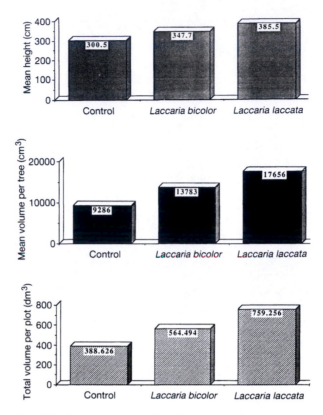

Fig. 16.8. Experiment 4: Ballandeix (France). Mean height (cm), mean volume per tree (cm³) and total volume per plot (dm³) of the three treatments 8 years after inoculation and 6 years after outplanting.

Experiment 5. (Artificial forest site 1) Reforestation with another species

Location: North Yorkshire UK – Wykeham. Elevation: 200 m. Annual precipitation: 1100 mm. Soil: Podzolic soil on sandstone. Vegetation: Site clear cut. Previous plantation of European larch. Fertilization: None. Host species: Douglas fir. Inoculum species: *T. terrestris* R 34, R 38, *L. proxima* 64, *L. laccata* S 238. (Walker, 1991; Jackson, 1991)

On this site, the vigour of natural ectomycorrhizal inoculum is higher than in the previous ones. One-year-old containerized Douglas fir seedlings were compared when mycorrhizal with four different fungal species (Fig. 16.9). The control seedlings were non-mycorrhizal at transplanting and the inoculated seedlings were mycorrhizal either with *Laccaria laccata* S 238, a British strain of *L. proxima* or with one of two British isolates of *Thelephora terrestris*.

Contamination of seedlings occurred after planting in many plots. In 1990, sporophores of *Laccaria* species were found in five control plots and in some plots

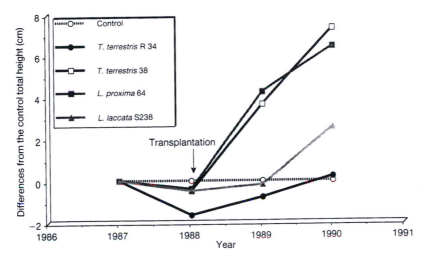

Fig. 16.9. Experiment 5: NYM 43 P 89 Wykeham, Yorkshire (UK). Growth (tree height expressed as differences from the control) of Douglas fir (*Pseudotsuga menziesii* (Mir)) containerized seedlings mycorrhizal with five different fungal strains two years after outplanting in a non-fertilized podzolic soil on sandstone.

where *T. terrestris* has been artificially introduced. *Paxillus involutus* was also recorded in some plots. However, *Laccaria proxima* fruit bodies occurred in seven out of the ten plots of the *proxima* treatment. Sporophores of *L. laccata* S 238 never appeared. Trees inoculated with *Thelephora* strain R 34 and *L. laccata* S 238 were no larger than the controls. *L. proxima* and *T. terrestris* B 34 produced significantly better growth than the other strains and from the control.

From the experience elsewhere it was known that the *Laccaria* strain S 238 from the United States was normally a very efficient symbiont on Douglas fir. It is possible that the inoculum, which has been produced in France and sent to Great Britain, had been contaminated by saprophytic fungi.

Experiment 6. (Artificial forest site 2) Reforestation with another species

Location: Galicia Spain – Caldas de Reyes. Elevation: 550 m. Annual precipitation: 1500 mm. Soil: Brown podzolic. Vegetation: Site clear-felled. Previous plantation of *Pinus pinaster*. Fertilization: None. Host species: Douglas fir. Inoculum species: *L. laccaria* S 238. (Alvarez, Parlade and Pera, unpublished data)

This site also contained a vigorous natural inoculum. Two-year-old bare-root Douglas fir seedlings were produced in France in two different nurseries, Morvan and Peyrat. The origin of the seeds (Washington zone 422) and the inoculated fungus (*L. laccata* S 238, USA), were the same in the two nurseries. The microflora was different in the two nurseries despite the fact that the naturally occurring fungal species were the same and were dominated by *Thelephora terrestris* and *Suillus* sp. In the Morvan nursery the natural microflora was more active. Irrespective of the nursery of origin the growth of Douglas fir seedlings associated with *Laccaria laccata* S 238 was the same (Figs 16.10 and 16.11). Differences from the control were small, the seedlings from the nursery (Peyrat) with the least effective natural inoculum being slightly favoured by inoculation.

These results show that, even with vigorous natural inoculum, the mycorrhizal status inherited from the nursery, continues to influence the growth of the trees

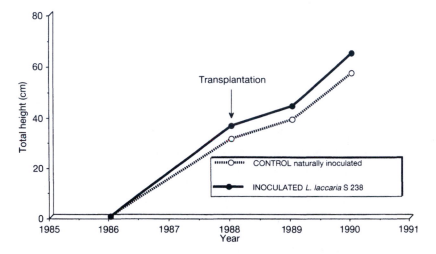

Fig. 16.10. Experiment 6: Caldas de Reyes Galicia (Spain). Growth (tree height) of Douglas fir (*Pseudotsuga menziesii* (Mir)) bare-roots seedlings (Morvan nursery) mycorrhizal with two different fungal strains two years after outplanting in a non-fertilized brown podzolic soil.

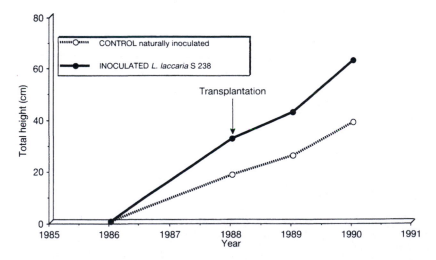

Fig. 16.11. Experiment 6: Caldas de Reyes Galicia (Spain). Growth (tree height) of Douglas fir (*Pseudotsuga menziesii* (Mir)) bare-roots seedlings (Peyrat Le Chateau nursery) mycorrhizal with two different fungal strains two years after outplanting in a non-fertilized brown podzolic soil.

after outplanting. This is true not only for the artificially inoculated fungi, but also for the natural microflora of the nursery.

Experiment 7. (Natural forest site) Reforestation with another species

> Location: Vosges – France – Brouvelieures. Elevation: 460 m. Annual pre-
> cipitation: 1000 mm. Soil: Podzolic soil on sandstone. Vegetation: Mixed
> forest, broadleaf trees and conifers (beech, oak, Norway spruce, and Scots
> pine). Fertilization: None. Host species: Douglas fir. Inoculum species:
> *Laccaria laccata* S 238. (Villeneuve *et al.*, 1991)

In this site, the mixed stand had sustained a rich natural ectomycorrhizal microflora. However, fungi had not co-evolved with Douglas fir and were not necessarily
compatible with this exotic species, of which two-year-old bare-root seedlings had
been planted. Plants naturally mycorrhizal with *T. terrestris* and an unidentified
Suillus, were compared with inoculated seedlings mycorrhizal with *Laccaria laccata*
S 38, from USA and slightly contaminated by the same *Suillus* as the control.

The inoculated seedlings grew significantly better than the controls during
the nursery phase and after outplanting (Fig. 16.12). At the end of the second
growing season after outplanting 20 seedlings of each treatment were randomly
selected. Seedlings were carefully hand excavated. Each root system was subdivided into three sections: 0–10 cm, 10–20 cm and >20 cm, based on the distance from the collar. The total number of root tips was determined and the
different types of fungi identified. The percentage of mycorrhizal infection according to distance from the collar was calculated (Figs 16.13 and 16.14). In the control plots, while the *Laccaria* mycorrhizas were not dominant, they were found on

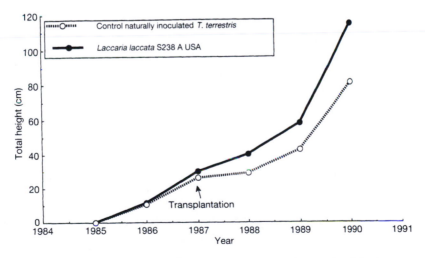

Fig. 16.12. Experiment 7: Brouvelieures (France). Growth (tree height) of
Douglas fir (*Pseudotsuga menziesii* (Mir)) bare-roots seedlings (Morvan nursery)
mycorrhizal with two different fungal strains two years after outplanting in a
non-fertilized podzolic sandstone.

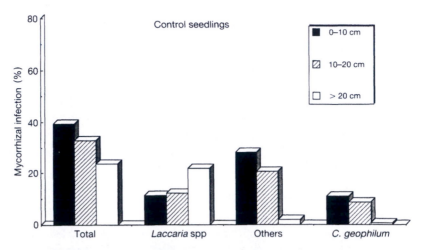

Fig. 16.13. Experiment 7: Brouvelieures (France). Colonization of 5-year-old Douglas-fir seedlings by different mycobionts according to radial distance from stems. Seedlings were previously inoculated with a naturally occurring fungus (*Thelephora terrestris*). Plants were 2 years old at the time of outplanting onto a clearcut mixed pine site.

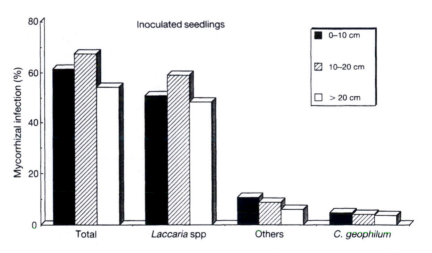

Fig. 16.14. Experiment 7: Brouvelieures (France). Colonization of 5-year-old Douglas-fir seedlings by different mycobionts according to radial distance from stems. Seedlings were previously inoculated with *Laccaria bicolor* (S238). Plants were 2 years old at the time of outplanting onto a clearcut mixed pine site.

new root tips. The other symbionts colonized the old roots after the establishment of *Laccaria*.

In the inoculated seedlings, the total mycorrhizal infection was higher, 60% compared with 40% in the control, and *Laccaria* mycorrhizas were dominant. We assume that this dominance is due to the competitiveness of the introduced *Laccaria* S 238 strain, which acted as a pioneer fungus and progressively colonized the newly formed root tips on all parts of the root system. This strain also persists very well on the old roots and reduces colonization by the native symbionts which are also less diverse than in the control.

Experiment 8. (Natural forest sites) Reforestation with the same natural species

> Location: Central Sweden – Acktjarnsgraven. Soil: Sandy glacial till. Vegetation: Vaccinio-pinetum (Scots pine) clear-felled 4 years before planta-tion and prepared by harrowing 1 year before plantation. Fertilization: None. Host species: *Pinus sylvestris*. Inoculum species: *L. laccata* (S 238), *Cenococcum geophilum, H. crustuliniforme*. (Stenström *et al.*, 1990)

At this site the natural inoculum is particularly vigorous. The response of 1-year-old containerized Scots pine seedlings to inoculation with three mycorrhizal fungi was determined (Fig. 16.15). After the nursery phase, the control seedlings were naturally infected by *T. terrestris*. The inoculated seedlings were mycorrhizal either with S 238 (USA), or with *Cenococcun geophilum* a French strain isolated from a plantation of Norway spruce, or with *H. crustuliniforme*, a Swedish strain isolated from a natural stand of Scots pine. All the introduced strains had a depressive effect, even S 238 which was so efficient on Douglas fir in most other sites. The inoculated fungi had not disappeared 5 years after outplanting and *L. laccata* S 238 fruit-body production was abundant around inoculated seedlings. It was there-fore assumed that the native fungi were sufficiently well adapted to Scots pine

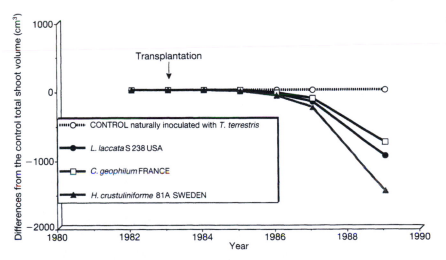

Fig. 16.15. Experiment 8: Acktjarnsgraven (Sweden). Growth (total shoot volume (cm^3) expressed as differences from the control) of Scots pine (*Pinus sylvestris*) containerized seedlings mycorrhizal with four different fungal strains, five years after outplanting in a forest soil glacial deposit till with a sandy composition.

to have a competitive superiority that reduced effectiveness of the introduced strains.

Conclusion

We now have the techniques for production of tree seedlings infected with competitive and efficient ectomycorrhizal fungi. These techniques are available for inoculation of bare-root seedlings in classical nurseries after soil fumigation, or for the production of infected containerized seedlings on artificial substrates. However, before ectomycorrhizal inoculation can be used on a large scale cheaper methods for the production of commercial ectomycorrhizal inoculum must be developed.

In western Europe there are now about 40 field experiments. This number is much lower than that in USA (Marx and Cordell, 1988), yet some interesting conclusions can be drawn.

In most experiments, several years after outplanting, there is an effect, positive or negative, of the mycorrhizal status inherited from the nursery phase. This result is very important by itself and makes evident the necessity of a better management of the ectomycorrhizal fungi during the nursery phase. The magnitude of the field response is influenced also by the quality of the natural inoculum in the nursery.

After transplantation to the forest site the most important factor determining the success or the failure of introduced inoculum appears to be the vigour of the natural ectomycorrhizal inoculum. In Europe this is normally high, even in the sites previously devoid of hosts. This is due to the fact that sporophores of epigeous basidiomycetes and ascomycetes have a seemingly limitless profligacy in the production of air-borne spores, which can be carried over very long distances.

The introduced fungi have to be adapted to the ecological conditions of the reforestation sites: climate, soil pH, fertility and other components of the ecosystems, in particular the native vegetation. More attention must be paid to possible antagonistic effects of other species such as grasses. To improve field responses quantitatively and to enlarge the number of sites where a positive response can be obtained efforts should be concentrated in two areas: the factors governing the competitive abilities and the factors governing the efficiency of the introduced fungi.

New methods such as RFLP and PCR, gene sequencing and DNA probes could become extraordinarily powerful tools for identifying fungal species or fungal strains associated with their hosts in field situations, even on a single mycorrhiza, and so assist in assays of distribution and competitive ability of introduced fungi. With these tools, competition studies, which are still in there infancy, could become a reality (Armstrong *et al.*, 1989; Gardes *et al.*, 1991; Martin *et al.*, 1991; Henrion *et al.*, 1991).

Competition between fungal strains is governed by intrinsic factors but also by interactions with other soil microorganisms. Work in Nancy (Duponnois and Garbaye, 1991) on the interactions between mycorrhizal colonization and rhizosphere bacteria could lead to improvements in competitive abilities of strains. We do not know why one strain is more efficient than another. The genetic selection of ectomycorrhizal fungi, which is now in progress in Quebec and in France could become a powerful tool to create more efficient strains. But what are the criteria to be followed in selection of the most efficient ectomycorrhizal fungi? What physiological criteria have to be employed? Could these criteria be achieved by gene transfer techniques which are now available for application to ectomycorrhizal fungi?

There are still large gaps between the levels of understanding achieved by ecologists, physiologists and molecular biologists. We have all to make efforts to reduce them.

Dedication

This paper is dedicated to the memory of Philippe Le Tacon who gave his life for science, July 12th 1991.

References

Armstrong, J.L., Fowles, N.L. and Rygiewicz, P.T. (1989) Restriction fragment length polymorphisms distinguish ectomycorrhizal fungi. *Plant and Soil* 116, 1–7.

Barrett, V., Dixon, R.R. and Lemke, P.A. (1990) Genetic transformation of a mycorrhizal fungus. *Applied Microbiology and Biotechnology* 33, 313–316.

Duponnois, R. and Garbaye, J. (1991) Some bacteria associated with *Laccaria laccata* ectomycorrhizas or sporocarps: effect on symbiosis establishment on Douglas fir *in vitro* and in glasshouse conditions. *Annales de Sciences Forestières*, 22 (3) 239–251.

Gardes, M., White, T.J., Fortin, J.A., Bruns, T.D. and Taylor, J.W. (1991) Identification of indigenous and introduced symbiotic fungi in ectomycorrhizae by amplification of nuclear and mitochondrial ribosomal DNA. *Canadian Journal of Botany* 69, 180–190.

Henrion, B., Le Tacon, F. and Martin, F. (1992) Rapid identification of genetic variation of ectomycorrhizal fungi by amplification of ribosomal RNA genes. *New Phytologist* (in press).

Jackson, R.M. (1991) Final report on contract '*Application of mycorrhizal technology in afforestation of marginally economic land.*' (MAIB.0086.UK (H)).

Le Tacon, F., Garbaye, J. and Carr, G. (1987) The use of mycorrhizas in temperate and tropical forests. *Symbiosis* 3, 179–206.

Le Tacon, F., Garbaye, J., Bouchard, D., Chevalier, G., Olivier, J.M., Guimberteau, J., Poitou, N. and Frochot, H. (1988) Field results from ectomycorrhizal inoculation in France. In: Lalonde, M. and Piche, Y. (eds), *Canadian Workshop on Mycorrhizae in Forestry*, 1–4 May 1988. CRBF, Faculté de Foresterie et de Géodésie, Université Laval, Ste-Foy, Québec, G1K 7P4.

Martin, F., Zaiou, M., Le Tacon, F. and Rygiewicz, R. (1991) Strain specific differences in ribosomal DNA from the ectomycorrhizal fungi *Laccaria bicolor* (Maire) Oron and *Laccaria laccata* (Scop ex Fr) Br. *Annales de Sciences Forestières* 48, 297–305.

Marx, D.H. (1980) Ectomycorrhizal fungus inoculations: a tool for improving forestation practices. In: Mikola, P. (ed.), *Tropical Mycorrhiza Research* Oxford University Press, Oxford, pp. 13–71.

Marx, D.H. and Cordell, C.E. (1988) Specific ectomycorrhizae improve reforestation and reclamation in the eastern United States. In: Lalonde, M. and Piche, Y. (eds), *Canadian Workshop on Mycorrhizae in Forestry*, 1–4 May 1988. CRBF, Université Laval, Sainte Foy, Québec, G1K 7P4, pp. 75–86.

Stenström, E., Ek, M. and Unestam, T. (1990) Variation in field response of *Pinus sylvestris* to nursery inoculation with four different ectomycorrhizal fungi. *Canadian Journal of Forest Research* 20, 1796–1803.

Villeneuve, N., Le Tacon, F. and Bouchard, D. (1991) Survival of inoculated *Laccaria bicolor* in competition with native ectomycorrhizal fungi and effects on the growth of outplanted Douglas-fir seedlings. *Plant and Soil* 135, 95–107.

Walker, C. (1991) Final technical report of project MA1B-0088-UK (AM). *Application of mycorrhizal technology in afforestation*.

17 Somatic Incompatibility – A Tool to Reveal Spatiotemporal Mycelial Structures of Ectomycorrhizal Fungi

A. Dahlberg

Department of Forest Mycology and Pathology, The Swedish University of Agricultural Science, P.O. Box 7026, S-750 07 Uppsala, Sweden

Introduction

Studies in fungal population ecology, which are dependent on the ability to identify fungal individuals, have recently emerged as a way to reveal new aspects of the biology of ectomycorrhizal fungi. The concept of the individualistic mycelia emerged in the late 1970s. Todd and Rayner (1980) showed that genetically different mycelia of the same fungal species can coexist as physiologically and spatially discrete individuals. Each genetically unique mycelium within a fungal population occupies a territorial domain, a genetic territory. Formerly, it was thought to be difficult or even impossible to define a fungal 'individual'. According to the concepts of the 'unit mycelium' and heterokaryosis, genetically different individuals of the same species could, via hyphal anastomosis, fuse and form a new functional unit. However, it is now realized that these concepts may have a much more restricted significance in natural populations than had previously been proposed (Gregory, 1984). Fungal individualism is based on somatic incompatibility, which is a genetic system that allows the recognition and subsequent acceptance or rejection of 'self' or 'non-self' mycelia. The underlying genetics behind somatic incompatibility expression is complex, polygenic or multiallelic (Rayner, 1991a). Somatic incompatibility seems to be widespread among fungi, myxomycetes, ascomycetes and basidiomycetes (Lane, 1981; Rayner, 1991a). So far, about 70 basidiomycetes, mainly wood-decomposing, saprophytic or pathogenic but also ectomycorrhizal (Table 17.1), have been thus studied in relation to population structure and mycelial interaction (Rayner, 1991a).

Terminology

The central unit in population ecology is the individual. The concept of a genet which is defined as a discrete genetic unit or assembly within a species, is generally preferred to the similar but less precise concepts of individual or clone (Braiser and

Table 17.1. List of ectomycorrhizal fungi studied and their expression of somatic incompatibility *in vitro*.

Showing varying degree of somatic incompatibility	
Laccaria bicolor	(T.W. Kuyper, pers. comm.)
Paxillus involutus	(pers. obs.)
Rhizopogon rubescens	(S. Miller, pers. comm.)
Rhizopogon subcaerulescens	(K. Bergeson and S. Miller, pers. comm.)
Suillus brevipes	(S. Miller, pers. comm.)
Suillus bovinus	(Dahlberg and Stenlid, 1990 and 1991; Sen, 1990)
Suillus granulatus	(Fries and Neuman, 1990)
Suillus luteus	(Fries, 1987)
Suillus tomentosus	(K. Bergeson and S. Miller, pers. comm.)
Suillus variegatus	(Fries, 1987; Sen, 1990)
Mentioned as being somatic incompatible	
Amanita muscaria	(Fries, 1987)
Hebeloma mesophaeum	(Fries, 1987)
Laccaria proxima	(Fries, 1987)
Mentioned as not showing somatic incompatibility	
Paxillus involutus	(Fries; 1987, D. Mitchell, pers. comm.)
Pisolithus tinctorius	(Fries, 1987)
Telephora terrestris	(Fries, 1987)
Suillus grevillei	(D. Mitchell, pers. comm.)

Rayner, 1987). A population is an aggregate of genets within one species ranging from very local to world-wide settings.

The general life cycle of higher basidiomycetes is divided into sexual and vegetative phases, the primary and secondary mycelia each being capable of independent growth. The primary mycelium, or thallus, originates by germination of basidiospores and is hence both haploid, monokaryotic, and homokaryotic. The secondary mycelium originates as a result of fusion between hyphae of two primary thalli or of a primary thallus with a secondary mycelium, with complementary mating type alleles. The secondary mycelium constitutes the main phase for basidiomycetes. If no secondary mycelium occurs, the primary homokaryotic mycelium constitutes the vegetative phase. Such species are said to be homothallic since they may form sporocarps without mating. In some species, both homo- and heterothallic subpopulations may be present (Hallenberg, 1991; Rayner, 1991b; for further discussion of mating systems in fungi, see Burnett, 1976). If the secondary mycelium contains two genetically identical nuclei, it is termed a homokaryotic mycelium or described as having a homothallic mating system. Fungi having heterothallic mycelia are out-crossing, while homothallic fungi are non-outcrossing (Rayner and Boddy, 1988). It is important to verify that the fungus under investigation for somatic incompatibility is sexually out-crossing. In non-outcrossing species, somatic incompatibility does not reveal spatial patterns with such resolution since the genotype of all basidiospores produced by a genet is identical with the original mycelium. It is therefore impossible by somatic incompatibility testing to distinguish between the spread of such species by spores and mycelia.

Somatic Incompatibility

Two different intraspecific systems for recognition and rejection of self and non-self mycelia operate in basidiomycetes. In the heterokaryotic phase, genetically identical mycelia, originating from the same genet, may fuse while non-identical mycelia reject each other. This prevents genetic exchange and maintains the integrity of the individual. This system is called somatic or vegetative incompatibility. It should not be confused with the other recognition system based upon sexual compatibility or mating that operates in the monokaryotic phase and causes mycelia with identical mating alleles to reject each other while those that are non-identical may fuse and form a dikaryon, so promoting genetic exchange. Ascomycetes similarly have vegetative and sexual incompatibility, designated vegetative groups, v-c groups, and mating types respectively (Rayner *et al.*, 1984). In contrast to basidiomycetes, the vegetative mycelium in ascomycetes is the homokaryotic primary mycelium.

General Procedure

The general procedure for studying somatic incompatibility in ectomycorrhizal fungi is outlined in Fig. 17.1 and described in more detail elsewhere (Dahlberg and Stenlid, 1990). First, select an outcrossing fungus which can be cultured and

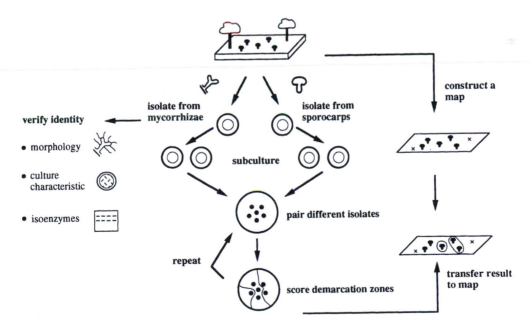

Fig. 17.1. General procedure for population studies in ectomycorrhizal fungi using either sporocarps or mycorrhiza.

show somatic incompatibility *in vitro*. Produce a map over the study site and isolate from the sporocarps. Pair selected isolates and score for somatic incompatibility reactions. At the mycelial level, somatic incompatibility commonly results in obvious demarcation lines between mycelial colonies of the same species, both in culture and in nature. The general procedure may also be used for isolates made from ectomycorrhizal root tips. Careful check of the identity of the mycorrhiza is necessary and for that purpose morphological description, culture characteristics and isoenzyme techniques may be used (Sen, 1990; Dahlberg and Stenlid, 1991). *In vivo* observations of somatic incompatibility of ectomycorrhizal fungi are lacking due to technical difficulties. However, in microcosms, somatic incompatibility has been observed between different genets of *S. bovinus* (R. Finlay, personal communication).

Results from *Suillus bovinus*

Combined results from nine populations of *Suillus bovinus* studied over the last 3 years in Sweden (Dahlberg and Stenlid, 1990, 1991) are presented.

The bolete *S. bovinus* is a common Scots pine symbiont in Scandinavia, occurring in young and old forests. It shows distinct reactions of somatic incompatibility. Sporocarps that were somatic compatible, belonging to the same genet, were located in aggregates close to each other, revealing meaningful patterns. The spatial pattern found is in accordance with the pattern found for saprophytes and pathogenic fungi (Stenlid, 1985; Kirby *et al.*, 1990).

The general patterns found are, that the number of genets per unit area decreases with increased age of forest, from 700–900 ha^{-1} in 10–20-year-old forest to 30–100 ha^{-1} in 100-year-old forest. This indicates that the rate of establishment by spores of *S. bovinus* genets decreases with the age of forest. Based on the number of genets per unit area in forests of different age, it is possible to estimate the median life expectancy for a genet, which is approximately 35 years in *S. bovinus*. A more accurate estimate would be to follow genets within populations for several years and calculate frequencies of birth and death. We are following up on populations studied in previous years to address such questions. The proportion of sporocarp-producing genets within a population is unknown. In a small study involving two sites, 4 and 7 m^2 respectively, where mycorrhizas were sampled around single genets of *S. bovinus* and tested for somatic incompatibility, no other genets than those appearing as sporocarps were observed.

The size and biomass per genet, indirectly reflected by number and biomass of sporocarps, increases with age of forest illustrating the ability to secondary spread by mycelial growth. In the older forests, the maximum size of genets, measured as the distance between the outermost compatible sporocarps, were 14–20 m, whereas it was 4–5 m in young forests. The spatial distributions of genets indicate that single genets may be hooked to root systems of more than one tree. This is in agreement with increasing awareness on the significance of mycorrhizal links between plants (Newman, 1988). Of 114 genets studied, only one had a distribution that spatially intermingled with other genets. This intermingling may be due to either fragmentation of the mycelium or mycelial growth at different layers in soil. Fragmentation of genets into physical and physiological independent but still genetical identical units, ramets, are likely to be a common feature in nature of fungi (Braiser and Rayner, 1987), but remains to be demonstrated to ectomycorrhizal fungi.

The relationship between the size of the *S. bovinus* genets studied and their production of sporocarps is not linear, which may indicate fragmentation. The rare incidence of intermingling of genets enables calculations of sporocarp production per genet. The biomass of sporocarps per genet was found to be 25–180 and 6–30 g DW, in old and young forest respectively. There may be relationships between the reproductive and vegetative biomass within genets of fungi, as has been shown in plants (Hartnett, 1990). For one *S. bovinus* genet, the proportions of biomass at the time of fruiting was calculated to be 30% in the sporocarps, 55% in the mycorrhizas and 15% in the extramatrical mycelia. In this particular genet there were approx 600 mycorrhizal clusters per produced sporocarp. An understanding of the relationship between sporocarps and mycelial distribution in soil is essential in order to interpret the pattern of sporocarp appearance.

Size strongly affects sexual maturation, fecundity and mortality, so for many organisms, size is more important than age in determining life history characteristics (Terence and Connell, 1987). Both size of genets and biomass of sporocarps within a genet may therefore be used in demographic studies for ectomycorrhizal fungi.

Limitations

The basic requirement for studying somatic incompatibility *in vitro* is the ability of the ectomycorrhizal fungi to be cultured. Only a limited number of species can be cultured, but those that do may help us to gain insight into patterns of population structure for ectomycorrhizal fungi.

Even though fungal individualism is likely to be the most widespread feature in nature, various mycorrhizal fungi may show different degrees or even absence of somatic incompatibility *in vitro* (Fries, 1987; cf Rayner and Boddy, 1988; K. Bergeson and S. Miller, personal communication). Therefore, special care to determine the best cultural medium for the species studied must be taken in order to ensure that any potential for somatic incompatibility is revealed. It is also important to distinguish true somatic incompatibility from other phenomena such as nutrient depletion and 'staling' (Sen, 1990).

Furthermore, the reaction and intensity of somatic incompatibility is dependent on the genetic relatedness between tested isolates. Dikaryotic mycelia formed by conjugation between haploids from the same sporocarp, siblings, may be so closely related genetically that the reaction of somatic incompatibility is only weak or absent. However, it appears that only a small fraction of formed siblings are so genetically related that they react as if they were identical (Fries, 1987; Hansen, Stenlid and Johansson, in preparation). Most basidiomycetes are believed to be sexually outcrossing, which further reduces the chance for such siblings to occur. Spores dispersed by wind are usually efficiently mixed. The situation for hypogeous basidiomycetes may deserve special attention in this respect. These factors combined with the fact that studies of populations have revealed meaningful patterns imply that somatic incompatibility is a useful tool in such studies.

Prospects for the Future

With awareness of fungal individualism and the tools for identifying genets, demographic studies can be performed on ectomycorrhizal fungi. Parameters of the dynamics of populations, such as natality, death, immigration and emigration, and life-history of single genets may be studied. Other techniques involving isoenzymes and DNA sequencing may also be used to identify genets. Selection of technique is determined by its availability, cost and the nature of the question being asked. We have an almost complete lack of information regarding population structure and dynamics in ectomycorrhizal fungi. Many of the theories developed in studies of population ecology in plants may be usefully applied to studies of ectomycorrhizal fungi. Availability of tools for identifying genets will enable studies in genetic exchange within populations and gene flow between populations to be carried out so improving our understanding of dynamics within species.

Acknowledgements

My thanks to Drs Jan Stenlid and Hugues Massicotte for commenting on earlier versions of the manuscript.

References

Braiser, C.M. and Rayner, A.D.M. (1987) Whither terminology below the species level in fungi? In: Rayner, A.D.M., Braiser, C.M. and Moore, D. (eds), *Evolutionary Biology of the Fungi*. Cambridge University Press, Cambridge, pp. 379–388.

Burnett, J.H. (1976) *Fundamentals of Mycology* 2nd edn. Edward Arnold. London.

Dahlberg, A. and Stenlid, J. (1990) Population structure and dynamics in *Suillus bovinus* as indicated by spatial distribution of fungal clones. *New Phytologist* 115, 487–493.

Dahlberg, A. and Stenlid, J. (1991) Size, distributions and biomass of genets in populations of *Suillus bovinus* revealed by somatic incompatibility. In: Dahlberg, A., 'Population and community structure and dynamics of ectomycorrhizal fungi in conifer forests.' PhD thesis, Swedish University of Agricultural Science. Uppsala.

Fries, N. (1987) Somatic incompatibility and field distribution of the ectomycorrhizal fungus *Suillus luteus* (Boletaceae). *New Phytologist* 107, 735–739.

Fries, N. and Neumann, W. (1990) Sexual incompatibility in *Suillus luteus* and *S. granulatus*. *Mycological Research* 94, 64–70.

Gregory, P.H. (1984) The fungal mycelium – an historical perspective. In: Jennings, D.H. and Rayner, A.D.M. (eds), *The Ecology and Physiology of the Fungal Mycelium*. Cambridge University Press, Cambridge, pp. 1–22.

Hallenberg, N. (1991) Speciation and distribution in *Corticiaceae* (Basidiomycetes). *Plant Systematics and Evolution* 117, 93–110.

Hartnett, D.C. (1990) Size-dependent allocation to sexual and vegetative reproduction on four clonal composites. *Oecologia* 84, 254–259.

Kirby, J.J.H., Stenlid, J. and Holdenrieder, O. (1990) Population structure and response to disturbance of the basidiomycete *Resinicium bicolor*. *Oecologia* 85, 178–184.

Lane, E.B. (1981) Somatic incompatibility in fungi and Myxomycetes. In: Gull, K. and Oliver, S.G. (eds), *The Fungal Nucleus*. Cambridge University Press, Cambridge, pp. 239–258.

Newman, E.I. (1988) Mycorrhizal links between plants: their functioning and ecological significance. *Advances in Ecological Research* 18, 243–270.

Rayner, A.D.M. (1991a) The challenge of the individualistic mycelium. *Mycologia* 83, 48–71.

Rayner, A.D.M. (1991b) Natural genetic transfer systems in higher fungi. *Transaction of the Mycological Society of Japan* 31, 75–87.

Rayner, A.D.M. and Boddy, L. (1988) *Fungal Decomposition of Wood – Its Biology and Ecology*. John Wiley and Sons, Chichester, UK.

Rayner, A.D.M., Cotes, D., Ainsworth, A.M., Adams, T.J.H., Williams, E.N.D. and Todd, N.K. (1984) The biological consequences of the individualistic mycelium. In: Jennings, D.H. and Rayner, A.D.M. (eds), *The Ecology and Physiology of the Fungal Mycelium*. Cambridge University Press, Cambridge, pp. 509–540.

Sen, R. (1990) Intraspecific variation in two species of *Suillus* from Scots pine (*Pinus sylvestris* L.) forests based on somatic incompatibility and isoenzyme analyses. *New Phytologist* 114, 603–612.

Stenlid, J. (1985) Population structure of *Heterobasidion annosum* as determined by somatic incompatibility, sexual incompatibility and isoenzyme pattern. *Canadian Journal of Botany* 63, 2268–2273.

Terence, P.H. and Connell, J.H. (1987) Population dynamics based on size or age? *The American Naturalist* 129, 818–829.

Todd, N.K. and Rayner, A.D.M. (1980) Fungal individualism. *Science Progress, Oxford* 66, 331–354.

18 Mixed Associations of Fungi in Ectomycorrhizal Roots[1]

F. Brand

Institut für Systematische Botanik, Menzinger Strasse 67, D-8000 München 19, Germany

Introduction

One of the most fundamental requirements for an understanding of ectomycorrhiza in natural forest ecosystems is to obtain characterization and identification of the different types occurring. Methods and results of such analyses are summarized in Agerer (1987). Recent studies have also revealed that, mixed associations of fungal species can consistently occur in individual ectomycorrhizal roots. This chapter reviews such phenomena.

Competition among Ectomycorrhizal Fungi in Forest Soil

Numerous species of mycorrhizal fungi are known to grow in forest soils. In beechwood soil, within a square of 10 × 10cm, populations of up to 15 different species of ectomycorrhizas have been found in the organic soil layers (Brand, 1991). It is likely that there is competition among the different fungi. Mycorrhizal fungi of forest trees have to compete with other species for their place on or in the tree roots, which are their sources of energy.

Patterns of competition can often be observed when examining the different ectomycorrhizal species which colonize laterals of a long root. Black mycelium of *Cenococcum geophilum* for example, often can be seen growing along roots and on the mantle surface of other ectomycorrhizas. Microscopic examination of long roots and of mycorrhizal mantles shows that various mycelia can grow along the root (Brand and Agerer, 1986). Especially in spring, and after periods of rain, root tips can burst through the mantle of their fungal partners, but are rapidly reinfected, sometimes by the same fungus, but often by a different species. Two different fungal mantles can then be observed on the same root tip, these being usually separated by a ring-shaped restriction of the root. Success in the recolonization of a 'free' root tip,

---

[1] Part XXXIX of the series 'Studies on ectomycorhizae; part XXXVIII: Agerer, Ch. 11, in this volume.

seems to depend both on mycelia present in close proximity to the root, and on environmental conditions such as moisture, pH, or other chemical factors. Detailed field studies as well as experiments under nature-like conditions are necessary to understand these mechanisms.

Ecto- and Endomycorrhizal Fungi in a Single Root Tip

Competition between fungi may occur within as well as on the surface of a mycorrhizal root. Intracellular infections in root cortex cells of ectomycorrhizas have frequently been observed and reported (summarized in Agerer, 1990; Brand, 1991). Usually they have been interpreted to be signs of increasing pathogenicity of an ectomycorrhizal fungus, or to be typical of ageing mycorrhizas. They also occur in typical 'ectendomycorrhizas' of conifers in nurseries. Endomycorrhizal infections in ectomycorrhizas may arise as a result of ingrowths of the Hartig net of an ectomycorrhizal fungus, but, according to recent studies, there also exist mixed associations in which two fungi share one root. One of the associates forms the typical structures of an ectomycorrhiza, the fungal mantle and Hartig net, while a second fungus forms haustoria in the cortical cells of the root. Mixed associations seem to be not uncommon in forest soils. Two different types have been observed and studied.

Mixed Associations of Gomphidiaceae with Ectomycorrhizas of *Suillus* and *Rhizopogon* Species

Relationships between intracellular hyphae of *Chroogomphus* or *Gomphidius* species within ectomycorrhizas formed by *Suillus* or *Rhizopogon* species in conifer roots have been examined. Several different mixed associations have been described by Agerer (1990, 1991) who refers to the phenomenon as 'co-growth' of, or possible parasitism by, the Gomphidiaceae. The haustoria of these fungi are found in cortical root cells covered by the ectomycorrhizal Hartig net. In some cases their hyphae can be detected by amyloid staining in the 'host' ectomycorrhizal mantle, Hartig net, and rhizomorphs (Agerer, 1990). *Chroogomphus* does not form rhizomorphs, but the amyloid hyphae are found within rhizomorphs of *Rhizopogon* and *Suillus* which are used as carriers enabling it to reach host-ectomycorrhizas (Agerer, 1990). According to Agerer this co-growth which is additional to an ability to form typical independent ectomycorrhizas, may provide Gomphidiaceae with the necessary energy resource to enable fruit-body formation.

Endo-infections by Ascomycete Fungi in Ectomycorrhizas Formed by Russulaceae

Mixed associations of an endophytic ascomycete have been observed in ectomycorrhizas of the Russulaceae on beech roots (Brand, 1987, 1991). These ascomycete infections were first detected in ectomycorrhizas of *Lactarius subdulcis* on *Fagus sylvatica* (Brand and Agerer 1986), then were seen in ectomycorrhizas of *Lactarius fuliginosus*, *L. rubrocinctus*, *Russula ochroleuca*, and *R. foetens* on beech, but also in the ectomycorrhiza of *L. scrobiculatus* on spruce (Brand 1987, 1991).

Beech ectomycorrhiza formed by *L. subdulcis* have now been extensively examined in forest soil and in culture. The orange-brown, smooth ectomycorrhizas can be found frequently in beechwoods. On clay soil in thick litter layers, they often are the only or the dominant ectomycorrhizal type, colonizing most of the root system and forming a considerable biomass. Mycorrhiza formed by *L. subdulcis* can be distinguished from other beech ectomycorrhizas by their large size (diameter of tips often 0.5–2 mm, that of other species 0.3–0.5 mm), by characteristic mantle structures, and by the formation of small numbers of orange-brown rhizomorphs attached to the litter substrate (Brand and Agerer, 1986). Collections from numerous beechwoods in Germany and the British isles have shown that even the smallest tips of *L. subdulcis* mycorrhizas consistently show intracellular infections by the second fungus. Often, stiff, glassy hyphae of this second fungus can be seen projecting from the mycorrhizal surface (Fig. 18.1).

The second fungus is known to be an ascomycete, because in transmission electron microscopy simple septa with Woronin bodies are observed. Its colourless, large, thick-walled hyphae form a reticulum between the ectomycorrhizal mantle and the root cortex (Fig. 18.2) and form lobed haustoria in inner root cortex cells which are not covered by the ectomycorrhizal Hartig net (Fig. 18.3). Thus, the two fungi 'share' the cortex tissue (Brand, 1987). Haustoria also can be found in the meristematic region of the beech root. In tips of 2 mm length, 200–300 root cells are typically infected by haustoria. The ascomycete hyphae which penetrate the ectomycorrhizal mantle form ornamented chlamydospores in the soil (Brand, 1987, 1991).

Beech mycorrhizal roots with *L. subdulcis* have been cultured by sterile synthesis (Brand and Agerer, 1986) and grown in unsterile root chamber cultures (Brand, 1991). Aseptically synthesized mycorrhizas, grown in glass tubes on sterilized beech litter by infection with agar cultures, were small and did not show ascomycete infections. In contrast, ectomycorrhizas produced in root chambers on two-layered substrate (pasteurized beech litter and peat; Brand, 1989) were larger and constantly did show ascomycete infections following inoculation with fresh root tips from the forest. The ascomycete hyphae were found to grow along rhizomorphs of *Lactarius sp.* infecting beech roots together with the ectomycorrhizal fungus. In all of the root chambers tested, ascocarps of the fungus *Leucoscypha leucotricha* (Humariaceae, Pezizales) appeared in the vicinity of the *Lactarius* mycorrhizas. Microscopic examination has confirmed that this is the fungal partner of *L. subdulcis* on *Fagus* (Brand, 1991). So far, however, it has not been possible to grow the ascomycete on agar, either from surface sterilized ectomycorrhizas, or from ascospores or chlamydospores.

In view of the frequency of *L. subdulcis* mycorrhiza in beechwoods, the mixed association with *L. leucotricha* must be a widespread and typical phenomenon, which may be obligate. There are no signs of a parasitic relationship,

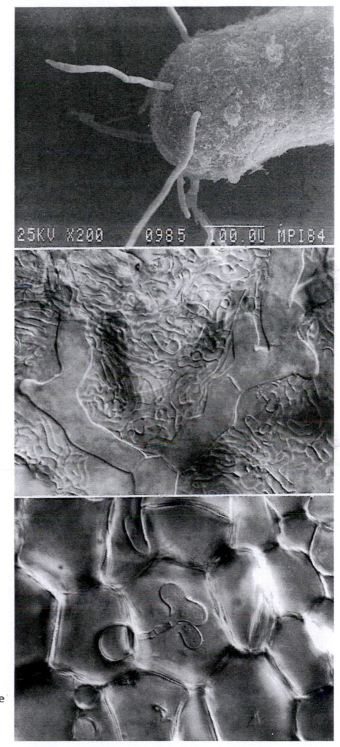

Figs 18.1–18.3
18.1. Ascomycete hyphae penetrating the fungal mantle of *Lactarius subdulcis* ectomycorrhiza on beech.
18.2. Ascomycete hyphae on the inner surface of the dissected ectomycorrhizal mantle of *Lactarius subdulcis* ectomycorrhiza on beech.
18.3. Ascomycete haustorium and intercellular hyphae in the inner root cortex of *Lactarius subdulcis* ectomycorrhiza on beech.

indeed the vigorous and widespread nature of this association suggests that the relationship is compatible and healthy. The larger size of these ectomycorrhizas appears to be caused by the ascomycete infection. This is supported by observations on beech ectomycorrhizas of *Russula ochroleuca* which are not obligately associated with *Leucoscypha*, but show much bigger size when infected (Brand, 1991).

Harley's 'Excised Beech Ectomycorrhiza' – A Mixed Mycorrhizal Association

It is of special interest that the mixed mycorrhizal association involving *Lactarius subdulcis*, *Leucoscypha leucotricha*, and *Fagus sylvatica* appears to have been specifically selected by Harley and co-workers over several decades for use in experiments on the physiology and nutrient uptake of ectomycorrhizas (see Harley, 1969; Harley and Smith 1983). The characteristics of this ectomycorrhiza, as described by Harley and McCready (1952a,b), closely resemble those produced by *L. subdulcis* (Brand and Agerer, 1986). In September 1988, former sampling plots of Harley and co-workers in the Chiltern hills were examined, and Professor Harley demonstrated roots of the type which had been used in his experiments. These were confirmed as being typical ectomycorrhizas of *L. subdulcis*. They were formed in profusion in the deep litter layers where they were associated with fruit-bodies of this fungus. All root tips subsequently examined were also infected by *Leucoscypha*, the number of haustoria per 2 mm root tip being in the range 250–310 (Brand, 1991).

This ectomycorrhizal type was selectively employed for the classic experiments on phosphate (e.g. Harley and McCready, 1952a,b) and carbohydrate (e.g. Lewis and Harley, 1965) metabolism of beech roots, in which the ectomycorrhizal mantle was routinely dissected from the root 'core' in order to examine the functions of the two tissues separately. Atkinson (1975) observed the occasional presence of a haustorium forming fungus in the cortical cells of these roots but appears to have been unaware of the extent or nature of the infection. Knowledge of the presence of an additional infection in these intensively studied ectomycorrhizal roots does not necessitate re-evaluation of the important physiological observations made using them, but it certainly provides a clearer picture of the complexity of the associations which occur in the field. As this tripartite association is so widespread in nature and can be re-assembled in the laboratory, further experiments are encouraged. In particular as recommended by Read (1987), studies of the extent and function of the associated mycorrhizal mycelia in natural substrates are needed.

References

Agerer, R. (ed.) (1987) *Colour Atlas of Ectomycorrhizae*. Einhorn-Verlag, Munich.

Agerer, R. (1990) Studies on ectomycorrhizae XXIV – Ectomycorrhizae of *Chroogomphus helveticus* and *C. rutilus* (Gomphidiaceae, Basidiomycetes) and their relationship to those of *Suillus* and *Rhizopogon*. *Nova Hedwigia* 50(1–2), 1–63.

Agerer, R. (1991) Studies on ectomycorrhizae XXXIV – Mycorrhizae of *Gomphidius glutinosus* and of *G. roseus* with some remarks on Gomphidiaceae (Basidiomycetes). *Nova Hedwigia* 53(1–2), 127–170.

Atkinson, M.A. (1975) The fine structure of mycorrhizas. DPhil thesis, University of Oxford, Oxford, UK.

Brand, F. (1987) A secondary fungal infection in mycorrhizae formed by *Lactarius* spp. with *Fagus sylvatica*. In: Sylvia, D.M., Hung, L.L. and Graham, J.H. (eds), *Mycorrhizae in the Next Decade*, University of Florida, Gainesville, Florida, p. 189.

Brand, F. (1989) Studies on ectomycorrhizae XXI – Beech ectomycorrhizas and rhizomorphs of *Xerocomus chrysenteron* (Boletales). *Nova Hedwigia* 48(3–4), 469–483.

Brand, F. (1991) Ektomykorrhizen an *Fagus sylvatica*. Charakterisierung und Identifizierung, ökologische Kennzeichnung und unsterile Kultivierung. *Libri Botanici* 2, IHW-Verlag, Eching.

Brand, F. and Agerer, R. (1986) Studies on ectomycorrhizae VIII – Mycorrhizae formed by *Lactarius subdulcis*, *L. vellereus* and *Laccaria amethystina* on beech. *Zeitschrift für Mykologie* 52(2), 287–320.

Harley, J.L. (1969) *The Biology of Mycorrhiza* 2nd edition. Leonard Hill, London.

Harley, J.L. and McCready, C.C. (1952a) Uptake of phosphate by excised mycorrhizas of beech. II. Distribution of phosphate between host and fungus. *New Phytologist* 51, 56–64.

Harley, J.L. and McCready, C.C. (1952b) Uptake of phosphate by excised mycorrhizas of beech. III. The effect of the fungal sheath on the availability of phosphate to the core. *New Phytologist* 51, 343–348.

Harley, J.L. and Smith, S.E. (1983) *Mycorrhizal Symbiosis*. Academic Press, London.

Lewis, D.H. and Harley, J.L. (1965) Carbohydrate physiology of mycorrhizal root of beech. I. Identity of endogenous sugars and utilization of exogenous sugars. *New Phytologist* 64, 224–237.

Read, D.J. (1987) Development and function of mycorrhizal hyphae in soil. In: Sylvia, D.M., Hung, L.L. and Graham, J.H. (eds), *Mycorrhizae in the next decade*. University of Florida, Gainesville, Florida, pp. 178–180.

Part Three

Mycorrhizas in Disturbed, Agricultural and Successional Ecosystems

19 Soil Disturbance in Native Ecosystems – The Decline and Recovery of Infectivity of VA Mycorrhizal Fungi

D.A. Jasper, L.K. Abbott and A.D. Robson

Soil Science and Plant Nutrition, The University of Western Australia, Nedlands, 6009, Australia

The Decline in Infectivity

The infectivity of VA mycorrhizal fungi in soil from Australian native ecosystems can be greatly decreased by soil disturbance. In such disturbance associated with mining, decreases in infectivity may range from one-third (Jasper *et al.*, 1987) to 100% (Jasper *et al.*, 1989a). This severe effect of disturbance in soil from Australian native ecosystems is not always observed elsewhere. For example, in ecosystems with a higher proportion of grasses, the soil disturbance may cause only slight delays in mycorrhiza formation (Visser *et al.*, 1984) or reduced propagule number with no reduction in root length colonized (Rives *et al.*, 1980). Typically these ecosystems in which the mycorrhizal infectivity is most tolerant of soil disturbance are those in which the soil also contains high spore numbers.

Presumably, high numbers of robust propagules such as spores and mycorrhizal roots should help ensure that the mycorrhizal infectivity is maintained during soil disturbance. Greater numbers of these robust propagules are likely to be found in more fertile soils with a high proportion of mycorrhizal plants (Jasper *et al.*, 1991). It has been estimated (Jasper *et al.*, 1991) that a productive pasture soil contained up to 10 or 25 times more infective propagules than soils from native forest or heathland respectively. Corresponding to this high number of propagules, the infectivity of the pasture soil was unaffected by soil disturbance (Table 19.1). By contrast, the infectivity of the fungi in the soils from the native ecosystems was reduced by half.

The abundance of robust propagules in some soils can be illustrated if the soils are serially diluted prior to a bioassay. Using this comparison, it can be seen that the soil from a grassland (Plenchette *et al.*, 1989) maintained original levels of infectivity despite being diluted to only 25% (Fig. 19.1). Soils from a sagebrush ecosystem (Rives *et al.*, 1980), and a pasture soil (Jasper *et al.*, 1991) behaved similarly in that the rate of reduction in colonization of bioassay plants with dilution was less than the rate of dilution, indicating each soil had an excess of robust propagules prior to dilution. By contrast, for the forest soil of Jasper *et al.* (1991)

151

Table 19.1. The effect of disturbing previously undisturbed soils, on the subsequent formation of VA mycorrhizas by clover bioassay plants (% root length colonized ± s.e.) (from Jasper et al., 1991).

Soil	VA mycorrhiza formation	
	Undisturbed	Disturbed
Forest	28 ± 2.9	15 ± 4.4
Heathland	11 ± 2.4	6 ± 1.5
Pasture	52 ± 2.8	50 ± 4.3

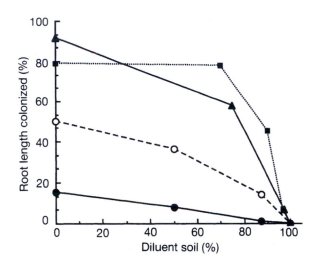

Fig. 19.1. The effect of diluting disturbed soils, on the subsequent formation of VA mycorrhizas by bioassay plants. (Symbols as follows: —●— Forest (Jasper et al., 1991), --○-- Pasture (Jasper et al., 1991), ····■···· Grassland (Plenchette et al., 1989) —▲— Sagebrush (Rives et al., 1980).)

colonization declined in direct proportion with the rate of dilution (Fig. 19.1), suggesting no excess of propagules in this soil after disturbance. Thus in this forest soil, the number of propagules was limiting VA mycorrhiza formation in the experimental situation.

The observations that the forest soil had a limited number of robust propagules, together with the marked decrease in mycorrhizal infectivity with soil disturbance suggests that in undisturbed soil, propagules such as hyphae are important. However, losses in infectivity, presumably due to destruction of hyphae, have occurred during summer periods when the soil is dry. This suggests that hyphae may survive, even in dry soil. To support this, Jasper et al. (1989b) have demonstrated that hyphae of *Acaulospora laevis* can remain infective in dry soil for 35 days. In a contrasting environment, the capacity of hyphae to survive extreme conditions has also been observed by Evans and Miller (1988) who found that hyphae could survive a period of winter freezing. In both cases the functional capacity of the hyphae was decreased with soil disturbance. Survival of hyphae in dry soil may not be a characteristic of all fungal species (D. Jasper, unpublished data). For example, hyphae of *Scutellospora calospora* can survive in dry soil regardless of the timing of commencement of spore production, but for *A. laevis*, hyphae do not maintain infectivity in dry soil if sporulation has commenced (D. Jasper, unpublished data).

The Return of Infectivity

An adequate level of infectivity of VA mycorrhizal fungi in soil to be revegetated after mining is likely to be an important component of successful re-establishment of a diverse, self-sustaining ecosystem. In this context, the severe decreases in infectivity as a result of soil disturbance in some Australian native ecosystems are of concern. Therefore it is important that we understand if, and how quickly, the level of infectivity returns to a pre-mining level.

Surveys of the VA mycorrhizal infectivity of soils revegetated after bauxite mining were conducted at mine sites at Weipa in north-eastern, and Jarrahdale in south-western Australia, using bioassays of undisturbed cores (D. Jasper, unpublished data). These studies revealed that the level of infectivity in the revegetated soils returned to equivalent to pre-mining levels after 2 years and 4 years respectively, correlating with the respective responses of infectivity in these soils to disturbance (Jasper *et al.*, 1987, 1989a).

Fig. 19.2. A schematic representation of the distribution of infectivity of VA mycorrhizal fungi, in 4-year-old revegetation of direct-returned soil after mining bauxite at Jarrahdale, Western Australia. (Dimensions of the study plot, 95 × 60 m. The value allocated to each square of 50 m^2 represents the level of colonization of bioassay plants grown in an intact core taken from the centre of the square.)

Characterization of the infectivity of VA-mycorrhizal infectivity in such areas of revegetation requires information on the distribution of infectivity, for example a measure of likely presence or absence of infectivity in each core. Such information would complement the calculated mean value for colonization that is typically derived from bioassays. An area of approximately 0.8 ha of 4-year-old revegetation at Jarrahdale in which the infectivity was approximately equivalent to undisturbed forest was sampled more intensively to establish this information. Intact cores were removed every 10 m on parallel transects that were 5 m apart. This sampling revealed extremely uneven distributions of mycorrhizal infectivity with up to half of the intact cores giving rise to no or less than 5% colonization of bioassay plants, while less than 4% of the cores supported colonization in excess of 50% of the roots of the bioassay plants (Fig. 19.2). This measurement of infectivity is restricted to the volume of soil in each core. The actual scale of the patches of high or low infectivity still needs to be determined.

Such uneven distribution is of concern given that the aim of revegetation is to achieve an evenly distributed but diverse range of plant species. If some plant species are heavily dependent on VA mycorrhizal fungi for growth and survival then their establishment may be restricted in those areas of soil where infectivity is absent or low. In the above study, uniform distribution of infectivity, equivalent to that found in undisturbed forest soil, was not achieved until 7 years after establishment. In contrast a mean level of colonization of bioassay plants equivalent to that in undisturbed forest was reached at only 4 years.

To ensure optimum diversity in revegetation at such sites where the normal rate of return of infectivity of VA mycorrhiza is low, it may be necessary to boost infectivity at such sites. This may be achieved by adjusting soil management techniques or, at some sites, by inoculation.

References

Evans, D.G. and Miller, M.H. (1988) Vesicular-arbuscular mycorrhizas and the soil-disturbance-induced reduction of nutrient absorption in maize. I. Causal relations. *New Phytologist* 110, 67–74.

Jasper, D.A., Robson, A.D. and Abbott, L.K. (1987) The effect of surface mining on the infectivity of vesicular-arbuscular mycorrhizal fungi. *Australian Journal of Botany* 35, 641–652.

Jasper, D.A., Abbott, L.K. and Robson, A.D. (1989a) The loss of VA mycorrhizal infectivity during bauxite mining may limit the growth of *Acacia pulchella* R.Br. *Australian Journal of Botany* 37, 33–42.

Jasper, D.A., Abbott, L.K. and Robson, A.D. (1989b) Hyphae of a vesicular-arbuscular mycorrhizal fungus maintain infectivity in dry soil, except when the soil is disturbed. *New Phytologist* 112, 101–107.

Jasper, D.A., Abbott, L.K. and Robson, A.D. (1991) The effect of soil disturbance on vesicular-arbuscular mycorrhizal fungi, in soils from different vegetation types. *New Phytologist* 118, 471–476.

Plenchette, C., Perrin, R. and Duvert, P. (1989) The concept of soil infectivity and a method for its determination as applied to endomycorrhizas. *Canadian Journal of Botany* 67, 112–115.

Rives, C.S., Bajwa. M.I., Liberta, A.E. and Miller, R.M. (1980) Effects of topsoil storage during surface mining on the viability of VA mycorrhiza. *Soil Science* 129, 253–257.

Visser, S., Griffiths, C.L. and Parkinson, D. (1984) Topsoil storage effects on primary production and rates of vesicular-arbuscular mycorrhizal development in *Agropyron trachycaulum*. *Plant and Soil* 82, 51–60.

20 Soil Disturbance and the Effectiveness of Arbuscular Mycorrhizas in an Agricultural Ecosystem

M.H. Miller and T.P. McGonigle

Department of Land Resource Science, University of Guelph, Guelph, Ontario N1G 2W1, Canada

Introduction

Disturbance of soil in maize (*Zea mays* L.) plots which had been untilled for a long period was first reported to result in a reduction of arbuscular mycorrhizal colonization and P absorption in young maize plants by O'Halloran *et al.* (1986). Subsequent growth-chamber studies provided evidence for a causal relationship between the reduction in P absorption and the reduction in mycorrhizal colonization (Evans and Miller, 1988). Fairchild and Miller (1988, 1990) established that the disturbance effect was not dependent on a lengthy undisturbed period. The effect was found to occur when soils which had been undisturbed during only two 3-week growth cycles were subsequently disturbed. Evans and Miller (1990) used compartmentalized growth units in which a mycorrhizal mycelium was established in the absence of plant roots to demonstrate that disruption of the extraradical mycelium was, at least in part, responsible for the reduction in colonization following disturbance. A similar conclusion was arrived at independently by Jasper *et al.* (1989).

McGonigle *et al.* (1990b) observed, in both a growth chamber and a field study, that P absorption by young maize plants increased progressively as the degree of soil disturbance decreased from severe disturbance (soil aggregates < 5 mm), to no disturbance. In these studies, however, disturbance did not affect the mycorrhizal colonization. This lack of an effect of disturbance on colonization lead to the conclusion that either the effect of disturbance on P absorption was not causally related to mycorrhizal colonization, or that disturbance reduced the effectiveness of the mycorrhizas through some mechanism other than, or in addition to, a reduction in colonization. We hypothesized (McGonigle *et al.*, 1990b) that disruption of the extraradical mycelium destroyed a system that was capable of acquiring nutrients for a newly connected developing root system, thereby obviating the need to establish a new extraradical mycelium.

This chapter reports on two studies: a field experiment to investigate further the effect of soil disturbance by tillage on arbuscular mycorrhizal colonization and

P absorption of young maize plants; and a glasshouse experiment designed to uncouple the effect of soil disturbance on colonization from that of the effectiveness of the extraradical mycelium as a nutrient acquisition system.

Field Study

Materials and methods

Shoot and root samples were collected weekly throughout the growing season from four replicates of each of four tillage treatments in a long-term experiment at the Ridgetown College of Agricultural Technology in southwestern Ontario. The following tillage treatments were sampled: (1) autumn mouldboard ploughing and spring disc harrowing (MP); (2) autumn chisel ploughing (CP); (3) ridge till (RT); and (4) no till (NT). All treatments were fertilized at the same rate based on requirements as indicated by soil test. Shoot dry mass was determined and the samples were analysed for P by the method of Thomas *et al*. (1967). Root samples were taken as cores 7.5 cm in diameter and 15 cm deep at standardized positions in the row. Roots were washed free of soil and stored in formyl acetic alcohol. Where the volume of roots in a sample was too great for subsequent procedures, the roots were cut into pieces so that the axes did not exceed 2 cm, dispersed in water, and a subsample was taken. Clearing and staining was by the method of Brundrett *et al*. (1984). The percentage of the root length containing arbuscules was determined by the method of McGonigle *et al*. (1990a).

Results and discussion

Arbuscular colonization of the roots, dry mass and P concentration of the shoots from the MP and RT treatments for the samples collected between 19 and 48 days after planting (DAP) are presented in Table 20.1. Arbuscular colonization increased from less than 10% at 19 DAP to as high as 60% at 48 DAP with the RT treatment. Colonization was greater on the RT than the MP treatment at each of these sampling times although the differences were not significant at $P = 0.05$ for 19 and 32 DAP. Colonization was similar for the RT and MP treatments at subsequent harvests (data not shown). Colonization of samples from the NT treatment was similar to those on RT, while colonization of the samples from the CP was intermediate between that on MP and RT treatments (data not shown). The differences in colonization found here to arise from disturbance are in contrast to the findings of a previous field study (McGonigle *et al*., 1990b) discussed earlier. No explanation can be offered at this time for these contrasting results. The data from this experiment are consistent with effects of disturbance on colonization observed in growth chamber experiments by Evans and Miller (1988) and Fairchild and Miller (1988, 1990).

Shoot dry mass and P concentration were not significantly different at 25 DAP. However the shoot P concentration at 32 DAP in the RT was greater than that in the MP treatment. This was accompanied by an increase in shoot dry mass. The differences between the two tillage treatments declined at later samplings and were

Table 20.1. Arbuscular colonization (AC) of roots, dry mass and phosphorus status of shoots from mouldboard plough (MP) and ridge-till (RT) treatments between 19 and 48 days after planting (DAP).

DAP	AC (%)		Shoot dry mass (g shoot⁻¹)		Shoot phosphorus			
					Conc. (mg g⁻¹)		Content (mg shoot⁻¹)	
	MP	RT	MP	RT	MP	RT	MP	RT
19	1.1a*	7.7a	0.008a	0.010a	n.d.†	n.d.	n.d.	n.d.
25	11a	42b	0.047a	0.048a	6.42a	6.12a	0.30a	0.29a
32	27a	39a	0.11a	0.21b	3.44a	4.77b	0.38a	1.00b
39	25a	53b	0.54a	0.60a	3.99a	5.22b	2.15a	3.10b
48	44a	60b	3.77a	4.38a	4.42a	4.41a	16.7a	19.3a

*Values for each parameter at each time followed by same letter are not significantly different at $P = 0.05$.
†n.d. = not determined.

not significant at 48 DAP. The increase in P content on the RT treatment between 25 and 32 DAP was much greater than that on the MP treatment (0.71 vs. 0.08 mg per shoot). This corresponds closely to the time period when the difference in arbuscular colonization was the greatest. This difference in P absorption is very similar to that observed in growth chamber studies (Fairchild and Miller, 1988, 1990).

Barry and Miller (1989) have reported that shoot P concentration at the 5–6-leaf stage can have a significant effect on final grain yield. As can be seen in Table 20.1, tillage can have a marked effect on shoot P concentration at this stage (32 DAP in this experiment). The effect of tillage appears to be due to an effect on the mycorrhizal symbiosis. Hence there would seem to be the potential to influence grain yield through management of this symbiosis.

Glasshouse Study

Materials and methods

The glasshouse study was designed to uncouple the effects of disturbance on colonization from that on the effectiveness of the extraradical mycelium as a nutrient acquisition system. To do so it was necessary to have a similar amount of root colonization in both the presence and absence of an extraradical mycelium.

In an attempt to achieve this situation, the procedure used by Fairchild and Miller (1988) to study the development of the disturbance effect was adopted. They had observed a much greater increase in colonization of maize roots in successive 3-week cycles of maize growth if the soil was not disturbed between cycles. This suggested that varying the cycle in which disturbance was imposed would affect colonization in the subsequent cycle.

A silt loam soil from the Elora Research Station having a very low $NaHCO_3$-extractable soil P value (3 p.p.m.) was used. Three phosphorus rates, 0, 75 and 150 $\mu gP g^{-1}$ of soil as $Ca(H_2PO_4)_2.H_2O$ were mixed thoroughly with the soil before potting in 4 litre pots at a bulk density of 1.1 mg cm^{-3}. Six maize seeds (Var. Pioneer 3949) selected for uniformity in weight (0.23–0.28 g per seed) were planted, and the soil was wetted to 25 gH_2O 100 g^{-1} dry soil which approximated to field capacity. Pots were watered periodically by weight to return the water content to the initial value. Pots were thinned to three shoots each after 7 days. The experiment was conducted between May and August 1990. Maize was grown for four 3-week cycles. Four pots of each P level were disturbed after each cycle (D), four were disturbed after the third cycle only (DL), while four were left undisturbed (U) throughout the studies. Disturbance consisted of pulverizing the soil to pass a 5-mm sieve, cutting roots into pieces approximately 2 cm in length and thoroughly mixing them into the soil before repotting. The upper and lower 7-cm depth fractions were disturbed separately and repotted in order.

At the conclusion of the fourth cycle, dry mass and P concentration of the shoots were determined. Roots were sampled by taking one 30-mm diameter soil core to the full 14-cm depth of the pot, midway between two shoots in each pot, washing the roots free of soil, and storing in formyl acetic alcohol. Subsampling, cleaning, staining and assessment of colonization were as described for the field experiment.

Results and discussion

Shoot dry mass (Fig. 20.1a) increased markedly with increasing P added in the D and DL disturbance treatments. In the U treatment shoot dry mass increased with the 75 $\mu gP g^{-1}$ soil, but there was no further increase at the 150 $\mu gP g^{-1}$ soil rate. At all soil P values, the greatest shoot dry mass occurred on the U treatment.

Shoot P concentration (Fig. 20.1b) was greater with the 75 $\mu gP g^{-1}$ soil than the zero rate in all disturbance treatments, but there was no further increase at the 150 $\mu gP g^{-1}$ soil rate. Shoot P concentration was greatest in the U treatment and least in the D treatment at all soil P rates. The effect of disturbance was least marked in the zero P added treatment.

Analysis of arbuscular colonization (Fig. 20.1c) provides some very interesting results. In the D and DL treatments, neither of which had an intact extraradical mycelium at the beginning of the fourth cycle, colonization increased as P supply increased to the 75 $\mu gP g^{-1}$ soil rate. There was a small but significant decline in the D treatment as the P rate increased to 150 $\mu gP g^{-1}$ soil. The colonization in the DL treatment was consistently greater than that in the D treatment, probably reflecting the greater level of inoculum present because of the lack of disturbance following previous cycles. In the U treatment, where there was an intact mycelium at the beginning of the fourth cycle, arbuscular colonization was greatest in the soil to which no phosphorus was added and declined progressively with increasing P rates, although the decline from 75 to 150 $\mu gP g^{-1}$ soil rate was not significant at $P = 0.05$. This pattern of colonization suggests a different influence of soil P

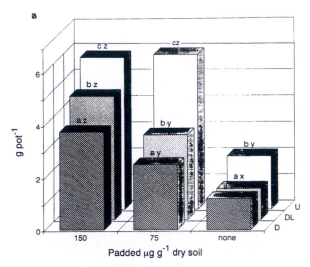

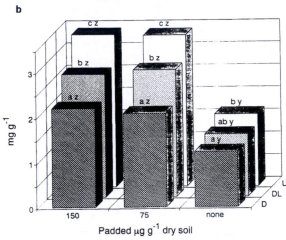

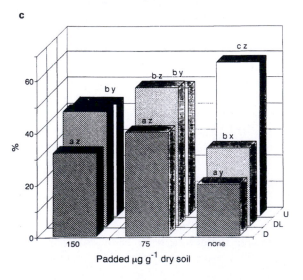

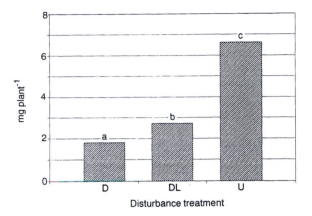

Fig. 20.2. Shoot P content at the intermediate P rate in three disturbance treatments (D – disturbed after each cycle, DL-disturbed after third cycle only, and U – undisturbed) following the fourth growth cycle in the glasshouse experiment. Bars with the same letter are not significantly different at $P = 0.05$.

on the infectivity of a disrupted compared with an undisrupted system composed of roots and extraradical mycelium.

The shoot P content from the three disturbance treatments on the intermediate P soil (75 μg added P g^{-1} soil) is shown in Fig. 20.2. At this P level, the colonization was least in the D treatment, whereas the DL and U treatments resulted in similar colonization (Fig. 20.1c). Neither the D or DL treatments had an intact extraradical mycelium at the beginning of the final growth cycle whereas the U treatment did. Thus we were successful in achieving a difference in colonization in the absence of an undisrupted mycelium (D versus DL) and a similar colonization in the absence and presence of an extraradical mycelium (DL versus U). Shoot P content was somewhat greater on the DL treatment compared with the D treatment, possibly reflecting the greater degree of colonization. However, shoot P content was much greater on the U than the DL treatment. Because these treatments produced comparable amounts of colonization, the increased P content can be attributed to the effectiveness of the pre-existing extraradical mycelium as a nutrient acquisition system. It is envisaged that the newly developing plant becomes attached to this already constructed mycelium, as the roots enter the undisturbed soil.

Fig. 20.1. Shoot dry mass (**a**), shoot P concentration (**b**) and arbuscular colonization of roots (**c**) in three disturbance treatments (D – disturbed after each cycle, DL – disturbed after third cycle only, and U – undisturbed) following the fourth growth cycle in the glasshouse experiment. Bars within a P rate with the same a, b, c letter and bars within a disturbance treatment with the same x, y, z letter are not significantly different at $P = 0.05$.

Summary and Conclusions

The data presented support the hypothesis that the increased absorption of P observed when soil is left undisturbed is due, at least in part, to the ability of the pre-existing extraradical mycelium to act as a nutrient acquisition system for the newly developing plant. Although it is risky to extrapolate this conclusion from a short-term glasshouse study to the field, the similarity of the effects in the field and controlled environment studies makes it tempting to do so. The implications of such an extrapolation are that the extraradical mycelium remains viable and retains its effectiveness as a nutrient acquisition system from one growing season to the next under the winter conditions in Ontario. Further studies are being conducted to test this hypothesis.

Acknowlegement

The authors wish to thank Doug Young, of Ridgetown College of Agricultural Technology, for allowing us to sample his tillage plots. The financial assistance of the Natural Sciences and Engineering Research Council of Canada is gratefully acknowledged.

References

Barry, D.A.J. and Miller, M.H. (1989) Phosphorus nutritional requirements of maize seedlings for maximum yield. *Agronomy Journal* 81, 95–99.

Brundrett, M.C., Piché Y. and Peterson, R.L. (1984) A new method for observing the morphology of vesicular-arbuscular mycorrhizae. *Canadian Journal of Botany* 62, 2128–2134.

Evans, D.G. and Miller, M.H. (1988) Vesicular-arbuscular mycorrhizas and the soil-disturbance-induced reduction of nutrient absorption in maize. I Causal relations. *New Phytologist* 110, 67–74.

Evans, D.G. and Miller, M.H. (1990) The role of the external mycelial network in the effect of soil disturbance upon vesicular-arbuscular mycorrhizal colonization of maize. *New Phytologist* 114, 65–71.

Fairchild, G.L. and Miller, M.H. (1988) Vesicular-arbuscular mycorrhizas and the soil-disturbance-induced reduction of nutrient absorption in maize. II Development of the effect. *New Phytologist* 110, 75–84.

Fairchild, G.L. and Miller, M.H. (1990) Vesicular-arbuscular mycorrhizas and the soil-disturbance-induced reduction of nutrient absorption in maize. III Influence of P amendments to soil. *New Phytologist* 114, 641–650.

Jasper, D.A., Abbott, L.K. and Robson, A.D. (1989) Hyphae of a vesicular-arbuscular mycorrhizal fungus maintain infectivity in dry soil, except when the soil is disturbed. *New Phytologist* 112, 101–107.

McGonigle, T.P., Miller, M.H., Evans, D.G., Fairchild, G.L. and Swan, J.A. (1990a) A new method which gives an objective measure of colonization of roots by vesicular-arbuscular mycorrhizal fungi. *New Phytologist* 115, 495–501.

McGonigle, T.P., Evans, D.G. and Miller, M.H. (1990b) Effect of degree of soil disturbance on mycorrhizal colonization and phosphorus absorption by maize in growth chamber and field experiments. *New Phytologist* 116, 629–636.

O'Halloran, I.P., Miller, M.H. and Arnold G. (1986) Absorption of P by corn (*Zea mays* L.) as influenced by soil disturbance. *Canadian Journal of Soil Science* 66, 287–302.

Thomas, R.L., Sheard, R.W. and Moyer, J.R. (1967) Comparison of conventional and automated procedures for nitrogen, phosphorus and potassium analysis of plant material using a single digest. *Agronomy Journal* 59, 240–243.

21 Development of Mycorrhizal Patches in a Successional Arid Ecosystem

M.F. Allen and E.B. Allen

Department of Biology, Systems Ecology Research Group, San Diego State University, San Diego, California 92182-0057, USA

Introduction

Most efforts at understanding the dynamics of mycorrhizal associations during recovery after disturbance have concentrated on the formation of mycorrhizas by the invading plants or on the influence of mycorrhizal fungi on plant colonization. The most basic model of succession consists of changes in mycotrophy during succession, wherein non-mycotrophic plants are followed by mycotrophic plants (e.g. Reeves *et al.*, 1979; Allen and Allen, 1980; Janos, 1980) or a subsequent variation of the model in which various mycotrophic stages can form the initial successional sere (Allen and Allen, 1990a). These models assume that succession is driven by the invasion of plants and mycorrhizal fungi. Specifically, early seral colonizing plants do not need mycorrhizas but are unable to compete successfully with later seral, mycorrhizal plants following the invasion by the fungi (Allen and Allen, 1984). A second set of models has concentrated on the architecture of a set of plants dispersed in patches and the ability of those differing structures to attract inoculum, whether dispersed by animals or by wind (Allen, 1988, 1991). This model assumes that early colonists are facultatively mycorrhizal plants that form mycorrhizal associations when the fungi disperse to the site. It is linked with the first model in that some early seral plants or planted species are facultatively mycorrhizal and, upon the successful invasion of the mycorrhizal fungi, form mycorrhizal associations that enhance this group of plants and provide inoculum for the later seral vegetation.

Our objective here was to examine a different question. Is there an inoculum network that establishes in a successional site and is formation of that network predictable? Specifically, as individual plants establish, do they initiate a mycelial network that is capable of intensifying the local inoculum and expanding beyond that plant, or is there simply a one-to-one relationship between fungus and host plant which promotes the growth of the individual plant and makes the fungus dependent upon that plant alone. The research reported here was carried out in southwestern Wyoming on a successional reclaimed coal surface mine. The vegetation is a shrub-steppe, which is described in detail elsewhere (Allen and MacMahon, 1985).

Succession of VA Mycorrhizal Fungi

We have previously demonstrated that VA mycorrhizal fungi invade rapidly, primarily by wind, although some animals can serve as vectors for the fungi (Warner *et al.*, 1987; Allen *et al.*, 1989). Moreover, the spore density and species diversity of mycorrhizal fungi was logarithmically related to the size of the individual plants (Allen, 1988, 1991). Two hypotheses may explain this relationship. Either the larger shrubs (or clumps of shrubs) have greater trapping capabilities, or large shrubs have a greater variety of habitats within a root system to enhance local spore production.

As a means of testing these hypotheses, we estimated the rate of production of spores and compared these with our estimates of spore deposition. Spore production was estimated by the change in spore counts in the subsurface soils during a growing season (unpublished data), a reasonable approach because the mine soils initially contained no inoculum. Spore deposition was estimated from direct counts (Allen, Friese and Hipps, in preparation). Following the initial

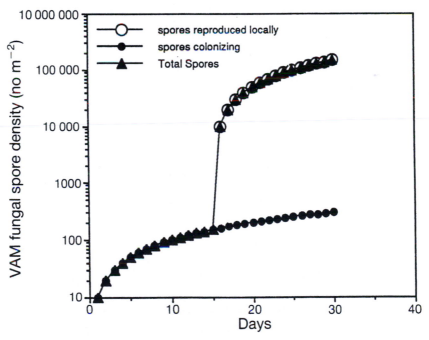

Fig. 21.1. Numbers of spores reproduced *in situ* versus the number of spores colonizing a group of established plants at the Kemmerer site during a single growing season. Shown are the estimated colonizing spores (●), the total spore counts (▲), and the numbers of spores estimated to have been produced (○) estimated by subtraction (see text).

formation of mycorrhizas by the primary colonists, the numbers of spores produced locally rapidly became more important than the number of immigrating
spores (Fig. 21.1). These data coupled with the observations by Friese and Allen
(1991) indicate that mycorrhizal fungal activity cannot be accounted for by incom-
ing inoculum. An initial establishment phase, probably initial spore trapping
followed by the development of a mycelial network and subsequent sporulation,
is essential to explain the densities of mycorrhizal spores in soil that is poor in
inoculum.

Based on these observations, how does a mycelial network develop? Friese and
Allen (1991) recently divided the VA mycelial network into distinctly differing
architectures. First, they found an absorbing hyphal matrix that was composed of
a dichotomously branching network which decreased in size as it expanded from
an infection unit. This network developed from an initial hypha of 8–10 μm
diameter with ~ 1 branching unit per day until a maximum of ~ 100 cm of hyphae
was produced. The hyphal tips appeared to be too small in diameter to continue
to allow further branching (~ 2 μm). Moreover, these tips appeared to be unable
to initiate any new infections. Importantly, these absorption networks were never
found except in the presence of arbuscules (presumably during periods of active P
and C transport) and were not found in the rhizosphere of non-mycotrophic plants
such as *Salsola kali* (unpublished observations). Secondly, a 'runner' hyphal network was observed that initiated new infections between roots. In contrast to
absorption hyphae, these runner hyphae were also observed in the rhizosphere of
S. kali (unpublished observations). Delta ^{13}C carbon analysis of the spores in the
rhizosphere of *S. kali* demonstrated that the fungus could gain carbon from 'non-
host' plants (Allen and Allen, 1990b). In summary, the absorbing hyphae formed
a fan-shaped network that could not initiate infections, while the runner hyphae
were coarser and could initiate infection, but are probably not important for
absorption.

To estimate the mycelial spread in a developing rhizosphere, we developed
a simple model based on the dominant mycotrophic species from our research
sites, *Artemisia tridentata* and *Agropyron* spp. We estimated root expansion of
these two species through the soil using data from Caldwell *et al.* (1985) and Allen
et al. (1989). The numbers of infecting units were determined using the size
classes from the same data sets. The expansion rates of the absorbing hyphae are
described using *Artemisia tridentata* and *Oryzopsis hymenoides* (which has a
rooting development similar to the *Agropyron* spp.) described in Friese and Allen
(1991). To summarize, the estimates of hyphae produced from this developmental
model provided a peak of 1100 cm of live hyphae per cm^3 of soil and a maximum total hyphal length of 2300 cm cm^{-3} soil (Fig. 21.2). These numbers were
calculated using only the absorbing hyphal network. These model values are in
the lower end of the range of our field estimates of viable VA mycorrhizal fungal
hyphae (1200 cm cm^{-3} soil) and of total hyphae (2800–5400 cm cm^{-3} soil).
While the model and field values are remarkably close, the values from the model
remain low. However, if accurate, it suggests that the absorbing hyphae account
for most of the hyphae present and that development of the hyphal network may
be supported by a single individual plant, as no anastomosis has been observed in
our system.

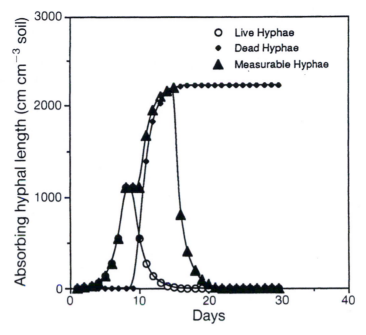

Fig. 21.2. The amount of VA mycorrhizal hyphae per unit soil volume based strictly on developing absorbing hyphal networks and the expanding rooting system (see text). Shown are the estimated live hyphae (○), the dead hyphae (●), and the total measurable hyphae (▲) assuming a predictable die back and disappearance.

However, we observed runner hyphae in the rhizospheres of both *Salsola kali* (Allen and Allen, 1986) which is capable if initiating new infections (Allen *et al.*, 1989) and of *Artemisia tridentata* (Allen and MacMahon, 1985) in the autumn, long after arbuscules disappeared and nutrient transport ceased. We modified the model to add a 'runner' hyphal component equal to the hyphal lengths observed in the two above conditions (an additional ~ 700 cm cm^{-3} soil). Using the modified model, we would predict that the length of viable hyphae in soil varies from a low of 700 cm cm^{-3} soil early in the growing season to a high of 2800 cm cm^{-3} soil and a total (live plus dead) peak of 3500 cm cm^{-3} soil (Fig. 21.3). These values are almost identical to the values that we have observed in the rhizosphere of *Agropyron smithii* in the field at the successional Kemmerer site (Allen and Allen 1986).

Importantly, these simple models, based on developmental features and data unrelated to the data from the successional sites, predict relatively accurately the total amount of hyphae in those sites. This suggests that a mycelial network is not just a mass of hyphae but a well organized aggregation designed for both resource acquisition for the plant and the search for new roots by the fungus. Moreover, the development of this network may be predictable in time and space.

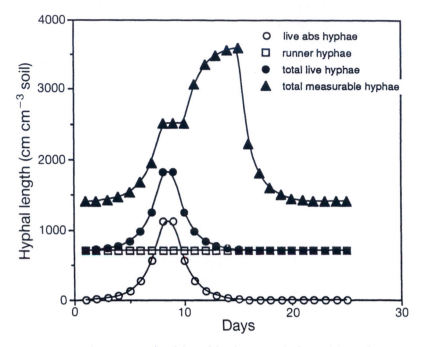

Fig. 21.3. Total VA mycorrhizal fungal hyphae in soil, derived from the previous model (Fig. 21.2) plus adding the estimated amounts of runner hyphae. Shown are the live absorbing hyphae (○), the runner hyphae (□), the total live hyphae, runner plus absorbing hyphae (●), and the total measurable hyphae (▲), assuming death and disappearance.

Synthesis and Conclusion

We can summarize by developing a conceptual model of the development of a mycorrhizal patch as shown in Fig. 21.4. A runner hyphal network initially develops from invading spores in association with whatever plant is present, be it a non-mycotrophic *Salsola kali*, or a mycotrophic shrub such as *Artemisia tridentata*. That network may or may not be involved in resource transport but is clearly capable of initiating new infections. Following the development of mycorrhizal infections, individual absorbing networks develop and disappear rapidly depending on the production of infectable roots during the growing season. Spores are then formed, which allows for the survival of the fungus during harsh times and further expands the mycelial network.

Clearly, more detail is needed to understand and quantify the mycelial network in soil and understand the relationship between spores, runner hyphae, and absorbing hyphae. However, these models suggest that the development of a mycelial network is essential to characterizing the amount of mycorrhiza present and understanding the dynamic interactions between plant and fungus. We would conclude that the effort to integrate mathematical models, laboratory experiments, and field observations will explain not only the story of the successional patterns

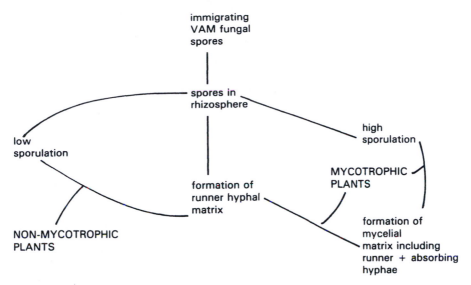

Fig. 21.4. A conceptual model of the development of a VA mycorrhizal fungal mycelial network in patch of plants from an arid ecosystem.

of a single site but also will provide a set of models that can be used to predict the dynamics of other newly disturbed sites.

References

Allen, E.B. and Allen, M.F. (1980) Natural re-establishment of vesicular-arbuscular mycorrhizae following stripmine reclamation in Wyoming. *Journal of Applied Ecology* 17, 139–147.

Allen, E.B. and Allen, M.F. (1984) Competition between plants of different successional stages: mycorrhizae as regulators. *Canadian Journal of Botany* 62, 2625–2629.

Allen, E.B. and Allen, M.F. (1986) Water relations of xeric grasses in the field: interactions of mycorrhizae and competition. *New Phytologist* 104, 559–571.

Allen, E.B. and Allen, M.F. (1990a) The mediation of competition by mycorrhizae in successional and patchy environments. In: Grace J.B. and Tilman G.D. (eds), *Perspectives on Plant Competition*. Academic Press, New York, USA, pp. 367–389.

Allen, M. F. (1988) Re-establishment VA of mycorrhizae following severe disturbance: comparative patch dynamics of a shrub desert and a subalpine volcano. *Proceedings of the Royal Society of Edinburgh* 94B, 63–71.

Allen, M.F. (1991) *The Ecology of Mycorrhizae*. Cambridge University Press, Cambridge, UK.

Allen, M.F. and Allen, E.B. (1990b) Carbon source of VA mycorrhizal fungi associated with Chenopodiaceae from a semi-arid steppe. *Ecology* 71, 2019–2021.

Allen, M.F. and MacMahon, J.A. (1985) Impact of disturbance on cold desert fungi: comparative microscale dispersion patterns. *Pedobiologia* 28, 215–224.

Allen, M.F., Allen, E.B. and Friese, C.F. (1989) Responses of the non-mycotrophic plant *Salsola kali* to invasion by vesicular-arbuscular mycorrhizal fungi. *New Phytologist* 111, 45–49.

Caldwell, M.M., Eissenstat, D.M., Richards, J.H. and Allen, M.F. (1985) Competition for phosphorus: differential uptake from dual-isotope-labeled soil interspaces between shrub and grass. *Science* 229, 384–386.

Friese, C.F. and Allen, M.F. (1991) The spread of VA mycorrhizal fungal hyphae in soil: inoculum type and external hyphal architecture. *Mycologia* 83, 409–418.

Janos, D.P. (1980) Mycorrhizae influence tropical succession. *Biotropica* 12, 56–64.

Reeves, F.B., Wagner, D. W., Moorman, T. and Kiel, J. (1979) The role of endomycorrhizae in revegetation practices in the semi-arid west. I. A comparison of incidence of mycorrhizae in severly disturbed vs. natural environments. *American Journal of Botany* 66, 1–13.

Warner, N.J., Allen, M. F. and MacMahon, J.A. (1987) Dispersal agents of vesicular-arbuscular mycorrhizal fungi in a disturbed arid ecosystem. *Mycologia* 79, 721–730.

22 Extraradical Hyphal Development of Vesicular-Arbuscular Mycorrhizal Fungi in a Chronosequence of Prairie Restorations

R.M. Miller and J.D. Jastrow

Environmental Research Division, Argonne National Laboratory, Argonne, Illinois 60439, USA

Introduction

Little quantitative information is available on the extraradical hyphal phase of the vesicular-arbuscular mycorrhizal association (Finlay and Söderström, 1989; Sylvia, 1990), especially at the community and ecosystem level. We have used an ecosystem restoration approach to study the development of roots and mycorrhizal colonization (Jastrow, 1987; Cook *et al.*, 1988) and the influences of vegetation and mycorrhizas on the development of soil structure (Jastrow, 1987; Miller and Jastrow, 1990). These studies were conducted in a chronosequence of tallgrass prairie restorations on a site that had been cultivated for rowcrops for at least 100 years. We showed an important association between root morphology and mycorrhizal colonization. In this report we present data on the development of the extraradical hyphal phase of the mycorrhizal association for the restoration chronosequence and for an adjacent cultivated field, and on the relationship between root growth and extraradical hyphae.

Methods and Materials

Fuller details are given in Jastrow (1987) and Cook *et al.* (1988).

Study area description

The study area included a chronosequence of tallgrass prairie restorations and an adjacent rowcrop field located at the US Department of Energy's National

Environmental Research Park at Fermi National Accelerator Laboratory (Fermilab), Batavia, Illinois, about 48 km west of Chicago. The chronosequence represents various phases of secondary succession typically associated with prairie restoration and included plots in their second to eleventh growing season since the last cultivation and planting to prairie. The rowcrop field had recently been in continuous corn. All sampled areas were located on similar silt loam and silty clay loam soil types.

Study design and methods

Ten sampling stations were established by using a stratified random design in each of four sampled prairie restorations (second, fifth, eighth and eleventh growing season) and in the rowcrop field. At each station, a randomly located $0.5 \, m^2$ circular quadrat was sampled during the last 2 weeks of June 1985: the aboveground vegetation was clipped and removed, and soil cores were taken to a depth of 20 cm.

Quantitation of extraradical hyphae

Hyphae were collected on a membrane filter by using a modification of the method of Hanssen et al. (1974). Two undried 10 g subsamples of each core (with roots, rhizomes and large organic debris removed) were placed in separate beakers of sodium hexametaphosphate solution ($35.7 \, g \, l^{-1}$), soaked overnight, sonicated for 20–25 s at 120 W and then diluted (c. 1:20). An aliquot of the diluted suspension was centrifuged at 1000 g, and the pellet was resuspended in 50% glycerol by using a vortex mixer. After centrifugation at 75 g for 30 s, the supernatant was filtered through a 20 μm mesh filter. The mesh filter was placed in a staining solution of lactic trypan blue, and hyphae were resuspended with a vortex mixer. After 1.5 h, the staining solution was quantitatively filtered through a white cellulose nitrate membrane filter with a 0.8 μm pore size.

When dry, filters were mounted on slides by using immersion oil and were viewed at 160×. Seventy fields of view, located by making stratified random movements of the filter with the microscope stage, were scored by the gridline intercept method and converted to hyphal lengths (Newman, 1966). Hyphae were recognized as mycorrhizal on the basis of morphology (e.g. Nicholson, 1959). Data for the two subsamples from each soil core were pooled.

Quantitation of root morphology

Total root lengths, the length of root colonized by mycorrhizal fungi and percentage colonization were quantified (Newman, 1966). Only fibrous roots (those < 1 mm in diameter) were assayed because roots with larger diameters were usually suberized or lacking cortex and were considered unlikely to be actively colonized. Total root lengths, colonized lengths, and percentage colonization were determined for two diameter size classes of fibrous roots, i.e. fine roots (0.2–1.0 mm diameter) and very fine roots (<0.2 mm diameter). To determine total and colonized lengths and percentage colonization for all fibrous roots, length data for the the two size classes were summed. Roots were scored as colonized if the cortex contained arbuscules, vesicles, or intraradical runner hyphae at the point of root intersection with the gridline. Specific root length (SRL) was calculated on the basis of dry weight as a measure of gross root morphology (Fitter, 1985).

Statistical analysis

The development of extraradical hyphae in the restoration chronosequence was modelled by using the NLIN procedure (SAS Institute, Inc., 1985). The relationships of extraradical hyphae to root parameters and to measures of intraradical mycorrhizal structures were determined by Pearson product-moment correlations. Data were transformed (logarithmic or arcsin square root) before analysis when appropriate.

Results and Discussion

The length of extraradical hyphae of mycorrhizal fungi (m cm^{-3} soil) increased with time since cessation of cultivation (Fig. 22.1), and could best be expressed by the exponential equation $y = \beta(1 - ce^{\alpha x})$, where β is the asymptotic maximum yield, α is a constant of proportionality and $c = (\beta - y_0)/\beta$. The parameter c was included in the model because in this study the regression does not pass

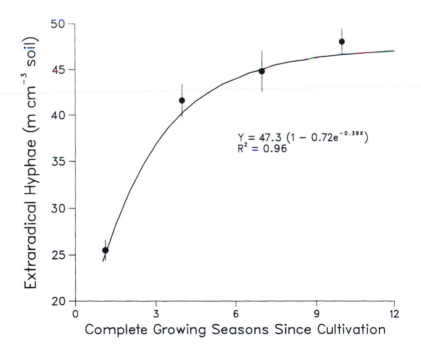

$$Y = 47.3 (1 - 0.72e^{-0.39x})$$
$$R^2 = 0.96$$

Fig. 22.1. The development of extraradical hyphae within the chronosequence of tallgrass prairie restorations ($n = 10$).

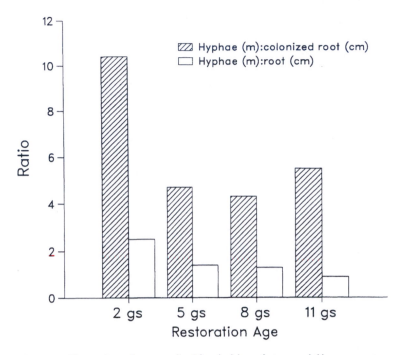

Fig. 22.2. The ratios of extraradical hyphal length to total fibrous root length and colonized fibrous root length for the sampled prairie restorations ($n = 10$). For X axis, numbers indicate restoration ages, i.e. number of growing seasons since last cultivation.

through the origin. The value of y_o (i.e., extraradical hyphal length when $x = 0$) was assumed to be $14 \, \text{m} \, \text{cm}^{-3}$ soil (the estimated value in the corn field at the beginning of the growing season), and this model predicted an asymptotic maximum yield for extraradical hyphae of $47.3 \pm 1.5 \, \text{m} \, \text{cm}^{-3}$ soil and an exponential growth rate to the asymptote of $0.39 \, \text{m} \, \text{cm}^{-3}$ per growing season ($R^2 = 0.96$).

Hyphal length as a portion of total fibrous root length declined over time (Fig. 22.2), but as a proportion of colonized fibrous root length, a relatively constant ratio was apparently achieved by the fifth season (Fig. 22.2). Hence, the extraradical hyphal phase may be responding in a coordinated way to both root growth and intraradical fungus growth.

Measurements of fine root length and colonized fine root length and of fibrous root mass and specific root length were all highly correlated with extraradical hyphal length (Table 22.1). In contrast, parameters associated with the very fine size class of roots, as a group, were not significantly associated with extraradical hyphae. In addition, mycorrhizal colonization, expressed as a percentage rather than as colonized root lengths, was not associated with hyphal length.

Extraradical hyphal length increased rapidly after cessation of cultivation (Fig. 22.1) and was related to the development of the root system within the restoration chronosequence (Table 22.1). The rate of hyphal development fitted an asymptotic exponential model; the approach to the asymptote was rapid, and

Table 22.1. Pearson product-moment correlations for extraradical mycorrhizal hyphae with root and intraradical mycorrhizal characteristics in the chronosequence of prairie restorations ($n = 40$).

Belowground parameters	r
Fine root length	0.71 ***
Very fine root length	0.27 n.s.
Colonized fine root length	0.57 ***
Colonized very fine root length	0.20 n.s.
Fine root colonization (%)	0.29 n.s.
Very fine root colonization (%)	0.01 n.s.
Fibrous root mass	0.64 ***
Root radius	0.10 n.s.
Specific root length	−0.46 ***

*** = $P \leq 0.001$; n.s. = $P > 0.05$.

by the fifth season, 90% of the asymptotic value of 47.3 m cm^{-3} soil was achieved (Fig. 22.1). After cultivation ceased, the extraradical mycorrhizal hyphae developed faster than the intraradical hyphal phase as demonstrated by the high ratio found for extraradical hyphal length to colonized root length in the second growing season (Fig. 22.2). Yet, as the asymptote for extraradical hyphal development was approached, the ratio appeared to decrease to a value of about four. The true ratio is actually higher, however, because the extraradical hyphal length is underestimated, due to the removal of roots during soil preparation for hyphal extraction, which results in the loss of extraradical hyphae attached to those roots (Miller and Jastrow, unpublished). The ratio of hyphal length to root length also declined with time. This decrease is most likely associated with a concurrent increase in root density (Cook *et al.*, 1988), especially in the very fine size class, for which proportionally less colonization occurs than for the fine roots.

We believe this report to be the first to determine the development of extraradical hyphae by using a chronosequence approach. Our findings indicate that hyphal development is associated with both root growth, as measured by biomass and length, and changes in gross root morphology as measured by specific root length. Furthermore, many of the trends for extraradical hyphae are a direct consequence of prairie species becoming established along the chronosequence. Both the asymptotic equilibrium for hyphal length and the establishment of a plant community dominated by prairie species occurred by the fifth growing season from planting (Jastrow, 1987). As a follow-up to this study we are investigating the association of root growth, and internal and external allocation of mycorrhizal fungi, in relation to changes in nutrient supply along the prairie chronosequence.

Acknowledgements

This research was supported by the US Department of Energy, Office of Energy Research, Office of Health and Environmental Research, Environmental Sciences

23 Interactions between Soil-dwelling Insects and Mycorrhizas during Early Plant Succession

A.C. Gange and V.K. Brown

Imperial College at Silwood Park, Ascot, Berks SL5 7PY, UK

Introduction

An extensive literature now exists indicating the beneficial effects that infection by vesicular-arbuscular mycorrhizas can have on plant growth and reproduction (Allen, 1991). However, a feature of this literature is that the vast majority of the experiments have taken place in controlled conditions. When studies have taken place in the field, the results are often inconclusive; reduction of mycorrhizal infection may result in reduced plant growth (e.g. Gange *et al.*, 1990), or have little or no effect (e.g. McGonigle, 1988).

There are a number of possible reasons for the apparent ineffectiveness of mycorrhizas in field situations, compared with the controlled conditions in 'pot trials' (Fitter, 1985). One of these reasons is the presence of grazing animals, especially insects (Brown and Gange, 1990). For example, Collembola (springtails) are often abundant in soil and a number of species are mycophagous. It has been demonstrated that these insects can reduce mycorrhizal infection in pot trials, leading to reduced plant growth compared with mycorrhizal controls (Warnock *et al.*, 1982). Furthermore, there are a number of much larger insect species that live in the soil, feeding on the subterranean parts of plants. It is highly likely that these will reduce mycorrhizal infection as an occupational hazard of root feeding (Rabatin and Stinner, 1991). This chapter describes an experiment which aims to reduce the natural levels of insects and/or mycorrhizas within a factorial design, to investigate whether soil insects may be one of the reasons for mycorrhizal ineffectiveness. The experiment took place in an early successional plant community, during the first 2 years of the colonization of bare ground.

Methods

An experimental site, measuring 450 m², was treated with a weedkiller in autumn 1987 to destroy perennial weeds. Following the death of the vegetation, the land was shallow ploughed in winter and hand raked in March of the following year. The site was divided into 2.5 × 2.5 m plots, separated by 1.5 m 'walkways'. There were four experimental treatments: (1) control, with natural levels of soil insects and mycorrhizas; (2) insecticide-treated, in which the soil insecticide Dursban 5G containing 5% w/w chlorpyrifos (DowElanco) was applied in granular form; (3) fungicide-treated, in which the contact fungicide Rovral containing 10% w/w iprodione (Rhône-Poulenc) was applied, also in granular form, and (4) insecticide and fungicide-treated. There were four replicates of each treatment, arranged in a randomized block design. Pesticide application began in March and continued at 6-weekly intervals for two growing seasons. The pesticides were applied just before rain, so that they were washed quickly into the soil and did not remain active on the surface for long. The advantages of using these compounds, together with tests of phytotoxicity, have been described previously (Brown and Gange, 1989; Gange et al., 1990, 1992).

The developing vegetation was sampled at 3-weekly intervals during the first growing season (April–October) and at monthly intervals during the second. On each occasion, a 38 cm linear steel grid containing ten point quadrat pins was placed randomly five times in each plot. The number of touches of living vegetation was recorded on each pin for all plant species. Data were condensed to provide information on the total touches (cover abundance) of the whole sward and the major life-history groups (annual forbs, perennial forbs and perennial grasses). These data were analysed according to dates by a split-plot analysis of variance, with pesticide treatments as the main effects. At the end of the season, soil samples were taken to determine the efficacy of the insecticide and fungicide treatments.

Results

Application of insecticide reduced insect populations. The mean number of pupal cases of one of the commonest herbivores, *Tipula oleracea*, was 9.2 ± 0.5 m^{-2} in control plots compared with 0.7 ± 0.1 m^{-2} in insecticide-treated plots. In addition, there were no insecticidal effects of fungicide application, with 8.7 ± 0.9 pupal cases m^{-2} found in plots receiving this treatment. The fungicide was successful in reducing mycorrhizal infection. In seven out of the 11 species whose roots were examined, there was a significant reduction in infection (Gange et al., 1990). In addition, there was no evidence of the fungicide controlling foliar pathogenic fungi, since *Spergula arvensis* was attacked (by mildew) in both treatment and control plots.

In the first year of succession, application of insecticide significantly increased the total vegetative cover ($F_{1,12} = 31.7$, $P < 0.001$), while that of fungicide decreased it ($F_{1,12} = 5.7$, $P < 0.05$) (Fig. 23.1a). There was no evidence of interaction between the compounds. However, in the second year, while insecticide again increased cover abundance ($F_{1,12} = 66.4$, $P < 0.001$) and fungicide decreased it ($F_{1,12} = 14.7$, $P < 0.01$) (Fig. 23.1b), there was also an interaction between the treatments ($F_{1,12} = 6.3$, $P < 0.05$). This is seen by first comparing the two lines of solid symbols in Fig. 23.1b, (when insects are present), i.e. control vs fungicide-treated. It can be seen that there is only a small reduction in cover

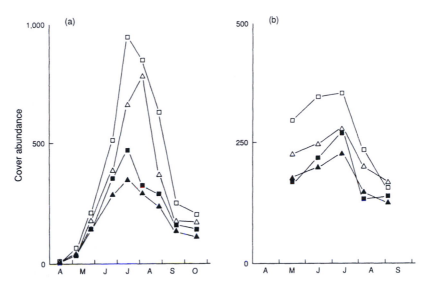

Fig. 23.1. Trends in cover abundance of a plant community during the **(a)** first and **(b)** second years of succession. (■) control, (□) application of soil insecticide, (▲) application of soil fungicide, (△) application of both compounds.

caused by the fungicide. However, if the two lines of open symbols are compared (when insects are reduced), there is a large reduction in cover caused by application of the fungicide. The effect of fungicide is therefore dependent on the insecticide, leading to the significant interaction term.

The annual forbs were the dominant life forms in the first year of the study, and changes in the total cover abundance were largely a reflection of changes in this life-history group (Fig. 23.2a). Thus, insecticide application significantly increased annual forb abundance ($F_{1,12} = 28.4$, $P < 0.001$), while fungicide decreased it ($F_{1,12} = 5.7, P < 0.05$). These forbs were very rare during the second year of the succession, and neither treatment had any significant effect on their abundance.

Perennial forbs began to appear in the community towards the end of the first year and during year two (Fig. 23.2b). In the second season, insecticide application increased perennial forb abundance ($F_{1,12} = 11.9$, $P < 0.01$) and fungicide reduced it ($F_{1,12} = 6.1$, $P < 0.05$) (Fig. 23.2b). There was a suggestion of an interaction between the treatments, with the reduction by fungicide being more pronounced when insects were controlled, but this was not significant, due to large variances in the data.

Perennial grasses also became established late in the first season. These responded in a similar manner to the perennial forbs in the second year, with insecticide application significantly increasing grass abundance ($F_{1,12} = 16.1$, $P < 0.01$), and fungicide decreasing it ($F_{1,12} = 15.0$, $P < 0.01$) (Fig. 23.2c). During this year, there was a dramatic increase in grass growth when insects were reduced and mycorrhizas present; however this effect virtually disappeared if mycorrhizas were reduced as well (Fig. 23.2c). There was therefore a significant

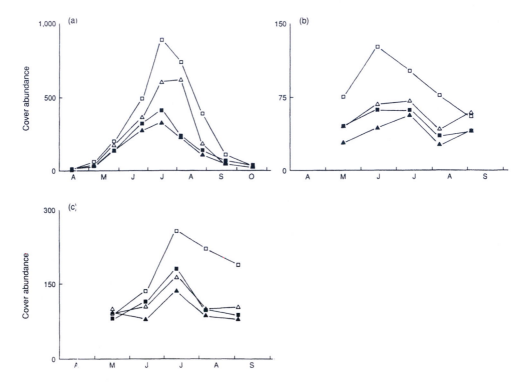

Fig. 23.2. Trends in the cover abundance of annual forbs during the first year of succession (**a**), perennial forbs during the second year of succession (**b**) and perennial grasses during the second year of succession (**c**). Conventions as in Fig. 23.1.

interaction term between the pesticides ($F_{1, 12} = 4.8$, $P < 0.05$) which explains the interaction seen in the overall community (Fig. 23.1b).

Discussion

A problem with field manipulation of mycorrhizas in past studies has been the choice of an effective fungicide. The systemic compound, benomyl, has most commonly been used (e.g. Fitter and Nichols, 1988), but this may not always be successful in reducing infection levels (Koide *et al.*, 1988). The work reported here confirms that of previous studies in demonstrating that granular formulations are likely to be the most effective pesticides for ecological studies in the soil (Brown and Gange, 1989; Gange *et al.*, 1990).

The results demonstrate that both subterranean insects and mycorrhizas are important in structuring an early successional community. Following the disturbance caused by ploughing, both insects and fungi might be expected to occur at relatively low densities (Evans and Miller, 1988; Brown and Gange, 1990). However, during the first year of the study, subterranean insects were found at a peak level of 137 m^{-2} and a number of annual forbs were found to be mycorrhizal (Gange *et al.*, 1990). Reducing herbivore numbers and fungal infection resulted in changes in the growth of a number of these annuals, with resulting changes in the overall plant community. However, it was during the second year of the study that the most interesting results were obtained. Reducing insect numbers caused an increase in total cover abundance, while reducing infection levels caused a decrease, similar to the results obtained in year one. However, it was found that mycorrhizal activity is highly dependent on the presence of insect herbivores, since a clear demonstration of mycorrhizal benefit was only obtained when insect numbers were reduced. This effect was due mainly to the performance of perennial grasses (e.g. *Agrostis stolonifera* and *Holcus lanatus*), the abundance of which was dramatically increased when insects were reduced but fungi present. When both insects and fungi were reduced, grass performance was no better than that of the controls, suggesting that the reduction of fungal activity by insects in the control plots was as effective as that achieved by fungicide in the treatment plots. A similar result was obtained by McGonigle and Fitter (1988), who applied benomyl and the insecticide chlorfenvinphos in a factorial design, to grassland dominated by *Holcus lanatus*. Insecticide application significantly reduced the abundance of mycophagous Collembola and it was suggested that this reduction allowed plants to benefit from the mycorrhizal infection. Such results may explain why other studies, which have attempted to reduce infections in the field, have not been very successful in terms of demonstrating mycorrhizal benefit (Fitter, 1986; Koide *et al.*, 1988). They have been done in the presence of subterranean insects, which can occur at densities of over 2000 m^{-2} in the sort of communities studied (Brown and Gange, 1990).

At the current time, we are attempting to determine which particular insects are responsible for reducing the response of plants to mycorrhizas. Mycophagous Collembola are clearly important (McGonigle and Fitter, 1988), but much larger insects such as Scarabaeid and Tipulid larvae, which are generally considered to feed on roots, may also play a vital role. Indeed, the effect recorded on perennial grasses in this study may be due to Scarabaeid larvae, since these insects have been shown to be positively associated with the roots of grasses (Gange *et al.*, 1991) and are known to be common in early successional communities.

Acknowledgements

We are grateful to the Natural Environment Research Council for financial support. The Dursban was generously provided by DowElanco and the Rovral by Rhône-Poulenc.

References

Allen, M.F. (1991) *The Ecology of Mycorrhizae*. Cambridge University Press, Cambridge.
Brown, V.K. and Gange, A.C. (1989) Differential effects of above- and below-ground insect herbivory during early plant succession. *Oikos* 54, 67–76.

Brown, V.K. and Gange, A.C. (1990) Insect herbivory below ground. *Advances in Ecological Research* 20, 1–58.

Evans, D.G. and Miller, M.H. (1988) Vesicular-arbuscular mycorrhizas and the soil-disturbance-induced reduction of nutrient absorption in maize. I. Causal relations. *New Phytologist* 110, 67–74.

Fitter, A.H. (1985) Functioning of vesicular-arbuscular mycorrhizas under field conditions. *New Phytologist* 99, 257–265.

Fitter, A.H. (1986) Effect of benomyl on leaf phosphorus concentration in alpine gasslands: a test of mycorrhizal benefit. *New Phytologist* 103, 767–776.

Fitter, A.H. and Nichols, R. (1988) The use of benomyl to control infection by vesicular-arbuscular mycorrhizal fungi. *New Phytologist* 110, 201–206.

Gange, A.C., Brown, V.K. and Farmer, L.M. (1990) A test of mycorrhizal benefit in an early successional plant community. *New Phytologist* 115, 85–91.

Gange, A.C., Brown, V.K., Barlow, G.S., Whitehouse, D.M. and Moreton, R.J. (1991) Spatial distribution of garden chafer larvae in a golf tee. *Journal of the Sports Turf Research Institute* 67, 8–13.

Gange, A.C., Brown, V.K. and Farmer, L.M. (1992) Effects of pesticides on the germination of weed seeds: implications for manipulative experiments. *Journal of Applied Ecology* 29, 303–310.

Koide, R.T., Huenneke, L.F., Hamburg, S.P. and Mooney, H.A. (1988) Effects of applications of fungicide, phosphorus and nitrogen on the structure and productivity of an annual serpentine plant community. *Functional Ecology* 2, 335–344.

McGonigle, T.P. (1988) A numerical analysis of published field trials with vesicular-arbuscular mycorrhizal fungi. *Functional Ecology* 2, 473–478.

McGonigle, T.P. and Fitter, A.H. (1988) Ecological consequences of arthropod gazing on VA mycorrhizal fungi. *Proceedings of the Royal Society of Edinburgh* 94B, 25–32.

Rabatin, S.C. and Stinner, B.R. (1991) Vesicular-arbuscular mycorrhizae, plant, and invertebrate interactions in soil. In: Barbosa, P., Krischik, V.A. and Jones, C.G. (eds), *Microbial Mediation of Plant–Herbivore Interactions*. John Wiley, New York, pp. 141–168.

Warnock, A.J., Fitter, A.H. and Usher, M.B. (1982) The influence of a springtail *Folsomia candida* (Insecta:Collembola) on the mycorrhizal association of leek *Allium porrum* and the vesicular-arbuscular mycorrhizal endophyte *Glomus fasciculatum*. *New Phytologist* 90, 285–292.

24 Are Mycorrhizal Fungi Present in Early Stages of Primary Succession?

J.N. Gemma and R.E. Koske[1]

Department of Plant Sciences and [1]Department of Botany, University of Rhode Island, Kingston, RI 02881 USA

The presence or absence of vesicular-arbuscular (VA) mycorrhizal fungi in secondary successional sites may determine the composition of the plant community that develops (Reeves *et al.*, 1979). Disturbed habitats lacking VA fungi have been found to be dominated in the early successional stages by facultative and non-mycotrophic plant species (Miller, 1979; Reeves *et al.*, 1979). Over time VA fungul propagules are introduced to the habitat (in association with facultative mycotrophs), allowing obligate mycotrophs to become established and eventually to dominate the late seral stages (Fig. 24.1a).

In primary succession of two sites in the western USA, a volcanic ash substrate from the Mt St Helens eruption (Allen, 1988) and an alpine glacial moraine (Trappe, personal communication) a similar lack of VA fungi in the early successional stages was reported. These data support the model of secondary and primary succession proposed by Janos (1980) (Fig. 24.1a).

Studies of primary succession in the Hawaiian Islands have suggested that mycorrhizal fungi may be present and important in the earliest stages of succession (Fig. 24.1b). Here we review that evidence and speculate on the role of mycorrhizas in the colonization of the Hawaiian Islands as they rose from the sea 30 million years ago.

The Hawaiian Islands contain many habitats in the early stages of primary succession. Lava fields are continuously formed by two active volcanos and sand dunes are numerous. The vegetation and substrates of these habitats have been examined for VA fungi using methods described elsewhere (Koske, 1988; Koske and Gemma, 1989; Gemma and Koske, 1990; Koske and Gemma, 1990; Koske *et al.*, 1990; Gemma *et al.*, 1992; Koske *et al.*, 1992).

Mycorrhizas in the Hawaiian Coastal Strand

VA fungi are a nearly constant component of the coastal strand vegetation. Over 70% of the 44 species of dune colonizing plants examined formed VA mycorrhizas

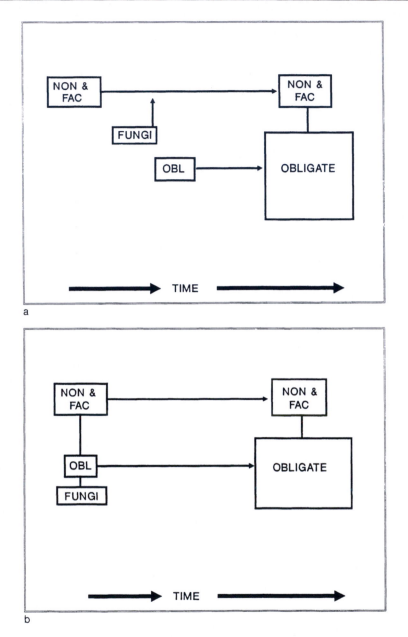

Fig. 24.1. Models of involvement of mycorrhizal fungi in primary succession. (**a**) Traditional scheme in which early colonizers are facultative (FAC) and non-mycotrophic (NON) species. Mycorrhizal fungi arrive later, establishing in association with species that are not obligate mycotrophs (OBL). Subsequently, obligate mycotrophs invade, eventually dominating the community (indicated by the relative size of boxes on the right). Scheme does not suggest that obligate mycotrophic species are derived from non- or facultative mycotrophs (and *vice versa*). (**b**) Model depicting mycorrhizal fungi arriving with earliest pioneer species, either being codispersed or arriving independently. Note that result of both models is the same: obligate mycotrophs dominate in later seral stages.

MYCORRHIZAS IN DUNE SITES

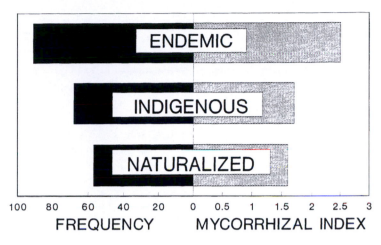

Fig. 24.2. Frequency (percentage of species) and intensity of mycotrophy in endemic, indigenous, and naturalized sand dune-inhabiting species in Hawaii. Mycorrhizal index is a measure of intensity of mycorrhizal formation in roots: 0 = none, 3 = maximal.

(Koske, 1988; Koske and Gemma, 1990; Koske *et al.*, 1992; Gemma *et al.*, 1992). Mycotroph infection was highest in the endemic species and least in naturalized species (Fig. 24.2).

The earliest stages of succession in coastal strands occur at the beach driftline where plant propagules (seeds, fruits, vegetative fragments) are deposited by wave action. The majority of seedlings growing on the driftline contained VA fungi. Propagules were found in both driftline materials and sand below the driftline debris as revealed by the mycorrhizal inoculum potential (MIP) procedure (Moorman and Reeves, 1979; Koske and Gemma, 1990). It appears that inoculum can be present in a primary successional site before the arrival of colonizing plant species.

Reproduction and disperal by vegetative fragmentation of rhizomatous plants is common in the dune habitat, and sprouting fragments may contain spores and hyphae of VA fungi (Gemma and Koske, 1989; Koske and Gemma, 1990). Such co-dispersal ensures the maintenance of the symbiosis in earliest stages of primary succession. Co-dispersal of plants and fungi occurs in rhizomatous grasses and other species in Hawaii and on the east and west coasts and Great Lakes of North America (Gemma and Koske, 1989), and VA fungi survive immersion in seawater (Koske and Gemma, 1990).

A scenario for the initial colonization of sand dunes emerges in which facultative mycotrophs and obligate mycotrophs, the latter with their co-dispersing symbionts, are disseminated by oceanic drift to distant shores. In addition, fungal propagules may arrive separately, in advance of their mycotrophic hosts, rendering the dune sites hospitable for obligate mycotrophs.

Mycorrhizas in Volcanic Substrates

Over 75 % of the 26 species colonizing volcanic substrates of various ages contained VA fungi (Gemma and Koske, 1990). Frequency of mycorrhizas and root colonization increase with the age of the substrate (Fig. 24.3); VA, and two other types of mycorrhiza (ericoid and orchid) were present in the 14-year-old site. Spores of VA fungi were recovered from all sites from the soil and occurred in roots and rhizomes of flowering plants and ferns.

As in the sand dunes examined above, there is a lack of dominance of non-mycotrophic species in the early seral stages. This apparently results from the arrival of VA fungi soon after the volcanic substrates cool (?co-dispersal) and the paucity of non-mycotrophic invaders in the general area (Gemma and Koske, 1990).

Mycotrophy in Hawaiian Flora

The incidence of mycotrophy in Hawaiian angiosperms and pteridophytes growing in a wide variety of habitats was examined. As predicted by the models in Fig. 24.1a and b, the majority of species in non-seral sites were obligate mycotrophs. However obligate mycotrophy was significantly more frequent in the angiosperms (86%) than in the native pteridophytes (60%) (Fig. 24.4).

One explanation for the difference in mycotrophy between angiosperms and pteridophytes may be found in the evolutionary histories of these two groups in

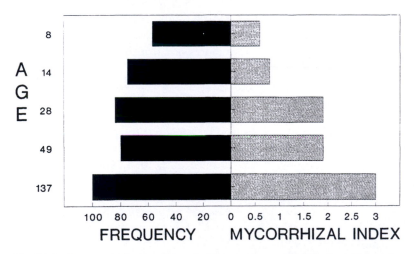

Fig. 24.3. Frequency (percentage of plant specimens sampled) and intensity of mycotrophy in endemic, indigenous, and naturalized species occurring in various types and ages of volcanic substrates in Hawaii. Sites 8 and 14 years old were lava flows; the 28-year-old sites was a cinder fall; the 49-year-old site was a geothermic soil; and the 137-year-old site was a volcanic soil. See caption for Fig. 24.2.

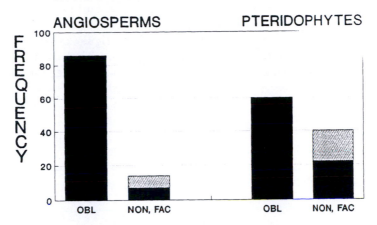

Fig. 24.4. Incidence of obligate, facultative (shaded lines), and non-mycotrophy (solid bars) in Hawaiian angiosperms and pteridophytes.

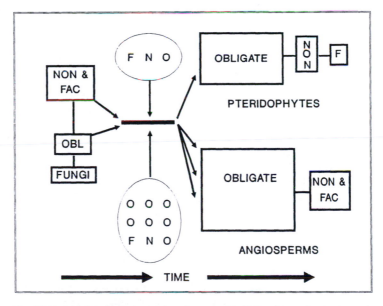

Fig. 24.5. Model depicting role of mycorrhizal fungi in colonization of the Hawaiian Islands by angiosperms and pteridophytes. A greater percentage of pteridophyte invaders appears to have been facultative (FAC or F) or non-mycotrophic (NON or N) species rather than obligately mycotrophic (O) pioneers that typified angiosperm invaders. Letters in ellipses refer to species arriving some time after the first colonizers were present. Extensive radiation by angiosperms (especially by the obligately mycotrophic species) is indicated by the arrows.

the Hawaiian Islands (Fig. 24.5). The high levels of obligate mycotrophy in the angiosperm flora suggest that the majority of the ancestral invaders were obligate mycotrophs. In the pteridophytes a sizable percentage of ancestors may have been facultative and non-mycotrophs. The present-day flora contains 956 native angiosperms 172 native pteridophytes (Fosberg, 1948, 1984). The prolific radiation in the angiosperms appears to have involved mainly obligate mycotrophs, whereas speciation in pteridophytes was more evenly accomplished by all three categories of mycotrophs.

The Hawaiian Islands represent 30 million years of continuous volcanic activity and sand dune formation. Consequently there has been continuous selection for adaptions that allow for the colonization of these primary successional sites. In the phosphate-fixing Hawaiian soils (Foote *et al.*, 1972) mycotrophy appears to have been a valuable attribute among the earliest colonizers.

References

Allen, M.F. (1988) Re-establishment of VA mycorrhizas following severe disturbance: comparative patch dynamics of a shrub desert and a subalpine volcano. *Proceedings of the Royal Society of Edinburgh* 94B, 63–71.

Foote, D.E., Hill, E.L., Nakamura, S. and Stephens, F. (1972) *Soil Survey of Islands of Kauai, Oahu, Maui, Molokai, and Lanai, State of Hawaii*. USDA Soil Conservation Service and University of Hawaii Agricultural Experiment Station, Honolulu.

Fosberg, F.R. (1948) Derivation of the flora of the Hawaiian Islands, In: Zimmerman, E.C. (ed.), *Insects of Hawaii*. Vol. 1. University of Hawaii Press, Honolulu, pp. 107–119.

Fosberg, F.R. (1984) Phytogeographic comparison of Polynesia and Micronesia. *In*: Radovsky, F.J., Raven, P.H. and Sohmer, S.H. (eds), *Biogeography of the Tropical Pacific*. Association of Systemics Collections and B.P. Bishop Museum, Honolulu, pp. 33–44.

Gemma, J.N. and Koske, R.E. (1989) Field inoculation of American beachgrass (*Ammophila breviligulata*) with VA mycorrhizal fungi. *Journal of Environmental Management* 29, 173–182.

Gemma, J.N. and Koske, R.E. (1990) Mycorrhizae in recent volcanic substrates in Hawaii. *American Journal of Botany* 77, 1193–1200.

Gemma, J.N., Koske, R.E. and Flynn, T. (1992) Mycorrhizae in Hawaiian Pteridophytes: Occurrence and evolutionary significance. *American Journal of Botany* 79, 843–852.

Janos, D.P. (1980) Mycorrhizae influence tropical succession. *Biotropica* 12, 56–64.

Koske, R.E. (1988) Vesicular-arbuscular mycorrhizae of some Hawaiian dune plants. *Pacific Science* 42, 217–229.

Koske, R.E. and Gemma, J.N. (1989) A modified procedure for staining roots to detect V-A mycorrhizas. *Mycological Research* 92, 486–488.

Koske, R.E. and Gemma, J.N. (1990) VA mycorrhizae in strand vegetation of Hawaii: evidence for long-distance codispersal of plants and fungi. *American Journal of Botany* 77, 466–474.

Koske, R.E., Gemma, J.N. and Englander, L. (1990) Vesicular-arbuscular mycorrhizae in Hawaiian Ericales. *American Journal of Botany* 77, 64–68.

Koske, R.E., Gemma, J.N. and Flynn, T. (1992) Mycotrophy in Hawaiian angiosperms: A survey with implications for the origin of the native flora. *American Journal of Botany* 79, 853–862.

Miller, F.M. (1979) Some occurrences of vesicular-arbuscular mycorrhiza in natural and disturbed ecosystems of the Red Desert. *Canadian Journal of Botany* 57, 619–623.

Moorman, T. and Reeves, F.B. (1979) The role of endomycorrhizae in revegetation practices in the semi-arid west. II. A bioassay to determine the effect of land disturbance on endomycorrhizal populations. *American Journal of Botany* 66, 14–18.

Reeves, F.B., Wagner, D., Moorman, T. and Kiel, J. (1979) The role of endomycorrhizae in revegatation practices in the semi-arid west. I. A comparison of incidence of mycorrhizae in severely disturbed vs. natural environments. *American Journal of Botany* 66, 6–13.

25 The Use of ^{15}N to Assess the Role of VA Mycorrhiza in Plant N Nutrition and its Application to Evaluate the Role of Mycorrhiza in Restoring Mediterranean Ecosystems

J.M. Barea, R. Azcón and C. Azcón-Aguilar

Departmento de Microbiología, Estación Experimental del Zaidín, CSIC. Prof. Albareda 1, 18008 Granada, Spain

Introduction

Previous reports (Carradus, 1966; Lundeberg, 1970; Lewis, 1976; Stribley and Read, 1974) support a now well-established role of both ericoid and ectomycorrhizal fungi in helping plants acquire N compounds from soils. In contrast, there is little published information on an equivalent role for VA mycorrhizas, though some conclusions have been reached. For example, VA fungi seem to be able: (1) to absorb both NO_3^- and NH_4^+ from the growth substrate (Bowen and Smith, 1981); (2) to increase the concentration and content of N in plants (Barea *et al.*, 1991a); (3) to assimilate ammonium, via glutamine synthetase activity (Smith *et al.*, 1985); and (4) to increase the N inflow to the root cells (Smith *et al.*, 1986). Besides, VA mycorrhizas have been shown to exert a P-mediated, therefore indirect, improvement of N_2 fixation in symbiotic systems (Barea and Azcón-Aguilar, 1983).

The use of ^{15}N-based methods offers the only possibility to evaluate a direct effect of any biological, physico-chemical, or ecological influence on N acquisition by plants, and is therefore the best procedure to ascertain whether VA infections enhance either N uptake from soil or N_2 fixation. The main objective of this study is to review, briefly, the published information on the topic. The fundamentals of the ^{15}N-aided methodologies used in research on VA mycorrhiza will be described first. Then, the possibilities of applying ^{15}N techniques to follow the impact of infection on N uptake, N_2 fixation and N transfer in the re-establishment of a plant cover, for the rehabilitation of degraded mediterranean ecosystem, will be discussed.

^{15}N-aided Methodologies in VA Mycorrhiza Research

Nitrogen has six isotopes, with mass numbers from 12 to 17, and only two of them (^{14}N and ^{15}N) are stable. The ratio ^{14}N: ^{15}N in natural materials is more or less constant, therefore, the use of N compounds artificially enriched in ^{15}N can be used as a tracer in soil fertility and plant nutrition studies (Zapata, 1990), or to assess biological N_2 fixation (Hardarson and Danso, 1990). Fig. 25.1 illustrates the fundamental principles of ^{15}N isotope application.

A basic assumption, widely accepted, is that 'when a plant is confronted with two or more sources of a nutrient element, the nutrient uptake from each of these sources is proportional to the amounts available in each source' (see Zapata, 1990). This implies that in a situation where a soil (where the natural abundance of ^{15}N : ^{14}N) and a fertilizer (with a % ^{15}N atomic excess) are the only sources of N available to the plant, the ^{15}N : ^{14}N ratio in plant must be rather constant. Additionally, the isotope ^{15}N can be used to measure the apparent N pool size in soil, a parameter which determines the plant available amount of N: i.e. the A_N value of the soil (see Fig. 25.1 and Zapata 1990). As an inherent property of the soil, and a yield-independent parameter, it would be constant for any one set of experimental conditions. However, if a given treatment is able to induce changes in the uptake patterns of an absorbing root system, the A_N value would be expected to change accordingly, because its apparent 'constancy' actually depends on the roots obtaining N from the same ^{15}N : ^{14}N pools in soil under the different treatments to be compared. VA mycorrhizas could bring about such a change (Fig. 25.1).

The ^{15}N methodology is the only direct procedure which gives a truly integrated measurement of N_2-fixation over a growing period. For quantitative estimates a non-fixing (reference) crop is needed to assess the ^{15}N : ^{14}N ratio, or the available N in the soil (A_N value). The advantages and difficulties of these techniques have been discussed by Hardarson and Danso (1990) and are outlined in Fig. 25.2. Changes in the ^{15}N : ^{14}N ratio, at the level of fourth decimal, can be used to assess N_2 fixation by the method of the 'natural abundance' (Högberg, 1990).

Root exudation and the decomposition of root nodule tissues may account for N enrichment of the rhizosphere in N_2-fixing plants (Heichel, 1987). There is evidence suggesting that such N, mostly deriving from N_2 fixation, can benefit a non-fixing plant growing nearby. This 'N transfer' process, can be evaluated using ^{15}N (Weaver, 1988). The fundamentals and application of the technique, are shown in Fig. 25.3.

VA Mycorrhizas and N Nutrition (^{15}N)

As far as we know the first report of N transport (^{15}N) from soil by VAM mycelium is that by Ames *et al.* (1983). Furthermore, Ames *et al.* (1984) suggested that VA mycorrhizal inoculation could induce a differential exploitation of different forms of soil N, thereby affecting the A_N value. By using P-compensated, non-VAM

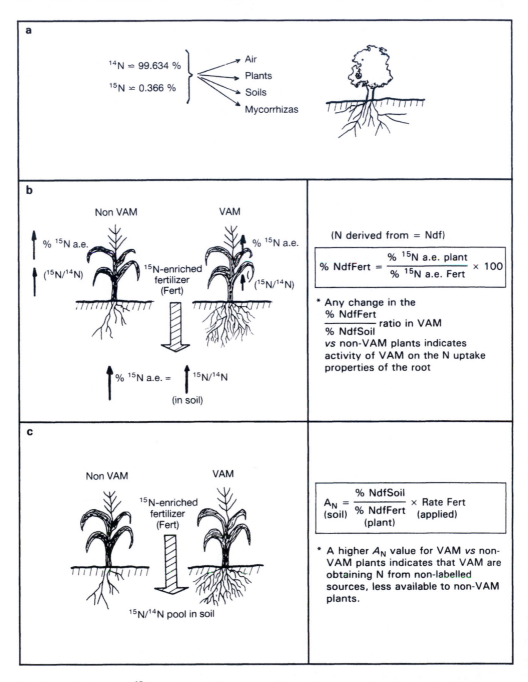

Fig. 25.1. The isotope ^{15}N, uptake and VA mycorrhizas. (**a**) natural abundance of stable N isotopes; (**b**)^{15}N atomic excess (a. e.) over the natural abundance; (**c**) the A_N value as an indicator of N utake from soil.

a

Non-fixing   N$_2$-fixing

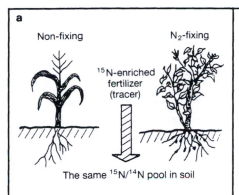

^{15}N-enriched fertilizer (tracer)

The same ^{15}N/^{14}N pool in soil

* The $\dfrac{\text{\% NdfFer}}{\text{\% NdfSoil}}$ is equal for $\begin{matrix}\text{Fixing}\\\text{Non-Fixing}\end{matrix}$
but the ^{15}N: ^{14}N in plant is lower in the 'Fixing' because 'dilution' with N from atmosphere (% Ndf Fix)

$$\text{\% NdfFix} = \left(1 - \frac{\text{\% }^{15}\text{N a.e. (Fixing)}}{\text{\% }^{15}\text{N a.e. (Non-Fixing)}}\right) \times 100$$

* The contribution of the N sources (% Ndf: Fix, Soil, Fert) can be measured

b

* The use of ^{15}N to assess if a given treatment, i.e. mycorrhizas, is acting on N$_2$ fixation or on N uptake from soil
* The experimental approach: to measure % NdfFix, Fert or soil in VAM plants *vs* P-given matched, non-VAM counterparts.

c

The same rate + enrichment of ^{15}N-added

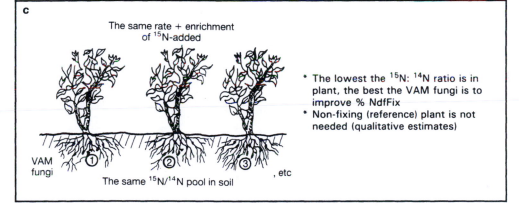

VAM fungi

The same ^{15}N/^{14}N pool in soil   , etc

* The lowest the ^{15}N: ^{14}N ratio is in plant, the best the VAM fungi is to improve % NdfFix
* Non-fixing (reference) plant is not needed (qualitative estimates)

Fig. 25.2. The isotope ^{15}N, N$_2$ fixation and VA mycorrhizas. **(a)** ^{15}N dilution techniques to measure N$_2$ fixation (Fix); **(b)** the use of ^{15}N to ascertain the VAM role in N$_2$ fixation; **(c)** ranking VAM fungi to improve NdFix (^{15}N).

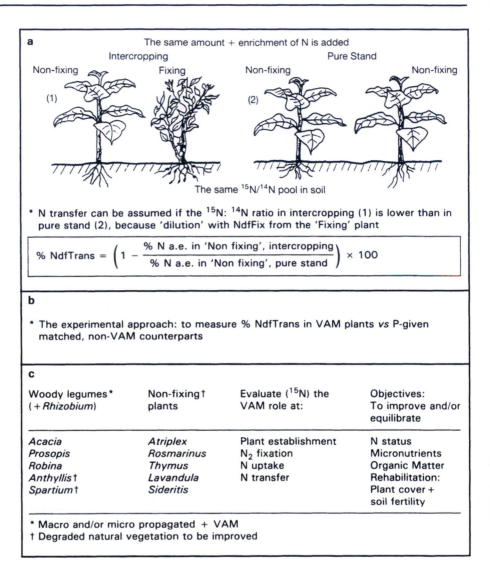

a

The same amount + enrichment of N is added

Intercropping   Pure Stand

Non-fixing   Fixing   Non-fixing   Non-fixing

(1)   (2)

The same $^{15}N/^{14}N$ pool in soil

* N transfer can be assumed if the ^{15}N: ^{14}N ratio in intercropping (1) is lower than in pure stand (2), because 'dilution' with NdfFix from the 'Fixing' plant

$$\% \text{ NdfTrans} = \left(1 - \frac{\% \text{ N a.e. in 'Non fixing', intercropping}}{\% \text{ N a.e. in 'Non fixing', pure stand}}\right) \times 100$$

b

* The experimental approach: to measure % NdfTrans in VAM plants *vs* P-given matched, non-VAM counterparts

c

Woody legumes* (+*Rhizobium*)	Non-fixing† plants	Evaluate (^{15}N) the VAM role at:	Objectives: To improve and/or equilibrate
Acacia	*Atriplex*	Plant establishment	N status
Prosopis	*Rosmarinus*	N_2 fixation	Micronutrients
Robina	*Thymus*	N uptake	Organic Matter
Anthyllis†	*Lavandula*	N transfer	Rehabilitation:
Spartium†	*Sideritis*		Plant cover + soil fertility

* Macro and/or micro propagated + VAM
† Degraded natural vegetation to be improved

Fig. 25.3. The isotope ^{15}N, N transfer and VA mycorrhizas. (**a**) estimation of N transfer from fixing to non-fixing plants; (**b**) the use of ^{15}N to ascertain the VAM role on N transfer; (**c**) an approach for sermiarid mediterranean degraded ecosystems.

matched controls, Barea *et al.* (1989b) demonstrated that the A_N values were significantly higher in VAM infected pots, suggesting that VAM hyphae can use N forms which are less available to non-VAM plants. This fact was further corroborated under different ecological conditions (Barea *et al.*, 1991a).

The isotope ^{15}N was also used to ascertain and measure the effect of VAM inoculation to improve N_2 fixation (Barea *et al.*, 1987, 1989b; Kucey and Bonetti, 1988). The qualitative method to estimate N_2 fixation (^{15}N) was used for a number of different purposes such as to demonstrate the role of VAM in a water stress situation (Azcón *et al.*, 1988), to detect microbial compatibilities to improve N_2 fixation (Azcón *et al.*, 1991) or to establish some relationships between N_2 fixation and N uptake (Azcón and Barea, 1992). Isotopic evidence for a VA mycorrhizal enhancement of N transfer has been only found in few instances (Kessel *et al.*, 1985; Haystead *et al.*, 1988; Barea *et al.*, 1989a, b).

The Use of ^{15}N to Evaluate the Role of VA Mycorrhizas in Restoring Mediterranean Degraded Ecosystems

These ecosystems can be taken as a model to apply VA mycorrhizal biotechnology to improve the status of a degraded plant cover under semi-arid conditions. The soils usually show a low level of organic matter, water, N and P availability, unbalanced micronutrient status, and low level of VA mycorrhizal propagules (Barea *et al.*, 1991b).

A number of approaches have been suggested aimed at preventing erosion and desertification in arid an semi-arid zones. The planting of woody legumes is the basis for a promising approach. These plants take advantage from their ability to fix N_2 in symbiosis with *Rhizobium* or *Bradyrhizobium* and to form mycorrhizas that may help the plant cope with stress situations such as nutrient deficiencies, drought, contamination and soil disturbance. Consequently, an experimental approach is proposed based on the establishment of plant material (autochthonous and adapted woody legumes) improved by the appropriate biotechnologies (plant micropropagation and symbiont inoculation) – these symbionts will be selected for functional compatibility with the plant (by following isotopic techniques) and ecological criteria. The two main goals will be obviously: (i) the establishment of a vegetation cover; and (ii) the rehabilitation of soil.

The information is discussed by Barea *et al.* (1990b) in terms of assessing the role of VA mycorrhiza in restoring soil fertility in arid and semi-arid zones, as based on the effect of the symbiosis in plant establishment and N and P cycling. The use of ^{15}N to evaluate the activity of VA infection in N_2-fixation, N-uptake and N-transfer can be critical, in the context of the significance of VA mycorrhiza in the rehabilitation of these ecosystems.

References

Ames, R.N., Reid, C.P.P., Porter, L.K. and Cambardella, C. (1983) Hyphal uptake and transport of nitrogen from two ^{15}N-labelled sources by *Glomus mosseae*, a vesicular-arbuscular mycorrhizal fungus. *New Phytologist* 95, 381–396.

Ames, R.N., Porter, L. K., St John, R.V. and Reid C.P.P. (1984) Nitrogen sources and A_N values for vesicular-arbuscular and non-mycorrhizal sorghum grown at three rates of ^{15}N-ammonium sulphate. *New Phytologist* 97, 269–276.

Azcón, R. and Barea, J.M. (1992) Nodulation, N_2 fixation (^{15}N) and N nutrition relationships in mycorrhizal or phosphate-amended alfalfa plants *Symbiosis* 12, 33–41.

Azcón, R., El-Atrach, F. and Barea, J.M. (1988) Influence of mycorrhiza vs. soluble phosphate on growth, nodulation, and N_2 fixation (^{15}N) in alfalfa under different levels of water potential. *Biology and Fertility of Soils* 7, 28–31.

Azcón R., Rubio, R. and Barea, J.M. (1991) Selective interactions between different species of mycorrhizal fungi and *Rhizobium meliloti* strains, and their effects on growth, N_2-fixation (^{15}N) and nutrition of *Medicago sativa* L. *New Phytologist* 117, 399–404.

Barea, J.M. and Azcón-Aguilar C. (1983) Mycorrhizas and their significance in nodulating nitrogen-fixing plants. In: Brady, N.C. (ed.), *Advances in Agronomy*. Academic Press, New York, pp. 1–54.

Barea, J.M., Azcón-Aguilar, C. and Azcón, R. (1987) Vesicular-arbuscular mycorrhiza improve both symbiotic N_2 fixation and N uptake from soil as assessed with a ^{15}N technique under field conditions. *New Phytologist* 106, 717–725.

Barea, J.M., Azcón, R. and Azcón-Aguilar, C. (1989a) Time-course of N_2-fixation (^{15}N) in the field by clover growing alone or in mixture with ryegrass to improve pasture productivity, and inoculated with vesicular-arbuscular mycorrhizal fungi. *New Phytologist* 112, 399–404.

Barea, J.M., El-Atrach, F. and Azcón, R. (1989b) Mycorrhiza and phosphate interactions as affecting plant development N_2-fixation, N-transfer and N-uptake from soil in legume-grass mixtures by using a ^{15}N dilution technique. *Soil Biology and Biochemistry* 21, 581–589.

Barea, J.M., Azcón-Aguilar, C. and Azcón, R. (1991a) The role of VA mycorrhizas in improving plant N acquisition from soil as assessed with ^{15}N. In: Flitton, C. (ed.), *The Use of Stable Isotopes in Plant Nutrition, Soil Fertility and Environmental Studies*. Joint IAEA/FAO Division, Vienna, pp. 209–216.

Barea, J.M., Salamanca, C.P. and Herrera, M.A. (1991b) The role of VA-mycorrhiza at improving N_2-fixation by woody legumes in arid zones. In: Werner, D. and Müller, P. (eds), *Fast Growing Trees and Nitrogen Fixing Trees*. Gustav Fisher-Verlag, Stuttgart, pp. 303–311.

Bowen, G.D. and Smith, S.E. (1981) The effect of mycorrhizas on nitrogen uptake by plants. In: Clark F.W. and Rosswall, R. (eds), *Terrestrial Nitrogen Cycles: Processes, Ecosystem Strategies and Management Impacts*. Ecological Bulletin N : 33, Swedish Natural Science Research Council, Stockholm, pp. 237–247.

Carradus, B.B. (1966) Absorption of nitrogen by mycorrhizal roots of beech. I. Factors affecting the assimilation of nitrogen. *New Phytologist* 65, 358–371.

Hardarson, G. and Danso, S.K.A. (1990) Use of ^{15}N methodology to assess biological nitrogen fixation. In: Hardanson, G. (ed.), *Use of Nuclear Techniques in Studies of Soil–Plant Relationships*. IAEA, Vienna, pp. 129–160.

Haystead, A., Malajczuk, N. and Crove, T.S. (1988) Underground transfer of nitrogen between pasture plants infected with vesicular-arbuscular mycorrhizal fungi. *New Phytologist* 108, 417–423.

Heichel, G.H. (1987) Legume nitrogen: symbiotic fixation and recovery by subsequent crops. In: Helsel, A.R. (ed.), *Energy in Plant Nutrition and Pest Control*. Elsevier Science Publishers B.V. Amsterdam, pp. 63–80.

Högberg, P. (1990) ^{15}N natural abundance as a possible marker of the ectomycorrhizal habit of trees in mixed African woodlands. *New Phytologist* 115, 483–486.

Kessel, Ch. Van, Singleton, P.W. and Hoben, H.J. (1985) Enhanced N-transfer from soybean to maize by vesicular-arbuscular mycorrhizal (VAM) fungi. *Plant Physiology* 79, 562–563.

Kucey, R.M.N. and Bonetti, R. (1988) Effect of vesicular-arbuscular mycorrhizal fungi and captan on growth and N_2 fixation of *Rhizobium*-inoculated field beans. *Canadian Journal of Soil Science* 68, 143–149.

Lewis, D.H. (1976) Interchange of metabolites in biotrophic symbioses between angio-sperms and fungi. *Perspectives in Experimental Biology* 2, 207–219.

Lundeberg, G. (1970) Utilization of various nitrogen sources, in particular bound soil nitrogen, by mycorrhizal fungi. *Studia Forestalia Suecica* 79, 1–96.

Smith, S.E., St John, B.J., Smith, F.A. and Nicholas, D.J.D. (1985) Activity of glutamine synthetase and glutamate dehydrogenase in *Trifolium subterraneum* L. and *Allium cepa* L: effects of mycorrhizal infection and phosphate nutrition. *New Phytologist* 99, 211–227.

Smith, S.E., St John, B.J., Smith, F.A. and Bromley, J.L. (1986) Effect of mycorrhizal infection on plant growth, nitrogen and phosphorus nutrition in glasshouse-grown *Allium cepa* L. *New Phytologist* 103, 359–373.

Stribley D.P. and Read, D.J. (1974) The biology of mycorrhiza in the Ericaceae. IV. The effect of mycorrhizal infection on uptake of ^{15}N from labelled soil by *Vaccinium macracarpon* Ait. *New Phytologist*, 73, 1149–1155.

Weaver, R.W. (1988) Isotope dilution as a method for measuring nitrogen transfer from forage legumes to grasses. In: Beck, D.P. and Materón, L.A. (eds), *Nitrogen Fixation by Legumes in Mediterranean Agriculture*. Icarda, Martinus Nijhoff Pbs., Dordrecht, pp. 359–365.

Zapata, F. (1990) Isotope techniques in soil fertility and plant nutrition studies. In: Hardarson, G. (ed.), *Use of Nuclear Techniques in Studies of Soil-Plant Relationships*. IAEA, Vienna, pp. 61–128.

26 Use of VA Mycorrhizas in Agriculture: Problems and Prospects

M.J. Daft

Department of Biological Sciences, University of Dundee, Dundee, DD1 1HN, UK

Introduction

In pot cultures endomycorrhizas can significantly enhance plant growth, particularly if the growth medium has been sterilized and contains an insoluble source of phosphorus (P) such as rock phosphate (Daft and Nicolson, 1966; Daft, 1991). Most soils contain mycorrhizal spores, hyphae and infected root segments. Their concentrations can vary considerably from habitat to habitat. This variation may result from physical conditions or biological influences. Long fallows can reduce subsequent crop growth and root colonization with vesicular-arbuscular mycorrhizas (Thompson, 1987). Soil fumigation, or the application of fungicides and herbicides can markedly reduce the natural mycorrhizal populations.

Harinikumar and Bagyaraj (1989) showed that different crops alter the production of infective propagules over three growing seasons. Not all vesicular-arbuscular mycorrhizal (VAM) species are equally effective in increasing plant growth. This suggests that a selection of suitable species for each habitat and particular crop is necessary for exploitation in the field. However competition with the indigenous mycorrhizal flora then becomes of importance. The effectiveness of the indigenous vesicular mycorrhizas (VAM) has been demonstrated by Dodd *et al.* (1983) in the Negev, Israel.

Interactions between rhizobia and VA mycorrhizas in chickpea, groundnut and lucerne have been discussed by Subba Rao *et al.* (1986) and Daft (1991). Vertebrate grazing did not reduce colonization (Wallace, 1987) and *Glomus manihotis* promoted plant growth in the presence of nematodes (Diederichs, 1987). Different mycorrhizal relationships may exist depending on the architecture of the root system, for example, tap and fibrous-rooted species (Anderson and Liberta, 1987). Another factor that compounds the situation is that of host genotype response. Krishna *et al.* (1985) found marked variations in levels of colonization in pearl millet genotypes and a similar situation occurs with groundnut and chickpeas (Daft, 1991).

Field trials with VA mycorrhizas are still beset with the problems of large scale production of inoculum, its storage and application to the crop. An alternative approach is to manipulate the indigenous mycorrhizal populations by cultural practices or by the application of soil amendments that increase the effectiveness of the potential inocula. Mycorrhizas are well known to be involved with the uptake of phosphate and appear to be good candidates for the assimilation of the immobile NH_4^+ ion (Raven, *et al.*, 1978). Urea, as a nitrogen source, has become more important in the tropics over the last 20 years (Spedding, 1983). Crop residues may be used to increase the organic matter content in soils, but if used in excess, nitrogen immobilization may take place.

The following experimental data illustrate the influence of rock phosphate, urea, straw and mycorrhizal infection on the growth of selected crop plants.

Results

Rock phosphate is commonly use in powered or granular form. The granular form is partially acidified and contains a higher concentration of soluble phosphate. A comparison of peanut plants grown either in the nodulated condition only or nodulated and infected by *G. clarum* and subsequently supplied with powdered or granular Kodjari and Florida rock phosphates, is given in Table 26.1. Dual infection increased shoot dry matter for both powdered and granular rock phosphates.

Table 26.1. Mean shoot dry weights and dependencies of peanut plants infected with *Rhizobium* (R) and *Glomus clarum* (G) and grown in two rock phosphates.

Form of rock phosphate	Mean shoot dry weight (g)		Mycorrhizal dependency
	R + G	R	
Kodjari			
Powdered	2.24	1.31	171
Granular	2.39	1.63	147
Florida			
Powdered	2.16	1.32	164
Granular	2.38	1.63	146

Table 26.2. Effects of urea on shoot and root dry weights (g) of maize plants infected with *Glomus clarum*.

	Urea concentration (mg kg^{-1} sand)			
	0	38	75	150
Shoot				
Control (Uninfected)	1.2	1.5	1.0	0.5
G. clarum	1.2	2.0	2.4	4.6
Root				
Control	0.5	0.7	0.5	0.3
G. clarum	0.4	0.6	1.0	1.8

Granular rock phosphate produced larger plants in both infected treatments and mycorrhizal dependency was greater when the powdered forms of each source of rock phosphate were applied.

When maize plants were supplied with various concentrations of urea (Table 26.2) the dry weights of the controls (uninfected) and those of plants infected by *G. clarum* plants were increased slightly by the addition of 38 mg urea kg^{-1} sand. At the two higher urea rates, infected and uninfected plants responded differently. At 75 mg urea the uninfected plants were similar to those without urea, while at 150 mg urea toxic effects were evident with two-thirds of the replicates dying. The infected plants gave increased dry weights, infected root lengths, percentage infections and spore production with increasing concentrations of urea.

The toxic effects of urea at high concentrations can be reduced by the incorporation of straw into the growth medium. When mycorrhizal maize was grown with straw (5 g kg^{-1} sand) incorporated into the medium, growth was inhibited. Combining 300 mg urea and the same amount of straw significantly increased the rate of growth and final yields of the infected plants. These results show that varying carbon:nitrogen ratios affect mycorrhizal growth responses differently.

In recent field trials in India (Lee, Bell, Wani and Daft, unpublished), we have combined the applications of rock phosphate, urea, straw and inoculation with *G. clarum* in studies of growth and yields of millet. The black vertisol soil contained less than 1 p.p.m. extractable phosphorus, had a pH of 8.7 and an indigenous spore population of less than 1 spore per gram of soil.

The most effective combined treatment was that which contained urea, straw, rock phosphate and *G. clarum*. This multiple treatment gave the greatest dry matter production as well as a significant increase of grain yield.

Discussion

In field work, sterilization, followed by reinoculation with a suitable VA fungus is a practical proposition. Many large scale nurseries employ this technique. However, over-inoculating non-sterilized fields can present problems. The indigenous mycorrhizal flora may mask the effects of an introduced species. The length of time over which applied inoculum remains effective is an important factor. It would appear that if large scale inoculation programmes are undertaken careful selection of the sites is necessary in order to obtain positive results. Another approach is to manipulate the indigenous flora by cultural practices so that the maximum benefit can be obtained by the crop plant. It is likely that combinations of various treatments will be required in order to maximize the potential of mycorrhizal infections.

References

Anderson, R.C. and Liberta, A.E. (1987) Variation in vesicular arbuscular mycorrhizal relationships of two sand prairie species. *American Midland Naturalist* 118, 56–63.

Daft, M.J. (1991) Influences of genotypes, rock phosphate and plant densities on mycorrhizal development and the growth responses of five different crops. *Agriculture, Ecosyctems and Environment* 35, 151–169.

Daft, M.J. and Nicolson, T.H. (1966) Effect of *Endogone* mycorrhiza on plant growth. 1. *New Phytologist* 65, 343–350.

Diederichs, C. (1987) Interaction between five endomycorrhizal fungi and the root-knot nematode *Meloidogyne javanica* on chickpea under tropical conditions. *Tropical Agriculture* 64, 353–355.

Dodd, J.C., Krikun, J. and Hass, J. (1983) Relative effectiveness of indigenous populations of vesicular-arbuscular mycorrhizal fungi from four sites in the Negev. *Israel Journal of Botany* 32, 10–21.

Harinikumar, K.M. and Bagyaraj, D.J. (1989) Effect of cropping sequence, fertilisers and farmyard manure on vesicular-arbuscular mycorrhizal fungi in different crops over three consecutive seasons. *Biology and Fertility of Soil* 7, 173–175.

Krishna, K.R., Shetty, K.G., Dart, P.J. and Andrew, D.J. (1985) Genotype-dependent variation in mycorrhizal colonisation and response to inoculation of pearl millet. *Plant and Soil* 86, 113–125.

Raven, J.A., Smith, S.E. and Smith, F.A. (1978) Ammonium assimilation and the role of mycorrhizas in climax communities in Scotland. *Botanical Society of Edinburgh Transactions* 43, 27–35.

Spedding, C.R.W. (ed.) (1983) *Fream's Agriculture*, 16th Edition. Royal Agriculture Society of England, John Murray, London.

Subba Rao, N.S., Tilak, K.V.B.R., and Singh, C.S. (1986) Dual inoculation with *Rhizobium* sp. and *Glomus fasciculatum* enhances nodulation, yield and nitrogen fixation in chickpea. *Plant and Soil* 95, 351–359.

Thompson, J.P. (1987) Decline of vesicular-arbuscular mycorrhizae in long fallow disorder of field crops and its expression in phosphorus deficiency of sunflower. *Australian Journal of Agricultural Research* 38, 847–867.

Wallace, L.L. (1987) Mycorrhiza in grassland: interactions of ungulates, fungi and drought. *New Phytologist* 105, 619–632.

27 Mycorrhizas, Forest Disturbance and Regeneration in the Pacific Northwestern United States

M.P. Amaranthus

Forest Service, Pacific Northwest Research Station, 3200 Jefferson Way, Corvallis, Oregon 97331, USA

Introduction

Much of the research on forest disturbance and soil biology has focused on ectomycorrhizas. Because most forest-tree species in the Pacific northwestern United States require ectomycorrhizas for nutrient and water uptake, the importance of understanding the relationship between disturbance, site conditions and mycorrhizas cannot be overstated. Numerous authors have reported reductions in mycorrhiza populations due to forest disturbance (Harvey *et al.*, 1976, 1980; Parke *et al.*, 1984; Perry *et al.*, 1982; Amaranthus *et al.*, 1987). However, the degree of reduction and its impact on forest regeneration varies widely and depends on many factors.

Factors Affecting Ectomycorrhizal Reduction

Type and severity of disturbance

Timber harvest and site preparation are the most widespread forest activities in the Pacific northwestern United States that alter both the aboveground and belowground environments. Wright and Tarrant (1958) found fewer ectomycorrhizas on Douglas fir seedlings growing in burned, compared with unburned, clearcuts. The greatest reductions were associated with the hottest burns. Parke *et al.* (1984) compared mycorrhiza formation in soils from burned and unburned clearcuts of 36 'difficult to regenerate' sites in northwestern California and southwestern Oregon. Douglas fir (*Pseudotsuga menziesii* (Mirb.) Franco) and ponderosa pine (*Pinus ponderosa* Laws.) seedlings grown in soils from burned and unburned clearcuts formed 40% and 20% fewer ectomycorrhizas respectively than seedlings grown in undisturbed forest soil. Amaranthus *et al.* (1987) found 90% fewer ectomycorrhiza and 44% less basal area growth on Douglas fir seedlings grown in clearcut and severely burned soils compared with undisturbed forest soil. However, studies on productive sites in the Oregon Cascade Mountains have not found mycorrhizal reductions following clearcutting and prescribed fire (Schoenberger and Perry, 1982; Pilz and Perry, 1984). It is difficult to generalize about effects of burning on ectomycorrhizal populations because they are highly dependent on duration and intensity of fire as well as soil and site conditions (Perry and Rose, 1983).

Because ectomycorrhizas predominate in the organic layers of the soil (Trappe and Fogel, 1977; Harvey *et al.*, 1976, 1979), the degree of organic matter lost from a site can influence mycorrhiza populations. Harvey *et al.* (1976) found up to 95% of the active ectomycorrhizas in humus and decaying wood in a mature Douglas fir – larch (*Larix*) forest. The importance of decaying wood to support ectomycorrhizal populations may be most critical following disturbance such as wildfire. Immediately following the 1987 wildfires in southern Oregon and northern California, decaying wood contained 25 times more moisture compared with mineral soil and was a centre of ectomycorrhizal activity for recovering vegetation (Amaranthus *et al.*, 1989). Decaying wood also acts as habitat for small mammals that are important in distributing fungal spores of several belowground mycorrhizal fungi (Maser *et al.*, 1978).

Forest practices that increase soil compaction and erosion adversely effect mycorrhizal formation and seedling establishment (M.P. Amaranthus, unpublished data). Increasing soil density can decrease fungal growth. In addition, decreasing soil structure restricts the movement of oxygen and water into soil and flushing out carbon dioxide. Populations of mycorrhizal fungi can be altered when levels of these basic elements become extreme. The diversity and abundance of these organisms is also reduced when topsoil is severely disturbed and eroded.

Ectomycorrhizal fungus diversity

Forest ecosystems within the Pacific northwestern United States generally contain an array of ecomycorrhiza-forming fungal species. The proportions of ectomycorrhiza types forming on seedlings often shifts following disturbance (Schoenberger and Perry, 1982; Perry and Rose, 1983; Pilz and Perry, 1984; Amaranthus and Perry, 1989b). Different environments reduced ectomycorrhiza formation by some fungus types and apparently stimulated formation for others. Clearcutting and site preparation did significantly reduce mycorrhizal formation and seedling performance on sites that contained only one or two ectomycorrhiza types (Perry *et al.*, 1982; Amaranthus *et al.*, 1987). Diversity likely provides a buffering capacity not found on sites with only one or few species.

Mycorrhizal diversity may also be tied to structural or habitat diversity within the forest ecosystem. Many mycorrhizal fungi require special habitats such as down logs. This habitat diversity may promote mycorrhizal diversity within the growing season. For example, in periods of adequate moisture, humus supports the highest level of ectomycorrhizal activity but during periods of soil drought soil wood becomes the most active site. Individual forest plants support a variety of mycorrhizal fungi. The diversity of mycorrhizal fungi formed by a given plant may increase its ability to occupy diverse belowground niches and survive a range of chemical and physical conditions.

Climatic conditions

Climate influences seedling growth and ectomycorrhiza formation (Harvey *et al.*, 1979; Pilz and Perry, 1984). The importance of early mycorrhiza formation in dry areas has been emphasized (Parry, 1953; Mikola, 1970). Dry climates may limit the activity of mycorrhizal fungi by decreasing the length of time for spore production, germination and optimal mycelial growth, which in turn can decrease the chances for planted seedlings to become colonized (Amaranthus and Perry, 1987). Seedlings in moist climates may be able to survive longer without mycorrhizas than those in dry climates, increasing their chances of becoming colonized. Moisture content also affects uptake of certain nutrients by mycorrhizas (Gadgil, 1972).

Seedlings growing in cold climates may also require rapid, early mycorrhizal colonization to take advantage of the short growing season and obtain the necessary nutrients and water to survive the long cold season and early frosts. In northwest California and southwest Oregon, Amaranthus and Perry (1987, 1989a, 1989b) found that mycorrhiza formation most strongly influences seedling survival and growth on sites limited by both moisture and temperature.

Biotic conditions

Little is known of the persistence and distribution of ectomycorrhizas in the absence of living hosts. It has been suggested (Hacskaylo, 1973) that ectomycorrhizal fungi do not persist long in the absence of host-supplied substrate. In many areas of the Pacific northwestern United States logged and burned sites are rapidly invaded by woody shrubs. These shrubs, especially members of the Ericaceae and Fagaceae, form mycorrhizas with many of the same fungi as do members of the Pinaceae (Molina and Trappe, 1982). These shrubs help preserve mycorrhiza fungus diversity during periods of rapidly changing aboveground community structure.

The plant community that occupies a site before disturbance can greatly influence the mycorrhizal fungus inocula potential for succeeding species. Amaranthus and Perry (1989b) planted Douglas fir seedlings at two locations: (1) a site cleared of whiteleaf manzanita (*Arctostaphylos viscida* Parry) and (2) a meadow cleared of annual grasses. In the first year seedling survival and growth was significantly greater on the manzanita site than the adjacent meadow with similar moisture and temperature conditions. By the third year Douglas-fir at the manzanita site was nearly 50 times on an area basis than seedlings in the cleared meadow. Douglas-fir seedlings outplanted at the manzanita site formed ectomycorrhizas more rapidly than those seedlings outplanted on the meadow. There were also dramatic shifts in the types of mycorrhizas found on seedlings grown in the two sites. Douglas fir seedlings, at the manzanita plots, contained siginificantly higher proportions of *Rhizopogon* sp. mycorrhizas. Certain *Rhizopogon* species have been demonstrated to decrease seedling moisture stress and improve seedling outplanted performance.

Rapid mycorrhiza formation, with fungi well adapted to site conditions, is the key to seedling establishment on sites difficult to regenerate.

Some woody shrub species may act as biological reservoirs, not only of mycorrhizal fungi but of other microflora as well. Significantly higher rates of nitrogen fixation – and increased seedling survival and growth – were found in association with the mycorrhizas of Douglas fir seedlings in a stand cleared of whiteleaf manzanita than in a meadow cleared of annual grass (Amaranthus *et al.*, 1990). *Azospirillium* spp., a nitrogen-fixing bacterium, was isolated from Douglas fir ectomycorrhizas at the manzanita site. Whiteleaf manzanita occupies particularly hot, dry sites with frequent fire. Because high nitrogen losses can accompany intense fire, natural mechanisms by which nitrogen is returned to the soil can be important to forest regeneration.

Effect of non-hosts over time

Past work strongly suggests that survival of propagules and input of spores are insufficient to maintain the ectomycorrhizal inocula potential in the absence of living hosts (Perry and Rose, 1983; Amaranthus and Perry, 1987). Invasion of sites by non-ectomycorrhizal plants over years can seriously affect reforestation, particularly in the case of ectomycorrhizal tree species growing on difficult sites where seedlings must establish ectomycorrhizas rapidly to survive. Ectomycorrhiza formation and seedling survival generally decreases as the length of time between disturbance and reforestation increases (Pilz and Perry, 1984; Perry *et al.*, 1989).

Learning from Natural Disturbance

The forests of the Pacific northwestern United States have evolved with frequent natural disturbance, particularly fire. Following such natural disturbances, biological remnants of the previous forest remain and provide the basis for recovery. These biological legacies include living trees and large inputs of organic matter. In much of the Pacific northwest United States natural fire regimes retain an abundance of living trees across the burned landscape. Photosynthate 'pumped' into the rhizosphere by living trees supports a wide array of mycorrhizal fungi that can actively colonize feeder roots of succeeding conifer seedlings. Thus seedlings 'plug into' the network of compatible ectomycorrhizal fungi supported by surviving living trees, thereby gaining opportunity for early ectomycorrhizal formation and establishment. Inputs of wood following the fire become an important site of mycorrhizal activity and a source of late season moisture for succeeding plants. Forest practices that mimic natural disturbance regimes, including maintaining living trees and providing inputs of woody material help maintain native populations of mycorrhizal fungi.

Conclusions

Studies from the Pacific northwestern United States indicate that reductions in mycorrhizal formation may affect the performance of outplanted seedlings, particularly on severely disturbed or environmentally limited sites where early growth is important. Maintaining native populations of mycorrhizal fungi on these sites aids regeneration. Why mycorrhizal formation is reduced following disturbance in some areas and not others is not completely understood. Great variability in the physical and biotic environment, disturbance history and intensity are likely contributing factors. One important factor is the presence of ectomycorrhizal host plants. Following disturbance, non-commercial ectomycorrhizal plants may serve as reservoirs of ectomycorrhizal fungal inocula while conifers are becoming established.

There is great opportunity to study the role of mycorrhizal fungi in forest recovery following natural disturbance. Natural disturbance provides a template to understand forest stability and resilience arising from complex interactions within the ecosystem. Trees, mycorrhizal fungi, and the entire ecosytem form a dynamic and coherent partnership whose relationship has coevolved over millennia. We can no longer consider trees as crops in isolation from the below-ground mutualists that promote forest regeneration.

References

Amaranthus, M.P. and Perry, D.A. (1987) Effect of soil transfer on ectomycorrhiza formation and the survival and growth of conifer seedlings in disturbed forest sites. *Canadian Journal of Forest Research* 17, 944–950.

Amaranthus, M.P. and Perry, D.A. (1989a) Interaction effects of vegetation type and Pacific madrone soil inocula on survival, growth and mycorrhiza formation of Douglas-fir. *Canadian Journal of Forest Research* 19, 550–556.

Amaranthus, M.P. and Perry, D.A. (1989b) Rapid root tip and mycorrhiza formation and increased survival of Douglas-fir seedlings after soil transfer. *New Forests* 3, 77–82.

Amaranthus, M.P., Perry, D.A. and Borchers, S.L. (1987) Reduction of native mycorrhizae reduces growth of Douglas-fir seedlings. In: Sylvia, D.M., Hung, L.L. and Graham, J.H. (eds), *Mycorrhizae in the Next Decade*. University of Flordia, Gainesville. p. 80.

Amaranthus, M.P., Parrish, D. and Perry, D.A. (1989) Decaying logs as moisture reservoirs following drought and wildfire, In: Alexander, E. (ed.), *Stewardship of Soil, Air, and Water Resources*. Proceedings Watershed 89, Juneau, Alaska. USDA Forest Service, Region 10. R10–MB–77 March 1989.

Amaranthus, M.P., Li, C.Y and Perry, D.A. (1990) Influence of vegetative type and madrone soil inoculum on associative nitrogen fixation in Douglas-fir rhizospheres. *Canadian Journal of Forest Research* 20, 368–371.

Gadgil, P.D. (1972) Effect of waterlogging on mycorrhizas of radiata pine and Douglas-fir. *New Zealand Journal of Forest Science* 2, 222–226.

Hacskaylo, E. (1973) Carbohydrate physiology of ectomycorrhizae. In: Marks, G.C. and Kozlowski T.T. (eds), *Ectomycorrhizae: Their Ecology and Physiology*. Academic Press, London, 207–230.

Harvey, A.E., Larsen, M.J. and Jurgensen, M.F. (1976) Distribution of ectomycorrhizae in a mature douglas-fir/larch forest soil in western Montana. *Forest Science* 22, 393–398.

Harvey, A.E., Larsen, M.J. and Jurgensen, M.F. (1979) Comparative distribution of ectomycorrhizae in soils of three western Montana forest types. *Forest Science* 25, 350–360.

Harvey, A.E., Larsen, M.J. and Jurgensen, M.F. (1980) Clearcut harvesting and ectomycorrhizae: survival of activity on residual roots and influence on a bordering forest stand in western Montana. *Canadian Journal of Forest Research* 10, 300–303.

Maser, C., Trappe, J.M. and Nussbaum, R.A. (1978). Fungal-small mammal interrelationships with emphasis on Oregon coniferous forests. *Ecology* 59, 799–809.

Mikola, P. (1970) Mycorrhizal inoculation in afforestation. *International Review of Forestry Research* 3, 123–196.

Molina, R. and Trappe, J.M. (1982) Lack of mycorrhizal specificity by the ericaceous hosts *Arbutus menziesii* and *Arctostaphylos uva-ursi*. *New Phytologist* 90, 495–509.

Parke, J.L., Linderman, R.G. and Trappe, J.M. (1984) Inoculum potential of ectomycorrhizal fungi in forest soils of southwest Oregon and northern California. *Forest Science* 30, 300–304.

Parry, M.S. (1953) Tree planting in Tanganyika: methods of planting. *East African Agricultural Journal* 18, 102–115.

Perry, D.A. and Rose, S.L. (1983) Soil biology and forest productivity: opportunities and constraints. In: Ballard R. and Gessel, S.P. (eds), *IUFRO Symposium on Forest Site and Continuous Productivity*, General Technical Report PNW-163. Portland, Or.: Pac. NW For. and Range Exp. Sta., USDA For. Serv., pp. 229–238.

Perry, D.A., Meyer, M.M., Egeland, D., Rose, S.L. and Pilz, D. (1982) Seedling growth and mycorrhizal formation in clearcut and adjacent undisturbed soils in Montana: a greenhouse bioassay. *Forest Ecology and Management* 4, 261–273.

Perry, D.A., Amaranthus, M.P., Borchers, J., Borchers, S.L. and Brainerd, R. (1989) Bootstrapping in ecosystems. *Bioscience* 39, 230–237.

Persson, H. (1982) *Changes in the tree and dwarf shrub fine-roots after clearcutting in a mature Scots pine stand*. Swedish Coniferous Forest Project, Technical Report 31.

Pilz, D.P. and Perry, D.A. (1984) Impact of clearcutting and slash burning on ectomycorrhizal associations of Douglas-fir. *Canadian Journal of Forest Research* 14, 94–100.

Schoenberger, M.N. and Perry, D.A. (1982) The effect of soil disturbance on growth and ectomycorrhizae of Douglas-fir and western hemlock seedlings: a greenhouse bioassay. *Canadian Journal of Forest Research* 12, 343–353.

Trappe, J.M. and Fogel, R.D. (1977) Ecosystematic functions of mycorrhizae. In: *The Belowground Ecosystem*. For Collins, Colo.: Colorado State University, Range Science Department Scientific Series, pp. 205–214.

Wright, E. and Tarrant, R.F. (1958) *Occurrence of mycorrhizae after logging and slash burning in the Douglas-fir forest type*. Research Note PNW-160. Portland, Or.: Pacific Northwest Forest and Ranges Experiment Station, USDA Forest Service.

28 Studies on the Effects of SO₂ and O₃ on the Mycorrhizas of Scots Pine by Observations Above and Below Ground

P.J.A. Shaw[1], J. Dighton[2] and J. Poskitt[2]

[1]Biology Laboratories, NPTEC, Kelvin Ave, Leatherhead, Surrey KT22 7SE:
[2]ITE Merlewood, Grange-Over-Sands, Cumbria LA11 6JU, UK

Introduction

The Liphook forest fumigation experiment was established in 1985 to investigate the long-term effects of realistic concentrations of SO₂ and O₃ on coniferous forest ecosystems in the UK. One major line of work entailed studying the responses of the mycorrhizal community to pollution stress, since under field conditions these fungi are crucial for the transfer of minerals from the soil solution to tree roots (Harley and Smith, 1983). Effects of acid precipitation and gaseous pollutants have been shown to reduce root growth and mycorrhizal development. Indirect effects of pollutants in reducing photosynthesis and hence carbon allocation to the root system, may also inhibit mycorrhizal development. The effects of these have recently been reviewed by Dighton and Jansen (1991). The work described below involved both counting fruiting bodies each autumn and taking root samples to identify mycorrhizal morphotypes. The results presented here cover root harvests and fruit body surveys from 1985 to 1990.

Methods

Site description

The experimental plots are located in Liphook forest on the Sussex/Hampshire border, at grid reference SU854298. The experimental area is on a 1:30 NNW facing slope where the soil was a humoferric podsol of the Shirrel Heath series (Jarvis *et al.*, 1984). Soil pH (distilled water) was approximately 4.1 in the organic layer and 4.0 in the bleached horizon. The area was accidentally burnt in 1976, and replanted with Corsican pine (*Pinus nigra* var. *laricio*) in 1982. In 1985 seven areas were cleared of vegetation and rotavated to 15 cm. Then the iron pan was broken by deep tining. These areas became the seven experimental plots. Each plot consisted of a 50 m diameter circle surrounded by gas release pipework. Gas was released at two heights to maintain a constant concentration across the width of an inner 30 m diameter circle (McLeod *et al.*, 1989). Seedlings of Scots pine, Sitka spruce and Norway spruce (196 per species per plot) were planted in April 1985 in the inner 30 m circles at 1 m spacing. The trees were planted in six sectors, with the two sectors of each species being diametrically opposite. Eight plants per species per plot (four per sector) were harvested annually for biometric measurements and root surveys.

The experimental design was an unreplicated factorial combination of two O_3 treatments and three SO_2 treatments, with an extra control making up the seven plots. O_3 treatments consisted of an ambient and a 'high' concentration, the latter being 1.5× ambient O_3. Continuous automatic monitoring allowed the O_3 target to be updated twice per hour. Ozone fumigation started in March 1988, but until December 1988 it was contaminated with N_2O_5 (Brown and Roberts, 1988). Subsequently a water scrubber was installed to control this problem.

The SO_2 treatments were based on a 2-year dataset collected from Bottesford, a rural town in the English Midlands (Martin and Barber, 1981). This dataset contains half-hourly means throughout the year for SO_2 and other gases, and was used to set corresponding half-hourly targets at Liphook.

The 'low' treatment was 1.5 × the Bottesford data, while the 'high' treatment was 3 × these data. The annual means for ambient, low and high SO_2 plots are approximately 4, 14 and 24 p.p.b. respectively. SO_2 fumigation commenced in May 1987 and has continued approximately uninterrupted since.

Root surveys

Following an annual harvest of trees for biometric analysis, the root systems of each cut plant were sampled for analysis of mycorrhizal morphotypes. In 1985 the whole root system was sampled, but in subsequent years this proved impractical and instead roots were collected from a 10 × 10 × 15 cm deep hole, 15 cm west of the cut stump. Samples were taken to ITE Merlewood, where roots were washed, fixed in 1.2% glutaraldehyde, and morphotypes defined by comparison with reference material. Subsequently mycorrhizas synthesized under pure culture conditions were compared with the reference samples, allowing a few types to be identified to species.

Fruitbody surveys

Each plot was checked at intervals of approximately 2 weeks from late August to November, and any fruitbodies found were recorded and marked (with white paint unless the cap was too viscid, in which case it was sliced with a scalpel). Most fungi were identified on the spot but doubtful specimens were returned to the laboratory for detailed examination. The diffuse crustose hymenium of *Thelephora terrestris* was noted but could not be quantified due to its growth habit.

Results

Mycorrhizal morphotypes

A total of ten morphotypes were identified on Scots pine, of which only four have been identified to genus. There were no significant treatment effects on any morphotype, and little change from year to year except for the appearance of some minor types in the later years, and a gradual decline in the frequency of *Suillus* type mycorrhizas. The results for all years (mean over all plots) are given in Table 28.1.

When the data were ordinated with Principal Components Analysis (PCA) the mean score on the first principal axis (28.1% of the total variance) was significantly correlated with year, indicating sucessional change in the data.

Table 28.1. Mean frequencies of mycorrhizal morphotypes in the Scots pine sectors of the Liphook plots. Results from all plots have been pooled.

Year	1985	1986	1987	1988	1989	1990
Morphotype						
Paxillus	22.0	10.1	16.9	25.4	21.7	17.4
Suillus	31.2	28.9	23.1	14.5	13.1	12.4
Cenococcum	5.9	0.6	3.5	1.8	5.1	8.4
Non-mycorrhizal	0.0	17.2	5.3	2.0	11.3	7.0
Type C	33.6	26.9	27.8	40.5	29.7	9.0
Type E	7.4	4.7	14.0	12.0	3.8	1.8
Type K	0.0	4.0	2.9	1.3	1.5	0.8
Type N	0.0	0.0	0.1	2.7	13.9	16.5
Type P	0.0	0.0	0.5	0.0	0.0	7.4
Type Q	0.0	0.8	0.5	0.0	0.0	0.2

Fruitbody surveys

No fruitbodies were found during 1985, reflecting the immaturity of the trees. In 1986 *Laccaria laccata*, *Paxillus involutus* and *Thelephora terrestris* emerged in all plots and under all tree species. From 1987 onwards almost all fruitbody production occurred under the Scots pine, and the subsequent data analyses refer solely to the pine sectors.

The initial community re-appeared along with *Suillus bovinus* and *Gomphidius roseus* (often in mixed clumps). This pattern was repeated in 1988, with addition of significant numbers of *S. variegatus*, *S. luteus* and *Boletus subtomentosus*. In 1989 *Suillus* species were the dominant fruitbdies while 1990 was marked by the appearance of significant numbers of *Cortinarius semisanguineus* and *Lactarius rufus*. The results for all years (pooled over all plots) are given in Table 28.2.

There was only one significant effect of treatment on a species; the mean count of *P. involutus* in each plot was significantly and positively related to the plot mean SO_2 concentration. However this is probably an artefact, since the high SO_2 plots

Table 28.2. Counts of ectomycorrhizal fruitbodies in the Scots pine sectors of the Liphook plots. Results from all plots have been pooled.

Year	1986	1987	1988	1989	1990
Fungal species					
Paxillus involutus	117	132	20	21	48
Laccaria laccata	567	2765	3868	182	297
Gomphidius roseus	0	99	431	896	289
Suillus bovinus	0	219	2982	5823	2492
Suillus variegatus	0	2	151	534	162
Suillus luteus	0	0	11	12	24
Boletus subtomentosus	0	7	18	17	7
Cortinarius semisanguineus	0	0	12	45	157
Lactarius rufus	0	0	0	2	61

held the highest numbers of *P. involutus* in 1986 before SO₂ fumigation started. When the data were ordinated by PCA the first axis (37% of the variance) detected the successional trend within the data. The eigenvector loadings were highest for the later successional species, and were significantly correlated with mean time to fruiting (calculated as summed fruitbody counts weighted by year, divided by the unweighted sum). The plot mean scores on the first axis were not significantly related to SO₂ treatment, but showed one of the control plots (plot 1) to have an anomalously advanced community. The mean plot score on the first PCA axis was significantly negatively correlated with the phosphorus concentration in soil water (measured by suction lysimeter at 30 cm depth).

No relationship was found between fruitbody emergence and morphotype distribution, either between plots or between years.

Discussion

The overriding trend in the fruitbody data is a change from a community dominated by *Laccaria laccata*, *Paxillus involutus* and *Thelephora terrestris* (all ubiquitous in the northern hemisphere) to a mixed community of 'early stage' fungi (Last *et al.*, 1987).

The occurrence of a succession of mycorrhizal species fruiting as a host tree matures is a well known phenomenon (Marks and Foster, 1967; Mason *et al.*, 1983; Last *et al.*, 1984; Dighton *et al.*, 1986). The majority of studies have involved visiting stands of widely differing ages on soils that are assumed to be reasonably similar, and the work presented above is unusual in being based on annual surveys of the the same site. Richardson (1970) found that pine forests should be visited at intervals of 3–5 days to be confident of recording all toadstools. This and the density of the foliage from 1988 onwards means that the fruitbody counts from Liphook cannot be regarded as absolute, but they provide a useful index of the population density.

The absence of a clear pollutant effect on the mycorrhizal community agrees with the observations of Termorshuizen and Shaffers (1987) that under Scots pine in the Netherlands mycorrhizal fruitbodies were not depressed by air pollution in young stands (5–13 years), although they were in mature stands (>50 years).

The first principal axis of the PCA on the Liphook fruitbody data picked up the successional trend in the data. The mean plot score on the first axis of this PCA provided an index of the speed of the fruitbody succession, and although this score was not related to SO_2 treatments it was significantly negatively correlated with the concentration of phosphate in the soil water. This suggests that low phosphate availability accelerated the fruitbody succession, possibly due to enhanced translocation of sugars to the mycorrhizas to increase the phosphate supply.

The work at Liphook was unusual in complementing the fruitbody surveys with annual monitoring of mycorrhizal infections on the tree roots. There were no significant effects of SO_2 on morphotype frequency nor on rate of appearance of new morphotypes. Similarly, there were no changes in morphotype composition corresponding to the fruitbody succession, implying that the changes in fruitbody emergence are due to alterations in the conditions that stimulate fruiting rather than differences in the intensity of mycorrhizal infection. The discrepancy between fruitbody and morphotype data is shown clearly by the data for *Suillus* mycorrhizas, which gradually declined in frequency over the experiment (Table 28.1), while fruitbodies of this genus peaked in the last two years of the experiment (Table 28.2). Similarly, an earlier analysis of Liphook data up to 1989 (Shaw, Dighton and Poskitt, 1992) noted the contradiction between the results for the fruitbodies of *P. involutus* (commonest in the SO_2 treated plots) and its mycorrhizas (which were non-significantly negatively correlated with SO_2 treatment). It is however possible that the Liphook morphotype data suffered from the small sample size (1500 cm^3 per tree) and concomitantly poor definition of the morph population.

By contrast, previous work has found effects of pollutants on mycorrhizal morphotypes. Dighton and Skeffington (1987) noted that acidic irrigation decreased the number of root tips of Scots pine with a brown coralloid morph ('Type F') later identified as *P. involutus*. Termorshuizen and Shaffers (1989) also noted that corralloid mycorrhizas were significantly negatively related to SO_2 concentrations in the Netherlands, although they did not identify their morphs to species.

Acknowledgements

We thank M.R. Holland for assistance during field sampling and A.R. McLeod for discussions. This work is published with the permission of National Power plc. The Liphook project now forms part of the Joint Environmental Programme of National Power pic and Powergen plc. Financial support was received from the commission of the European Communities for part of this study.

References

Brown, K.A. and Roberts, T.M. (1988) Effects of ozone on foliar leaching in Norway spruce *Picea abies* L. Karst): confounding effects due to N$_2$O$_5$ production during ozone generation. *Environmental Pollution* 55, 55–73.

Dighton, J. and Skeffington, R.A. (1987) Effects of artificial acid precipitation on the ectomycorrhizae of Scots pine seedlings. *New Phytologist* 107, 191–202.

Dighton, J. and Jansen, A.E. (1991) Atmospheric pollutants and ectomycorrhizas, more questions than answers. *Environmental Pollution* 73, 179–204.

Dighton, J., Poskitt, J.M. and Howard, D.M. (1986) Changes in the occurrence of basidiomycete fruitbodies during forest stand development with specific reference to mycorrhizal species. *Transactions of the British Mycological Society* 87 163–171.

Harley, J.L. and Smith, S.E. (1983) *Mycorrhizal Symbiosis*. Academic Press, London.

Jarvis, M.G., Allen, R.H., Fordham, F.J., Hazelden, J., Moffat, A.J. and Sturdy, R.G. (1984) *Soils And Their Use In South East England*. Bulletin 15 of the Soil Survey of England and Wales, Harpenden.

Last, F.T., Mason, P.A., Ingleby, K. and Fleming, L.V. (1984) Succession of fruitbodies of sheathing mycorrhizal fungi associated with *Betula pendula*. *Forest Ecology and Management* 9, 229–234.

Last, F.T., Dighton, J. and Mason, P.A. (1987) Successions of sheathing mycorrhizal fungi. *Trends in Ecology and Evolution* 2, 157–161.

McLeod, A.R., Brown, K.A., Skeffington, R.A. and Roberts, T.M. (1989) Studies of ozone effects on conifers and the formation of oxides of nitrogen during ozone generation. In: Bucher, J.B. and Bucher-Wallis, I. (eds), *Air Pollution and Forest Decline*. Proceedings of the 14th International Meeting on Air Pollution effects on Forest Ecosystems, IUFRO, Interlaken, Switzerland Oct 2–8 1988. Birmensdorf 1989, pp. 181–187.

Marks, G.C. and Foster, R.C. (1967) Succession of mycorrhizal associations on individual roots of radiata pine. *Australian Forestry* 31, 193.

Martin, A. and Barber, F.R. (1981) Sulphur dioxide, oxides of nitrogen and ozone measured continously for two years at a rural site. *Atmospheric Environment* 15, 567–578.

Mason, P.A., Wilson, J., Last, F.T. and Walker C. (1983) The concept of succession in relation to the spread of sheathing mycorrhizal fungi on innoculated tree seedlings growing in unsterile soil. *Plant and Soil* 71, 247–256.

Richardson, M.J. (1970) Studies on *Russula emetica* and other agarics in a Scots pine plantation. *Transaction of the British Mycological Society* 55, 217–229.

Shaw, P.J.A., Dighton, J. and Poskitt, J. (1992) Studies on the mycorrhizal community infecting trees in the Liphook forest fumigation experiment. *Forest Ecology and Management* (in press).

Termorshuizen, A.J. and Schaffers, A.P. (1987) Ocurrence of carpophores of ectomycorrhizal fungi in selected stands of *Pinus sylvestris* in the Netherlands in relation to stand vitality and air pollution. *Plant and Soil* 104, 209–217.

Termorshuizen, A.J. and Schaffers, A.P. (1989) The relation in the field between fruitbodies of mycorrhizal fungi and their mycorrhizas. *Agricuture, Ecosystems and Environment* 28, 509–512.

29 Sequences of Sheathing (Ecto-) Mycorrhizal Fungi Associated with Man-made Forests, Temperate and Tropical

F.T. Last[1], K. Natarajan[2], V. Mohan[2] and P.A. Mason[1]

[1]*Institute of Terrestrial Ecology, Bush Estate, Penicuik, Midlothian, Scotland EH26 OQB, UK:* [2]*Centre of Advanced Study in Botany, The University, Guindy Campus, Madras, 600025, India*

Having recorded the late summer/autumnal above-ground production of fruit-bodies of fungi associated with the formation of the sheathing ectomycorrhizas of *Betula pendula* and *B. pubescens* planted on a former agricultural site, Mason *et al.* (1982) and Deacon *et al.* (1983) developed the concept of mycorrhizal succession. They noted that fruitbodies were arranged systematically both in time and space. Those of *Hebeloma* (*crustuliniforme, sacchariolens*), *Laccaria* (*proxima, tortilis*), *Thelephora terrestris* and *Inocybe* spp., found during the first 4 years – early-stage fungi – became less abundant thereafter and were substituted or augmented by those of species of *Lactarius* (*glyciosmus, pubescens*), *Leccinum* (*scabrum, versipelle*), *Cortinarius* and *Russula* (*betularum, grisea*) – late-stage fungi. With the exception of *Laccaria tortilis*, fruitbody production tended to be maximal in arcs whose radii, from bases of trees, increased from year to year.

These observations of deciduous *Betula* spp. have been repeated on a variety of evergreen trees planted on 'new' sites at diverse locations (temperate and high altitude tropical). Chu-Chou (1979) in New Zealand observed numerous fruitbodies of *Hebeloma crustuliniforme*, *Laccaria* sp. and *Rhizopogon* spp. in association with seedlings of *Pinus radiata* in forest nurseries. While *H. crustuliniforme* stopped producing fruitbodies when the seedlings were transplanted those of *Laccaria* sp. and *Rhizopogon* spp. were joined and later superseded by fruitbodies of species of *Inocybe* and *Suillus* which, in turn, were joined by those of *Amanita muscaria* and *Scleroderma verrucosum*. Termorshuizen and Schaffers (1987), observing stands of *Pinus sylvestris* in the Netherlands that were either 5–10 or 50–80 years old, found fruitbodies of *Laccaria proxima*, *Inocybe* (*brevispora, umbrina*) and *Suillus bovinus* in all stands 5–10 years old but in none of those 50–80 years old, whereas fruitbodies of *Lactarius hepaticus* and *Russula emetica* were concentrated in the older stands. Now Mohan (1991) has detected a similar sequence with *Pinus patula* growing at 2100 m at lat. 11 °N (southern India). (Table 29.1).

The evidence from diverse sources suggests that communities of mycorrhizal fungi change with time after planting trees on 'new' sites. But how accurately do

Table 29.1. Numbers of fruitbodies of different mycorrhizal fungi associated with stands of *Pinus patula* (Mohan, 1991).

Genera of fungi	Age of stand (yr)					
	6	10–11	14–15	15–16	16–17	20–21
Thelephora	9×10^3	2×10^3	Nil	Nil	Nil	Nil
Laccaria	1×10^3	2×10^4	1×10^2	1×10^2	8×10^2	5×10^2
Rhizopogon	1×10^3	1×10^3	8×10^2	7×10^2	9×10^2	1×10^3
Lycoperdon	33	19	7×10^2	7×10^2	5×10^2	3
Suillus	8	5×10^2	4×10^2	6×10^2	7×10^2	7×10^2
Scleroderma	Nil	Nil	1	6	4×10^2	5×10^3
Amanita	Nil	Nil	1×10^3	3×10^3	3×10^3	4×10^3
Tricholoma	Nil	Nil	Nil	2×10^2	2×10^2	3×10^2
Russula	Nil	Nil	7×10^2	Nil	Nil	5×10^2

fruitbodies above-ground reflect the occurrence of mycorrhizas in soil? Their value has been disputed but accumulating evidence suggests that there is a good correspondence. Observations made by Warcup (Mason *et al.*, 1983) have been substantiated (Fig. 29.1) while Jansen (1991) obtained a highly significant correlation between numbers of mycorrhizas in soil and the cumulative abundance of fruitbodies of mycorrhizal fungi. Because of the methods used it is accepted that correlations can only exist for fungi that form observable fruitbodies, so excluding *Cenococcum geophilum*, E-strain fungi, those that form vesicular arbuscular mycorrhizas (VAM) at early stages with eucalypts (Hilton *et al.*, 1989), etc.

There are many potential reasons for sequences of mycorrhizal fungi, involving the availability of inocula and the ability of available inocula to infect. This aspect was highlighted by observations made by Deacon *et al.* (1983) who transferred birch seedlings, previously grown in aseptic conditions, to soil cores taken beneath fruitbodies of *Laccaria laccata*, *Inocybe lanuginella*, *Hebeloma* (*sacchariolens*, *populinum*), *Lactarius pubescens* or *Leccinum* (*subleucophaeum*, *rigidipes*) associated with *Betula* spp. growing on former agricultural land. The cores contained mycorrhizas attributable to the different fungi. The mycorrhizas that developed on seedlings in *Laccaria*-, *Inocybe*-and *Hebeloma*-soils were virtually all attributable to these fungi whereas no *Leccinum*, and very few *Lactarius*, mycorrhizas developed in *Leccinum*-and *Lactarius*-soils.

In summarizing work with spores added to horticultural and agricultural substrates Fox (1986) concluded that species of *Hebeloma*, *Inocybe* and *Laccaria* (*proxima*, *tortilis*) could colonize birch seedlings in unsterile conditions, but species of *Amanita*, *Cortinarius*, *Elaphomyces*, *Lactarius*, *Leccinum*, *Russula*, *Scleroderma* and *Suillus* could not despite the latter group being able to infect in axenic conditions. Mohan (1991) in southern India has obtained comparable results. Inoculating unsterile soils (forest and grassland) with spores of *Thelephora*, *Laccaria* or *Rhizopogon* resulted in the formation of mycorrhizas on *Pinus patula* attributable to these fungi, whereas *Amanita*, *Lycoperdon*, *Scleroderma* and *Suillus* failed. However when the latter fungi were inoculated to partially

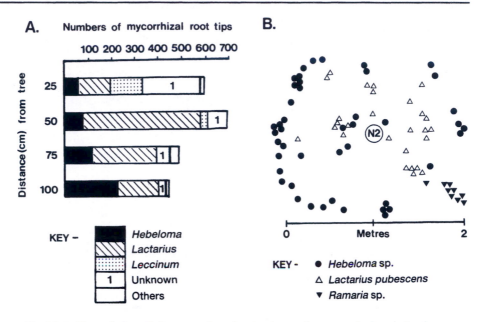

Fig. 29.1. The relation, 7–8 years after planting into a former agricultural site, between the occurrences of **(A)**, below-ground mycorrhizas (Deacon et al., 1983) and **(B)**, above-ground fruitbodies of mycorrhiza-forming fungi associated with Betula pubescens (Last et al., 1984).

sterilized forest soil they, like the others, formed appreciable numbers of mycor-rhizas. Marx (1980) reported that mycorrhizas developed on pine seedlings (P. taeda, P. echinata) in soil inoculated with spores of Pisolithus tinctorius and Thelephora terrestris but not with those of species of Amanita and Lactarius. Comparable tests have been made on Pinus banksiana with mycelial inocula (Danielson et al., 1984). Where cultures of E-strain fungi, Hebeloma sp., Laccaria proxima and T. terrestris were used, nearly all short roots became mycorrhizal with mycorrhiza of the inoculant fungi: no mycorrhizas attributable to the inoculant fungus developed when Amphinema byssoides, Hydnum imbricatum, Suillus (tomentosus, umbonatus) or Tricholoma (flavovirens, pessundatum) were tested.

The evidence shows that added spore and mycelial inocula of early-stage fungi can colonize roots of seedlings growing in unsterile substrates (horticultural/agricultural) whereas those of late-stage fungi cannot. This difference did not exist when seedlings were planted into undisturbed soil with established trees sup-porting mycorrhizas and extramatrical hyphae. Late-stage fungi spread from established trees to form mycorrhizas on introduced seedlings. Why is it therefore that a period of colonization by early-stage fungi has to elapse before late-stage fungi can establish mycorrhizas on seedlings in horticultural/agricultural sub-strates? Why is it that early-stage fungi can colonize without delay yet are later displaced? In addition to temporal changes and changes in a horizontal plane Reddell and Malajczuk (1984) examined the vertical distribution of mycorrhizas of Eucalyptus marginata in a sandy podzolic soil. They found that two types of

Table 29.2 Vertical distribution of mycorrhizas of *Eucalyptus marginata* in forests on sandy podzolic soil (Reddell and Malajczuk, 1984).

Horizon	Depth (cm)	No. of mycorrhizas		
		White*	Brown*	Black†
Litter	0–2	21	54	0
	2–4	320	110	0
	4–6	160	54	12
Mineral soil	6–8	67	6	130
	8–10	0	0	52

* Basidiomycete types
† *Cenococcum geophilum*

mycorrhizas were concentrated in upper litter horizons while a third, *Cenococcum geophilum*, was numerous in lower mineral horizons (Table 29.2). There was a very strong correlation between numbers of white and brown mycorrhizas and the amount of litter in soil samples. While Baar *et al.* (personal communication) have not sorted mycorrhizas into different types they found, after removing litter and humus from the floor of *Pinus sylvestris* forests, that the assemblages of mycorrhizal fruitbodies changed to include those of *Laccaria* (*bicolor, proxima*) and *Inocybe lacera*, typical early-stage fungi, instead of *Lactarius hepaticus* and *Xerocomus badius*, typical late-stage fungi, and *Paxillus involutus*. Other observations also suggest that litter contents of soils have different effects on different fungi. Gardner and Malajczuk (1988), noted correspondence between fruitbody production and the occurence of mycorrhizas on *Eucalyptus* spp. growing on a ridged rehabilitated mine site, and found that the proportions of fruitbodies on mineral soil ridges, attributable to *Laccaria* sp. (an early-stage fungus) and *Cortinarius sanguineus* (a late-stage fungus), were 58 and 28% respectively whereas in litter-filled troughs the percentages were reversed. Tyler (1985) found, in woodlands on a range of soil types that *Laccaria laccata* fruited mainly at sites with <13% organic matter, whereas *Amanita citrina*, *Cortinarius* spp., and others fruited mainly, but not exclusively, at sites with >30% organic matter. When Fleming *et al.* (1986) grew birch seedlings in cores of forest soil that had had time to develop a litter layer and in which mature birches were associated with fruitbodies of late-stage fungi (*Amanita*, *Leccinum*, *Tricholoma* and *Cortinarius*), a significant proportion of seedling mycorrhizas were attributable to these fungi, except *Amanita muscaria*, even when the cores of soil were incubated in glasshouse conditions.

Although many biotic and abiotic soil factors (e.g. soil moisture availability and bulk density, cation exchange capacity and rates of decomposition) change when trees mature, much of the evidence suggests that activities of early-and late-stage fungi are intimately linked with litter accumulation (Dighton and Mason, 1985). It seems that early-stage fungi are ideally suited to soils, mineral or peat, with little or no litter whereas late-stage fungi establish themselves in substrates where litter has accumulated. In recent years many inoculation experiments have been widened to include unsterile, in addition to sterile, substrates. It is now desirable to include soils with or without litter at different stages of decomposition, expecting that late-stage fungi, unlike early-stage, would succeed in soils with, but not in those without, litter. Warcup (1991) has suggested that the abundance of ascomycete mycorrhizas with eucalypt seedlings is increased directly or indirectly by the removal of litter. The seemingly strong role of substrate quality brings to mind procedures for growing the cultivated mushroom *Agaricus bisporus* and its critical dependence on *Pseudomonas putida* a microbe that profoundly affects its vegetative growth, in addition to stimulating fruitbody production (Hayes *et al.* 1969). It accelerates radial hyphal extension, decreases the frequency of hyphal branching and alters branching angles (Rainey, 1991) – characteristics associated with domain extension and exploration. Is it conceivable that the efficacy of populations of 'mycorrhization helper bacteria' (MHB), discussed by Stenström *et al.* and Duponnois and Garbaye at this symposium, is (1) of more importance to some mycorrhizal fungi than others and (2) critically dependent on substrate quality.

References

Chu-Chou, M. (1979) Mycorrhizal fungi of *Pinus radiata* in New Zealand. *Soil Biology and Biochemistry* 11, 557–562.

Danielson, R.M., Visser, S. and Parkinson, D. (1984) The effectiveness of mycelial slurries of mycorrhizal fungi for the inoculation of container-grown jack pine seedlings. *Canadian Journal of Forestry Research* 14, 140–142.

Deacon, J.W., Donaldson, S.J. and Last, F.T. (1983) Sequences and interactions of mycorrhizal fungi of birch. *Plant and Soil* 71, 257–262.

Dighton, J. and Mason, P.A. (1985) Mycorrhizal dynamics during forest tree development In: Casselton, D., Wood, D.A. and Frankland, J.C. (eds), *Development Ecology of Higher Fungi*. University Press, Cambridge, pp. 117–139.

Fleming, L.V., Deacon, J.W. and Last, F.T. (1986) Ectomycorrhizal succession in a Scottish birchwood. In: Gianinazzi-Pearson, V. and Gianinazzi, S. (eds), *Physiological and Genetical Aspects of Mycorrhizae*. INRA, Paris, pp. 259–64.

Fox, F.M. (1986) Groupings of ectomycorrhizal fungi of birch and pine, based on establishment of mycorrhizas on seedlings from spores in unsterile soils. *Transactions of the British Mycological Society* 87, 371–380.

Gardner, J.H. and Malajczuk, N. (1988) Recolonisation of rehabilitated bauxite mine sites in Western Australia by mycorrhizal fungi. *Forest Ecology and Management* 24, 27–42.

Hayes, W.A., Randle, P.E. and Last, F.T. (1969) The nature of the microbial stimulus affecting sporophore formation in *Agaricus bisporus* (Lange) Sing. *Annals of Applied Biology* 64, 177–187.

Hilton, R.N., Malajczuk, N. and Pearce, M.H. (1989) Larger fungi of Jarrah forest: an ecological and taxonomic survey. In: Dell, B., Havel, J.J. and Malajczuk, N. (eds), *The Jarrah Forest*. Kluwer Academic Publishers, Dordrecht, pp. 89–109.

Jansen, A.E. (1991) The mycorrhizal status of Douglas Fir in the Netherlands: its relation with stand age, regional factors, atmospheric pollutants and tree vitality. *Agriculture, Ecosystems and Environment* 35, 191–208.

Last, F.T., Mason, P.A., Ingleby, K. and Fleming, L.V. (1984) Succession of fruitbodies of sheathing mycorrhizal fungi associated with *Betula pendula*. *Forest Ecology and Management* 9, 229–234.

Marx, D.H. (1980) Ectomycorrhizal fungus inoculation: a tool for improving forestation practices. In: Mikola, P. (ed.), *Tropical Mycorrhiza Research*. Clarendon Press, Oxford pp. 13–71.

Mason, P.A., Last, F.T., Pelham, J. and Ingleby, K. (1982) Ecology of some fungi associated with an ageing stand of birches (*Betula pendula* and *B. pubescens*). *Forest Ecology and Management* 4, 19–39.

Mason, P.A., Wilson, J., Last, F.T. and Walker, C. (1983) The concept of succession in relation to the spread of sheathing mycorrhizal fungi on inoculated tree seedlings growing in unsterile soils. *Plant and Soil* 71, 247–256.

Mohan, V. (1991) Studies on ectomycorrhizal association in *Pinus patula* Schlect. and Cham. plantations in the Nilgiri Hills, Tamil Nadu. Unpublished PhD thesis, University of Madras.

Rainey, P.B. (1991) Effect of *Pseudomonas putida* on hyphal growth of *Agaricus bisporus*. *Mycological Research* 95, 699–704.

Reddell, P. and Malajczuk, N. (1984) Formation of mycorrhizae by jarrah (*Eucalyptus marginata* Donn ex Smith) in litter and soil. *Australian Journal of Botany* 32, 511–520.

Termorshuizen, A.J. and Schaffers, A.P. (1987) Occurrence of carpophores of ectomycorrhizal fungi in selected stands of *Pinus sylvestris* in the Netherlands related to stand vitality and air pollution. *Plant and Soil* 104, 209–217.

Tyler, G. (1985) Macrofungal flora of Swedish beech forest related to soil organic matter and acidity characteristics. *Forest Ecology and Management* 10, 13–29.

Warcup, J.H. (1991) The fungi forming mycorrhizas on eucalypt seedlings in regeneration coupes in Tasmania. *Mycological Research* 95, 329–332.

30 Mycorrhizal Succession and Morel Biology

F. Buscot

Bundesforschungsanstalt für Landwirtschaft, Institut für Bodenbiologie, Bundesalle 50 D 3300 Braunschweig, Germany

Introduction

In stable forest ecosystems, morels form storage and over-wintering sclerotial organs which surround main roots of trees, and provide the nutrients for the vernal fructification (Buscot and Roux, 1987; Buscot, 1989; Buscot and Bernillon, 1991). In the association between the late fructifying *Morchella esculenta* and *Picea abies*, the fine rootlets emerging from such sclerotia formed vital mycorrhizas with the morel, whereas other mycorrhizas formed with a heterobasidiomycete in their vicinity were senescent and invaded by vital morel hyphae (Buscot and Kottke, 1990). No mycorrhiza synthesis with morel in pure culture has been achieved, but sclerotia formation around tree roots was stimulated when the trees were mycorrhizal with other fungi (unpublished observations). Thus one can hypothesize that morels are not able to form mycorrhizas as primary, but only as secondary, fungal partners of a root system, by succeeding to mycorrhizas previously formed by other fungi. This chapter concerns the association between *Morchella elata* and *Picea abies*, in which early stages of mycorrhiza formation by the morel and its interactions with other mycorrhizas were observed.

Materials and Methods

Two non-mature ascocarps of *M. elata* (Fr.) Boudier were collected with the surrounding soil in February 1990 in a 60-year-old plantation of *P. abies* near Herrenberg (Germany). The material was washed and prepared for light and transmission electron micoscopy as described by Buscot and Kottke (1990). Mycorrhizas were characterized according to the method of Haug *et al.* (1987), by analysing their morphology, anatomy and ultrastructure. This analysis also served to evaluate the vitality of the mycorrhizas and their relation with the morel.

Results

Description of the mycorrhizal types

The subterranean mycelial structure connected with the ascocarps of *M. elata* was dense and voluminous, and unlike *M. esculenta* (Buscot and Kottke, 1990), it also surrounded clusters of about 70 mycorrhizal rootlets, among which six different types could be distinguished (Table 30.1). The single ascomycete types could not be identified, but its colour and cytological features confirmed that it was not the morel. Types C and D corresponded to types 15 and 3 of Haug *et al.* (1987). No well developed morel mycorrhizas could be observed, but on six ageing mycorrhizas of type D the beginning of secondary mycorrhiza formation by the morel occurred (Fig. 30.1).

Table 30.1. Anatomical features of the six mycorrhizal types surrounded by the sclerotial subterranean structure formed by *Morchella elata*.

Mycorrhizal type	Colour	Structure of the hyphal mantle	Depth of the Hartig net	Fungal partner
A	Rust to fuscous black when ageing	5–8 μm thick polygonal plectenchyma, opaque interhyphal matrix	Developed till the central cylinder	Ascomycete
B	Olivaceous black	4–6 μm thick compact prosenchyma	Developed till the central cylinder	Basidiomycete
C	Translucide, clay pink to fuscous black when ageing	Outer 7–8 μm thick loose, and inner 12–20 μm thick, compact prosenchyma	Developed till the central cylinder	Basidiomycete
D	Distally white and proximally brown	3–8 μm thick compact prosenchyma	Developed around 2–3 cortical cell layers	Basidiomycete
E	Cigar brown	6–8 μm thick polygonal synenhyma, opaque interhyphal matrix	Developed around 1–2 cortical cell layers	Basidiomycete
F	White to buff when ageing	8–15 μm thick compact prosenchyma, clear interhyphal matrix	Developed around 2–3 cortical cell layers	Basidiomycete

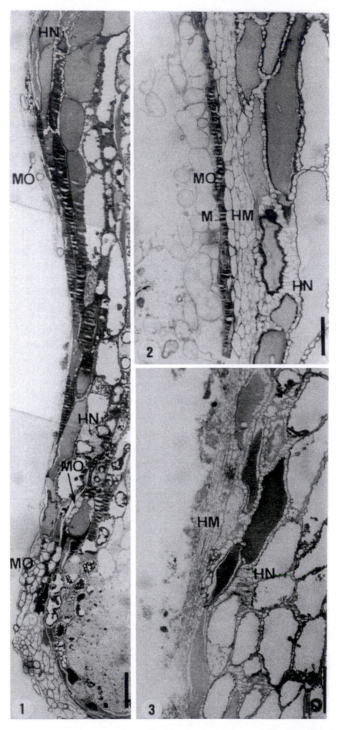

Figs 30.1–30.3. 30.1. Formation of secondary mycorrhizas by *Morchella elata*. The morel mycelium (MO) has a loose contact (small arrows) with the ageing type D mycorrhiza, but forms a hyphal mantle and a Hartig net (big arrow) at the distal part

Vitality of the mycorrhizas and relation to the surrounding morel hyphae

Three degrees of attachment between the mycorrhizas and the surrounding morel mycelium were distinguished: (1), *Diffuse contact* (Fig. 30.3); (2), *Tight attachment*, with formation of a secondary hyphal mantle by *M. elata*, but without penetration into the original mycorrhizas (Fig. 30.2); (3) *Succession*, with formation of secondary mycorrhizas by *M. elata*, and localized transition between the two symbionts (Fig. 30.1). These degrees of attachment did not represent different stages of a single association, but different types of association.

The mycorrhizas of types A, C and D were vital or senescent, whereas all of type B were vital and all of types E and F were senescent (Table 30.2). Thus, many of the mycorrhizas were active at the time of the observation. Obviously all of them were completely surrounded by the morel sclerotial structure and therefore had no contact with the soil.

Except in the cases of succession on ageing mycorrhizas of type D, there was no correlation between the vitality of the mycorrhizas and their degree of attachment to the morel mycelium (Table 30.2). On the contrary, the degree of attachment appeared to be specific of the mycorrhizal type.

Table 30.2. The vitality of six mycorrhizal types and their relation to the mycelium of *Morchella elata* surrounding them in February.

Mycorrhizal type	Vitality	Relation to the surrounding morel mycelium
A	Vital to senescent	Tight attachment, formation of a concentric, secondary hyphal mantle by the morel
B	Vital	Loose contact
C	Vital to senescent	Loose contact
D	Vital to senescent	Loose contact when vital, succession of the morel as symbiont when ageing
E	Senescent	Loose contact
F	Senescent	Loose contact

Figs 30.1–30.3. Contd.
(HN$_1$, Hartig net of type D mycorrhiza; scale bar, 10 μm). **30.2.** Hyphae of the subterranean, sclerotial organ of *Morchella elata* (MO) grow in close contact with type A mycorrhizas, forming an external and concentric, secondary hyphal mantle around it. The organisms remained separated by an opaque matrix (M); (HM and HN, hyphal mantle and Hartig net of the type A mycorrhiza respectively; scale bar, 5 μm). **30.3.** The contact of the subterranean, sclerotial organ of *Morchella elata* to the mycorrhizas of type F is so loose that its hyphae completely detach from these mycorrhizas during their preparation (HM and HN, hyphal mantle and Hartig net of a type F mycorrhiza respectively; scale bar, 5 μm).

Discussion

The investigations of this system reinforce the idea that morels only form mycorrhizas when they succeed ageing mycorrhizas formed by other fungi. There must be a specific recognition procedure for this succession which only occurred with one of the mycorrhizal types surrounded by the morel mycelium. As morels also can develop entirely saprophytically (Ower, 1982), the biological significance of the phenomenon of secondary mycorrhizas should be studied thoroughly in order to elucidate whether such fungi are evolving to strict symbionts or whether they are losing this potential. This model of secondary mycorrhizas should also be useful to investigate the recognition procedures during mycorrhizal synthesis.

References

Buscot, F. (1989) Field observations on growth and development of *Morchella rotunda* and *Mitrophora semilibera* in relation to forest soil temperature. *Canadian Journal of Botany* 67, 589–593.

Buscot, F. and Bernillon, J. (1991) Mycosporins and related compounds in field and cultured mycelial structures of *Morchella esculenta*. *Mycological Research* 95, 752–754.

Buscot, F. and Kottke, I. (1990) The association of *Morchella rotunda* (Pers.) Boudier with roots of *Picea abies*. *New Phytologist* 116, 425–430.

Buscot, F. and Roux, J. (1987) Association between living roots and ascocarps of *Morchella rotunda*. *Transactions of the British Mycological Society* 89, 249–252.

Haug, I., Kottke, I. and Oberwinkler, F. (1987) Licht- und Elektronenmikroskopische Untersuchungen von Mykorrhizen der Fichte (*Picea abies* (L.) Karst.) in Vertikal-profilen. *Zeitschrift für Mykologie* 52, 373–392.

Ower, R. (1982) Notes on the development of the morel ascocarp: *Morchella esculenta*. *Mycologia* 74, 142–144.

Part Four
Mycorrhizas in Heathland Ecosystems

31 The Role of Ericoid Mycorrhizas in the Nitrogen Nutrition and Ecology of Heathland Ecosystems

J.R. Leake

Department of Animal and Plant Sciences, The University of Sheffield, Sheffield S10 2UQ, UK

Introduction

Ericoid mycorrhizas have been shown to mobilize nitrogen from a wide range of organic sources, including protein, peptide, amino acids, chitin and lignin (Bajwa, *et al.*, 1985; Bajwa and Read, 1986; Haselwandter, *et al.*, 1990; Leake and Read, 1990a). Since rates of organic matter mineralization are very slow in heathlands, and almost all of the soil nitrogen is organically bound, the direct involvement of ericoid mycorrhizas in the recycling of this vital, growth-limiting, nutrient is of great importance for the nutrition of individual plants and for the nitrogen cycling dynamics of these ecosystems.

While some progress has been made in our understanding of the nature and properties of some of the enzymes which catalyse the hydrolysis and assimilation of products from organic polymers by mycorrhizal fungi (Spinner and Haselwandter, 1985; Leake and Read, 1989, 1990b) little is known of the regulation of production of enzymes within the soil. Organic, heathland soil has been found to contain high concentrations of 'free' amino acids, particularly in autumn (Abuarghub and Read, 1988), but their effects on proteinase production by ericoid mycorrhizas have not been investigated. This chapter reports a comparative analysis of the effects of amino acids, ammonium, protein hydrolysate and protein on proteinase production by the ericoid endophyte *Hymenoscyphus ericae*.

Materials and Methods

The seasonal and profile distribution of free amino acids in *Calluna* heathland soil, reported by Abuarghub and Read (1988) indicated that 13 amino acids are present in significant quantities. These amino acids can be grouped into three categories – basic, neutral and acidic. To investigate their effects on proteinase production by *Hymenoscyphus ericae* the fungus was grown in 20 ml Norkrans basal medium (Norkrans, 1949) (pH adjusted to 3.5) from which mineral N and yeast extract were omitted, but with one of each of the amino acids at a concentration of 32 mgN l^{-1}. For comparative purposes, parallel cultures were established in media containing ammonium ($(NH_4)_2SO_4$), casein, hydrolysate, and protein (bovine serum albumin) each of which provided a nitrogen concentration of 32 mg l^{-1}.

The activity of the enzyme in culture filtrates was measured at five harvests over 21 days of growth, by incubating 50–500 µl of culture filtrate with 50 µl of fluorescein isothiocyanate labelled bovine serum albumin (FITC-BSA) (1 mg ml^{-1}) in 1 ml of 0.2 M glycine HCl buffer (pH 2.2) and distilled water to give a final assay volume of 2.05 ml. After incubation for 3 h at 37 °C in a shaking water bath the reaction was terminated by addition of 1 ml 10% trichloroacetic acid. After centrifugation at 3000 g, for 7 min, 50–200 µl volumes of supernatant were diluted in 4 ml 0.4 M boric acid: NaOH buffer (pH 9.7) and measured by a Perkin-Elmer LS30 automated luminescence spectrometer. Excitation and emission wavelengths were 495 nm and 516 nm respectively. At each harvest there were four replicate cultures for each treatment. Mycelium was separated from the culture solution by filtration on Whatman No. 1 paper, transferred to pre-weighed pieces of aluminium foil and oven dried at 80 °C for 24 h before weighing. Enzyme activities are expressed as crude fluorescence units released in 3 h ml^{-1} of culture filtrate (total activity), and as fluorescence units released per ml of culture filtrate per mg dry weight of mycelium in 3 h (specific activity).

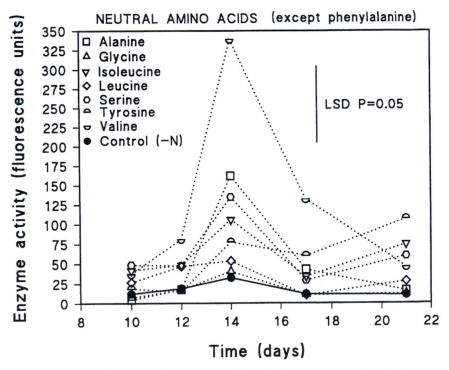

Fig. 31.1. Proteinase activity in culture filtrates of *H. ericae* provided with each of six neutral amino acids as sole source of N.

Results

Neutral amino acids

With the exception of phenylalanine, the neutral amino acids gave only a modest increase in proteinase production relative to the control treatment which lacked N (Fig. 31.1). While valine appeared to cause some stimulation of proteinase production it was not significantly greater than that found in cultures supplied with ammonium as the sole N source (c.f. Fig. 31.5). In every treatment, with the sole exception of that containing tyrosine, maximum enzyme production had occurred by the 14th day, and thereafter it declined. Enzyme activity per mg of dried mycelium (specific activity) indicated that none of the neutral amino acids, excepting phenylalanine, were very effective as inducers of proteinase production, glycine, alanine and tyrosine significantly reducing the quantity of enzyme produced per mg of mycelium with respect to the control treatment (− N) (Fig. 31.2). Phenylalanine was a very effective inducer of proteinase, its effects are described later.

Basic and acidic amino acids

With the exception of histidine, the basic and acidic amino acids failed to stimulate proteinase production by *H. ericae* (Fig. 31.3). Indeed enzyme production in the

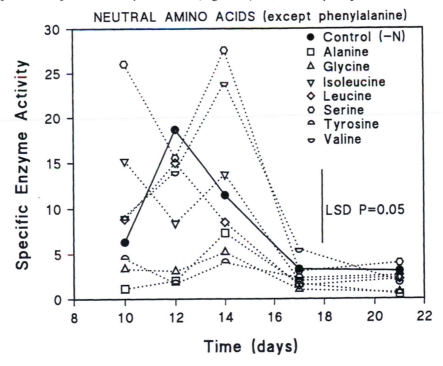

Fig. 31.2. Specific enzyme activity (fluorescence release per ml of culture filtrate per mg of mycelium dry weight per hour) of endophyte grown with each of six amino acids as sole N source.

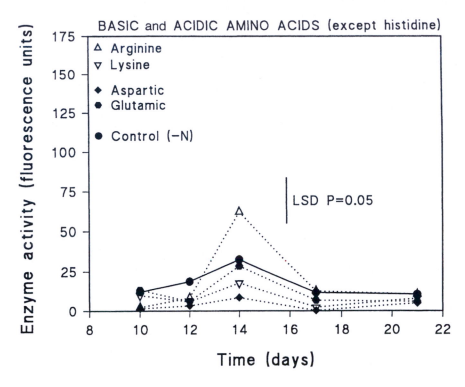

Fig. 31.3. The proteinase activity in cultures of *H. ericae* grown with basic and acidic amino acids (except histidine).

lysine, aspartic acid and glutamic acid treatments were consistently lower than in the control, which lacked N, despite the mycelium dry weight being much higher when the fungus was provided with N. The specific enzyme activities of cultures grown with these amino acids confirmed that they tend to repress rather than induce proteinase production, aspartic acid being a particularly effective inhibitor (Fig. 31.4). Overall, the effects of the basic and neutral amino acids, excepting histidine, on proteinase production were very similar.

Histidine, phenylalanine, protein, protein hydrolysate and ammonium

The effect of histidine was remarkable. It stimulated a massive increase in proteinase production, far in excess of that found with any of the other amino acids tested, and much greater than that found in the casein hyrolysate or protein treatments (Fig. 31.5). This effect was apparent by the first harvest (10 days) at which time there was only 3 mg dry weight of mycelium in each flask, a tenth of the biomass present at 21 days (data not shown). The role of histidine as an inducer of proteinase is more striking when the results are expressed as specific activity (Fig. 31.6). At the first harvest (10 days) the specific activity of endophyte grown on histidine was an order of magnitude higher than the most active cultures of the other treatments. Its subsequent decline reflected the increasing biomass of the fungus throughout the experiment and the fall in enzyme activity after the 14th

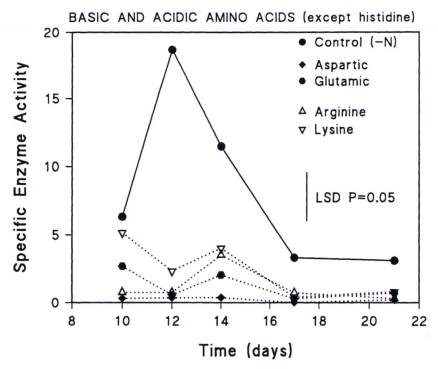

Fig. 31.4. Specific enzyme activity of cultures of *H. ericae* grown with basic and acidic amino acids (except histidine).

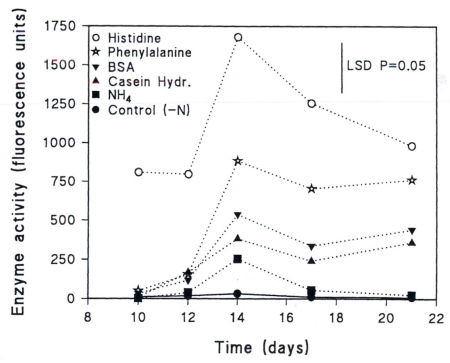

Fig. 31.5. The proteinase activity of cultures of *H. ericae* grown with histidine, phenylalanine, protein (BSA), protein hydrolysate (casein hydrolysate) and ammonium as sole nitrogen sources.

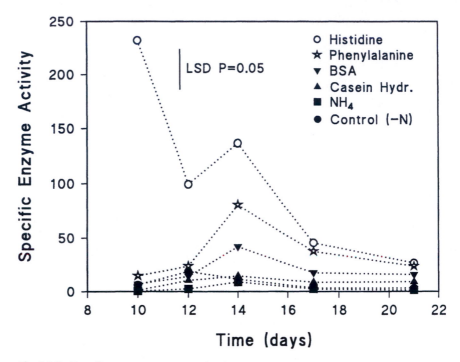

Fig. 31.6. Specific enzyme activity of cultures of the endophyte grown with a range of mineral and organic N sources.

day, possibly due to depletion of the inducer from the culture solution, since this provided the sole source of nitrogen. Phenylalanine was also an effective inducer, but less so than histidine. None the less, it stimulated much greater enzyme production than was found on protein or protein hydrolysate. The latter two treatments induced higher proteinase activities than most of the amino acids, and have been noted as very effective inducers in earlier studies (Leake and Read, 1990b). Ammonium also stimulated enzyme production, more so than many of the amino acids.

Supplementary experiment

To investigate further the role of histidine as an inducer of proteinase production by *H. ericae*, the endophyte was grown with histidine as the sole source of N, providing a concentration range from 0 to $64 \, mgN \, l^{-1}$. The materials and methods were as described for the earlier experiments, except that the number of harvests was increased. The results revealed that as little as $4 \, mgN \, l^{-1}$ as histidine triggered induction of proteinase by 6 days, and that maximum enzyme activity occurred at the 8th day (Fig. 31.7). Full induction was achieved by histidine concentrations of 16, 32 and $64 \, mgN \, l^{-1}$, which indicated that the induction mechanism is fully activated by concentrations of between 8 and $16 \, mgN \, l^{-1}$. Further increasing the concentration of the amino acid sustained the maximum enzyme

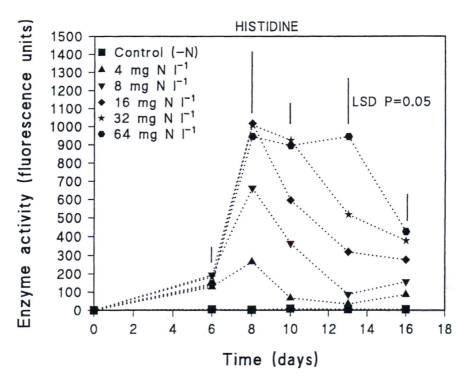

Fig. 31.7. Induction of proteinase production in *H. ericae* by histidine supplied at concentration of 4–64 mg l^{-1}.

activity at about 1000 crude units for a longer period, but did not increase peak activity at 8 days. The decline in enzyme activity after 8 days is almost certainly due to depletion of the amino acid from the culture media, since it provided the sole source of N for the fungus. The early decline in enzyme activity in the treatments containing the lower concentrations of the inducer lend support to this view.

Discussion

Since protein molecules are too large to pass through cell membranes intact, their hydrolysis products must be employed for regulation of proteinase production in relation to the supply of available substrates. Amino acids are the major products of proteolysis which are assimilated by mycorrhizas. Induction of proteinase production by some amino acids provides a convenient means of ensuring that enzyme production is linked to substrate availability. The results indicate, however, that most of the amino acids present in heathland soil are either ineffective as inducers, or may even repress enzyme production. The exceptional response to histidine is therefore interesting since it suggests that this compound may have a special role in controlling proteinase production by *H. ericae*. This view is reinforced by the unusual seasonal dynamics of histidine in *Calluna* soil reported by Abuarghub and Read (1988). It was quantitatively the most important 'free' amino acid in the soil in autumn, which was when the pool of 'free' amino acids attained maximum concentration.

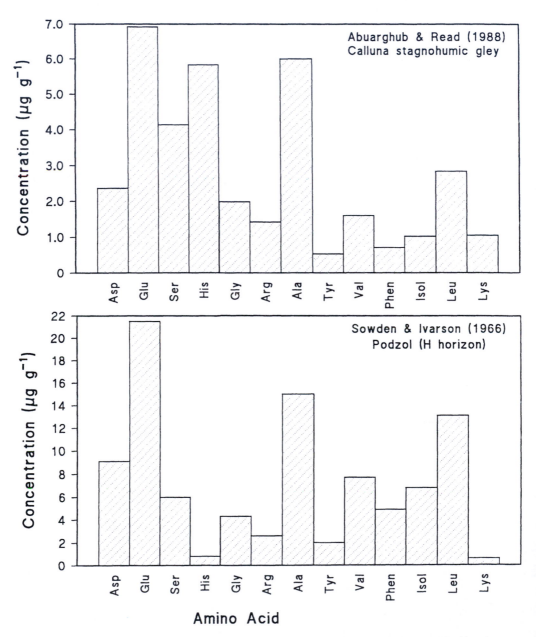

Fig. 31.8. The mean annual concentration of individual 'free' amino acids in heathland soil reported by Abuarghub and Read (1988), and the concentrations of the same compounds detected in extracts of a podzol by Sowden and Ivarson (1966).

Whereas phenylalanine is also an effective inducer, it is doubtful whether it is has a significant effect on ericoid mycorrhizas, since it has only been detected at very low concentrations in heathland soil (Abuarghub and Read, 1988).

The source of the histidine is unknown. One possibility is that it is produced by the host plant and released in senescent tissues, signalling the availability of fresh substrates for proteolysis. The concentrations of histidine found in *Calluna* soil do appear to be unusually high, approaching 8 mgN l^{-1} in November. When expressed as average concentrations in the top 10 cm of soil over the year, the results of Abuarghub and Read (1988) indicate that it is present in similar concentrations to glutamic acid and alanine (Fig. 31.8). An analysis of the spectrum of amino acids in a forest podzol by Sowden and Ivarson (1966), while using different extractant and detection methods, found a very similar overall pattern of relative concentrations of 12 of the 13 amino acids detected by Abuarghub and Read (1988). The major difference between the results of these two studies was that the concentration of histidine in Sowden and Ivarson's study was the second lowest of the amino acids, while in the *Calluna* soil it was third most important (and by far the most important in late autumn). This raises the exciting possibility that histidine occurs at uniquely high concentrations in heathland soil, perhaps released by the host plant as a signal compound which regulates proteinase production by ericoid endophytes. Such a mechanism of regulation would be very effective since the peak annual concentration of histidine coincides with the period when the greatest quantities of fresh litter and other substrates for proteolysis are released into the rooting zone.

The extent of the specificity of the ericoid endophytes response to histidine is unknown. In particular it would be of considerable interest to determine whether other mycorrhizal fungi respond in a similar way to the compound. The effect of histidine on proteinase production by ectomycorrhizal fungi is currently being investigated.

References

Abuarghub, S.M. and Read, D.J. (1988) The biology of mycorrhiza in the Ericaceae. XII. Quantitative analysis of individual 'free' amino acids in relation to time and depth in the soil profile. *New Phytologist* 108, 433–441.

Bajwa, R. and Read, D.J. (1986) Utilization of mineral and amino-nitrogen sources by the ericoid mycorrhizal endophyte *Hymenoscyphus ericae* and by mycorrhizal and non-mycorrhizal seedlings of *Vaccinium*. *Transactions of the British Mycological Society* 87, 269–277.

Bajwa, R., Abuarghub, S.M. and Read, D.J. (1985) The biology of mycorrhiza in the Ericaceae. X. The utilization of protein and the production of proteolytic enzymes by the mycorrhizal endophyte and by mycorrhizal plants. *New Phytologist* 101, 469–486.

Haselwandter, K., Bobleter, O. and Read, D.J. (1990) Utilization of lignin by ericoid and ectomycorrhizal fungi. *Archives of Microbiology*, 153, 352–354.

Leake, J.R. and Read, D.J. (1989) The biology of mycorrhiza in the Ericaceae. XIII. Some characteristics of the extracellular proteinase activity of the ericoid endophyte *Hymenoscyphus ericae*. *New Phytologist* 112, 69–76.

Leake, J.R. and Read, D.J. (1990a) Chitin as a nitrogen source for mycorrhizal fungi. *Mycological Research* 94, 993–995.

Leake, J.R. and Read, D.J. (1990b) Proteinase activity in mycorrhizal fungi. II. The effects of mineral and organic nitrogen sources on induction of extracellular proteinase in *Hymenoscyphus ericae* (Read) Korf & Kernan. *New Phytologist* 116, 123–128.

Norkrans, B. (1949) Some mycorrhiza forming *Tricholoma* species. *Svensk Botanisk Tidskrift* 43, 309–327.

Sowden, F.J. and Ivarson, K.C. (1966) The 'free' amino acids of soil. *Canadian Journal of Soil Science* 46, 109–114.

Spinner, S. and Haselwandter, K. (1985) Proteins as nitrogen sources for *Hymenoscyphus* (= *Pezizella*) *ericae*. In: Molina, R. (ed.), *Proceeding of the 6th North American Conference on Mycorrhizae* Forest Research Laboratory, Oregon state University, Corvallis, Oregon, p. 422.

32 Mycorrhizal Aspects of Improved Growth of Spruce when Grown in Mixed Stands on Heathlands

E.A. Ryan and I.J. Alexander

Dept of Plant and Soil Science, University of Aberdeen, Aberdeen AB9 2UE, UK

Introduction

When Sitka spruce (*Picea sitchensis* (Bong.) Carr.) is planted on poor upland peats and heathlands in northern Britain satisfactory growth can only be achieved with repeated applications of nitrogen (N) fertilizer. Sitka spruce planted in mixture with Scots pine (*Pinus sylvestris* L.), lodgepole pine (*P. contorta* Doug.) or Japanese larch (*Larix kaempferi* (Lamb.) Carr.) has improved N-nutrition, comparable to that of N-fertilized Sitka spruce monoculture (Garforth, 1979; Taylor, 1985). This 'mixtures effect' occurs between 6 and 12 years after planting when the trees have reached the critical height of 2.5 m (Dickson, 1977). Research to date (Carey *et al.*, 1986, 1988) has confirmed that this improved growth is solely due to improved N nutrition but has failed to explain how the extra N is accessed. Heather (*Calluna vulgaris* (L.) Hull), when present accentuates the problem but is not the sole cause of N deficiency in pure stands of spruce (Taylor and Tabbush, 1990).

Read and his co-workers (Stribley and Read, 1980; Bajwa and Read, 1985; Leake and Read, 1987, 1989) have demonstrated that the ericoid mycorrhizal fungus, *Hymenoscyphus ericae* (Read) Korf & Kerman, can utilize amino acids, peptides and proteins and that the assimilated nitrogen is rapidly transferred to infected plants. Abuzinadah and Read (1986a,b, 1988) have shown that certain ectomycorrhizal fungi also have this ability. In particular they found that *Suillus bovinus* (L. ex Fr.) a mycorrhizal associate of pine but not of spruce, could degrade protein. Many other *Suillus* spp. only fruit in nature in association with pine or larch (Trappe, 1962; Watling, 1971; Moser, 1983). Alexander and Watling (1987) did not record any *Suillus* spp. in their survey of the macrofungi of pure Sitka spruce plantations. The possibility arises that the 'mixtures effect' may be related to the presence, in the mixed stands, of ectomycorrhizal fungi capable of exploiting organic N in the soil. In this chapter the composition of the mycorrhizal community of Sitka spruce in a pure stand and of both Sitka spruce and Scots pine in a mixed stand are examined. The aims were to see if there was a difference in the mycorrhizal associates between pure and mixed stands, and to find whether *Suillus* spp. if present, had infected spruce. The proteinase activity of a range of ectomycorrhizal fungi isolated from pure and mixed stands and the utilization of the cleavage products of fungal proteinase were also examined.

Materials and Methods

The communities of mycorrhizas

An Old Red Sandstone site at Culloden Forest near Inverness, North East Scotland with a severely indurated gley soil and a shallow (< 10 cm) acid humus layer was used for this study. Sample plots (40 × 20 m with a 5 m buffer strip) had been planted in 1969. There were four pure Sitka spruce plots and four plots with a mixture of Sitka spruce and Scots pine. The mixed plots were planted with an alternate 3 × 3 triplet arrangement of Scots pine and Sitka spruce at 2-m spacing.

Twenty 4-cm diameter cores of the top 6 cm of soil were taken from the central area of each of the four pure Sitka spruce plots. Each core was 50 cm from the base of a tree. A further 40 cores from each of the four mixed plots were collected, one core from beneath the centre tree of each of 20 Sitka spruce triplets and one core from beneath the centre tree of each of 20 Scots pine triplets.

Each set of 20 cores was randomly divided into four groups of five cores. The cores were broken up, washed through a 2 mm sieve and collected on a 200 μm sieve. All the material on the 200 μm sieve was fixed in 3% glutaraldehyde. A random subsample of 100 tips of pine or spruce as appropriate was picked out, orientated in water agar in groups of 25 and post fixed in 3% glutaraldehyde, dehydrated in an alcohol series and embedded in LR white resin. Sections (4 μm) were cut and stained with toluidine blue.

Sections were scored for ectomycorrhizal infection and were allocated to types on the basis of the morphology of the sheath and Hartig net, hyphal characteristics and staining reactions. In the spruce root samples those few pine roots which had been embedded in error could be readily identified and excluded from the counts.

Protein breakdown

Isolates of *Suillus variegatus* (Swartz ex Fr.), *S. luteus* (L. ex Fr.) O. Kuntze, *S. bovinus*, *Paxillus involutus* (Batsch) Fr., and *Lactarius rufus* (Scop.) were obtained from fruiting bodies collected from the field site. An isolate of *Tylospora fibrillosa* (Burt) Donk (Taylor and Alexander, 1990) was obtained from surface sterilized spruce roots. Bovine serum albumin (BSA) MW 66000 was chosen as a test protein substrate because of its high solubility at low pH and its purity (98–99% albumin). Plugs of each test isolate 4 mm in diameter were cut from the edge of a 14-day-old culture on malt-extract-free MMN (Marx, 1969) agar. Each plug was transferred to the centre of a previously boiled and sterilized cellophane disc placed on MMN agar and left for 48 h at 20°C to allow hyphae to regenerate. Inoculated discs were then transferred to 90 mm Petri dishes containing 20 ml of Norkrans (1950) basal mineral medium containing 9.112 mg ml^{-1} glucose and 1.6 mg ml^{-1} BSA with the pH adjusted to 2.5. This gave a C:N ratio of 15:1 (Abuzinadah and Read, 1986a). The cellophane discs were balanced on three 1.5% water agar plugs to allow the fungi to grow at the liquid–air interface. The concentration of protein in the culture medium was measured at the outset and then every 5 days (fruiting body isolates) or 7 days (root isolate) using the Coomassie-blue protein assay. There were four replicates for each fungus at each harvest.

Proteinase activity

Isolates of *S. variegatus*, *S. luteus*, *S. bovinus*, *P. involutus*, *L. rufus*, *T. fibrillosa* and *Hymenoscyphus ericae* (Read 101) were grown using the same method as for the BSA assay. At days 7, 14, 21 and 28, dry weights and proteinase activity were measured using the modified Twining (1984) fluorimetric proteinase assay described by Leake and Read (1989). Mycelium was harvested by filtering the culture medium and mycelium through preweighed Whatman GF/B glassmicrofibre paper and dried at 80°C for 24 h. There were four replicates for each fungus at each harvest.

Utilization of cleaved protein

Suillus variegatus and *H. ericae* but not *L. rufus*, were found to have active proteinase at low pH. The ability of *L. rufus* to use protein cleaved by either *S. variegatus* or *H. ericae* was investigated. *S. variegatus* and *H. ericae* were grown on the BSA Norkrans medium previously described at pH 3.9. The BSA concentration was monitored until it all had been degraded. An aliquot of culture medium was filtered through a Whatman GF/B microfilter and the total N content measured (Allen, 1974). The cleaved culture media were then filter sterilized and dispensed into 90 mm Petri dishes in 20 ml aliquots. A further series of dishes were set up containing 20 ml of Norkrans basal mineral medium with either no N, BSA or ammonium chloride added to give the same N content as the cleaved *S. variegatus* or *H. ericae* media. Each dish was inoculated with a 2-day-old *L. rufus* plug growing on cellophane prepared as above. Dry weight was measured at days 10, 20, 30 and 40.

Results

Mycorrhizal associates on Sitka spruce roots

Twelve mycorrhizal types were recognized, which could be placed in four groups, *Tylospora fibrillosa* (Taylor and Alexander, 1990), Russulaceae spp. (Agerer, 1986), Ascomycetous (Weiss and Agerer, 1988), and others (this group comprised fewer than 12% of the total tips and was made up of nine types none of which accounted for more than 4%).

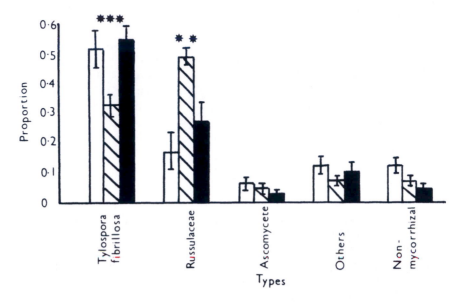

Fig. 32.1. Proportion of different mycorrhizal types on Sitka spruce roots collected beneath Sitka spruce in pure (□), and mixed stands (■), and Sitka spruce roots collected beneath Scots pine in mixed stands (▨). Vertical bars show standard error. Asterisks show proportions which are significantly different from the others ***P < 0.001, **P < 0.05.

The proportions of *Tylospora fibrillosa* and Russulaceae mycorrhizas were significantly affected by the source of the roots. Spruce roots from the pure plots or from beneath spruce trees in the mixed plots, had a higher proportion (50–55%) of *T. fibrillosa* mycorrhizas than those collected beneath pine in the mixed plots (33%) (Fig. 32.1). Conversely spruce roots from beneath pine in the mixed plots had a higher proportion of Russulaceae mycorrhizas (48%) than those collected under spruce (16–26%) in mixed or pure plots. No mycorrhizas with the characteristic anatomy of *Suillus* spp. associates were found on spruce roots.

Mycorrhizal associates on Scots pine roots

Thirteen mycorrhizal types were recognized. These could be placed into six groups Russulaceae (Agerer, 1986), *Suillus* spp. (Palm and Stewart, 1984), *Tylospora fibrillosa* (Taylor and Alexander, 1990), *Piceirhiza obscura* (Gronbach and Berg, 1988), a sheathless type, and others (this group comprised less than 13% and was made up of seven types none of which accounted for more than 5%).

Again, the proportion of *T. fibrillosa* and Russulaceae mycorrhiza groups were significantly affected by the source of the roots. Pine roots from beneath pine trees had a higher proportion (43%) of Russulaceae mycorrhizas than those beneath spruce (8%). Pine roots from under spruce had a higher proportion of *T. fibrillosa* mycorrhizas (17%) than under pine (4%). There was no significant difference in the proportion of *Suillus* mycorrhizas under Sitka spruce or Scots pine (Fig. 32.2).

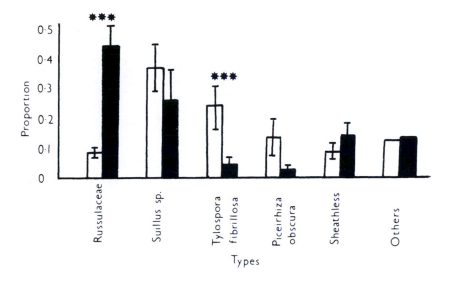

Fig. 32.2. Proportion of different mycorrhizal types on Scots pine roots collected from beneath Scots pine (■) or Sitka spruce (□) in mixed stands. Vertical bars show standard error. *** = difference significant, $P < 0.001$.

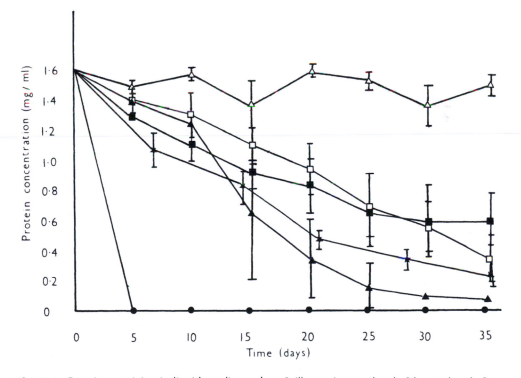

Fig. 32.3. Protein remaining in liquid medium when *Suillus variegatus* (–●–), *S.luteus* (–■–), *S. bovinus* (–▲–), *Paxillus involutus* (–□–), *Lactarius rufus* (–△–) and *Tylospora fibrillosa* (–★–) are grown with bovine serum albumin as the only nitrogen source. Vertical bars show standard deviation ($n = 4$).

Protein breakdown

S. variegatus had the greatest degradative ability. The isolate broke down all of the BSA within 5 days (Fig. 32.3). *L. rufus* broke down relatively little protein. *S. bovinus*, *S. luteus*, *P. involutus* and *T. fibrillosa* occupied an intermediate position.

Proteinase activity

At each sampling (Fig. 32.4) *S. variegatus* had the greatest proteinase activity. *S. variegatus* and *H. ericae* both showed their greatest proteinase activity at day 7. *S. variegatus* had a significantly higher proteinase activity than *H. ericae* at day 14 and 28. *L. rufus* had little or no proteinase activity.

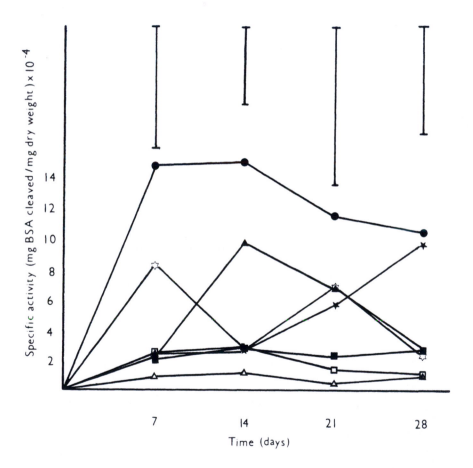

Fig. 32.4. Specific activity of the proteinase from *Suillus variegatus* (–●–), *S.luteus* (–■–), *S.bovinus* (–▲–), *Paxillus involutus* (–□–), *Lactarius rufus* (–△–), *Tylospora fibrillosa* (–★–) and *Hymenoscyphus ericae* (–☆–). Vertical bars show l.s.d. ($P = 0.05$).

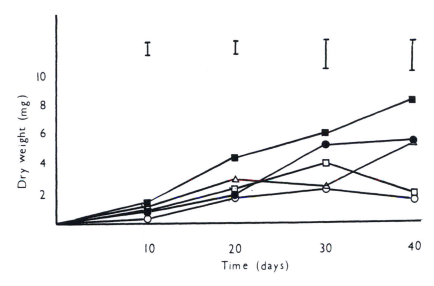

Fig. 32.5. Growth of *Lactarius rufus* on various nitrogen sources, no nitrogen (–○–), ammonium (–△–), bovine serum albumin (BSA) (–□–), BSA cleaved by *Hymenoscyphus ericae* (–●–), BSA cleaved by *Suillus variegatus* (–■–). Vertical bars show l.s.d. (*P* = 0.05).

Utilization of cleaved protein by *Lactarius rufus*

The *S. variegatus* conditioned and *H. ericae* conditioned media contained 53% and 51% respectively of the original N added as BSA. The best growth of *L. rufus* was on *S. variegatus* conditioned medium, and it was significantly better than growth on ammonium-N at days 30 and 40 (Fig. 32.5). *L. rufus* also grew on the *H. ericae* medium but growth was significantly better on the *S. variegatus* media at days 20 and 40. At day 40 *H. ericae* media supported similar growth to the ammonium control.

Discussion

There was little difference in the overall level of mycorrhizal infection of spruce in mixed (95%) or pure (88%) plots. It is therefore unlikely that the level of mycorrhizal infection affected the N status of the trees. Sitka spruce did not form mycorrhizas with *Suillus* spp. at Culloden even when admixed with *Suillus*-infected pine, so the improved N nutrition can not be ascribed to spruce roots becoming infected with fungi normally restricted to pine. However the composition of the mycorrhizal community on the spruce roots rooting under a pine canopy was clearly different from that rooting under a spruce canopy, in that the proportion of Russulaceae mycorrhizas increased and that of *T. fibrillosa* decreased. The reasons for this are unknown but there are a number of factors which could be different under the pine canopy including the quantity and quality of throughfall and litter. The change in the mycorrhizal associates of spruce when in mixture may be a component of the mixtures effect.

Within 5 days *S. variegatus* cultures broke down all the BSA added. The high proteinase activity of this fungus, at least equivalent to that of *H. ericae*, was confirmed by the FITC-linked BSA assay. If this proteinase remains active in soil in the presence of *Calluna* and polyphenol–protein complexes it could be an important factor in the success of pine on *Calluna*-dominated heathland and moorland. It may be significant that *S. variegatus* is recorded as an associate of pine 'especially . . . on poor soil' (Lange and Hora, 1965).

L. rufus, on the other hand, neither broke down BSA nor showed significant proteinase activity, confirming the conclusion of Abuzinadah and Read (1986a) that *L. rufus* is a 'non-protein' fungus. *L. rufus* fruited abundantly in the mixed stands, where mineral N levels in the soil are known to be higher (Carey *et al.*, 1988), and it seems likely that many of the Russulaceae type mycorrhizas were formed by this fungus. *L. rufus* was capable of utilizing the cleavage products which resulted from the proteolytic activity of *S. variegatus* for growth, and it is possible that this is a route by which the 'extra' N is absorbed by spruce in mixed stands. *L. rufus* also grew on the cleavage products resulting from the proteolytic activity of *H. ericae*. Apparently this is not an important pathway for N transfer to spruce in the field, possibly because of phytotoxic compounds associated with *Calluna* or its residues, or because *Calluna* mycorrhizas are extremely efficient competitors for the products of proteolysis.

Acknowledgements

This research was supported by EEC contract No. MAIB–0026–UK(BA) and by a grant from the Scottish Forestry Trust.

References

Abuzinadah, R.A. and Read, D.J. (1986a) The role of proteins in the nitrogen nutrition of ectomycorrhizal plants. I. Utilization of proteins and peptides by ectomycorrhizal fungi. *New Phytologist* 103, 481–493.

Abuzinadah, R.A. and Read, D.J. (1986b) The role of proteins in the nitrogen nutrition of ectomycorrhizal plants. III. Protein utilization by *Betula*, *Picea* and *Pinus* in mycorrhizal association with *Heboloma crustiliniforme*. *New Phytologist* 103, 507–514.

Abuzinadah, R.A. and Read, D.J. (1988) Amino acids as nitrogen sources for ectomycorrhizal fungi. *Transactions of the British Mycological Society* 91, 473–479.

Agerer, R. (1986) Studies on Ectomycorrhizae III. Mycorrhiza formed by four fungi in genera *Lactarius* and *Russula* on spruce. *Mycotoxin* 27, 1–59.

Alexander, I.J. and Watling, R. (1987) Macrofungi of Sitka spruce in Scotland. *Proceedings of the Royal Society of Edinburgh* 93B, 107–115.

Allen, S.E. (ed.) (1974) *Chemical Analysis of Ecological Materials*. Blackwell Scientific, Oxford.

Bajwa, R. and Read D.J. (1985) The biology of mycorrhiza in the Ericaceae. IX. Peptides as nitrogen sources for the ericoid endophyte and for non-mycorrhizal plants. *New Phytologist* 101, 459–467.

Carey, M.L., Maher, H., Horgan, T., Hendrik, E., Heskub K. and Griffen, E. (1986) Maintenance and enhancement of forest productivity through manipulation of the nitrogen cycle. European R and D programme in the field woods as a renewable raw material. 1982–85. Reference No. 08–012 Bos. Report to Commission of the European Communities Daiz, Brussels, 126 pp.

Carey, M.L., McCarthy, R.G. and Miller, H.G. (1988) More on nursing mixtures. *Irish Forestry* 45, 7–20.

Dickson, D.A. (1977) Nutrition of Sitka spruce on peat problems and speculations. *Irish Forestry* 34, 31–46.

Garforth, M.F. (1979) Mixtures of Sitka spruce and Lodgepole pine in South Scotland: history and future management. *Scottish Forestry* 33, 15–27.

Gronbach, E. and Berg, B. (1988) *Piceirhiza obscura*. In: *Colour Atlas of Ectomycorrhizae*, Plate 20. Einhorn-Verlag, Schwabisch Gmund.

Lange, M. and Hora, F.B. (1965) *Collins Guide to Mushrooms and Toadstools*. Collins, London and Glasgow.

Leake, J.R. and Read, D.J. (1987) Metabolism of phyto-and fungitoxic phenolic acids by the ericoid mycorrhizal fungus. In: Sylvia, D.M., Hung, L.L. and Graham, J.H. (eds), *Proceedings of the 7th North American Conference on Mycorrhizae* Institute of Food and Agricultural Sciences, University of Florida, Gainesville, Florida. p. 332.

Leake, J.R. and Read, D.J. (1989) The biology of mycorrhiza in the Ericaceae. XIII. Some characteristics of the extracelluar proteinase activity of the extracelluar proteinase activity of the ericoid endophyte *Hymenoscyphus ericae*. *New Phytologist* 112, 69–76.

Marx, D.H. (1969) The influence of ectotrophic mycorrhizal fungi on the resistance of pine roots to pathogenic infections. I. Antagonism of mycorrhizal fungi and soil bacteria *Phytopathology* 59, 153–163.

Moser, M. (1983) *Keys to Agarics and Boleti*. Roger Phillips, London.

Norkrans, B. (1950) Studies in growth and cellulolytic enzymes of *Tricholoma*. *Symbiolae Botanica Upsalensis* 11, 1–26.

Palm, M.E. and Stewart, E.L. (1984) In vitro synthesis of mycorrhizae between presumed specific and nonspecific *Pinus* and *Suillus* combinations. *Mycologia* 76(4), 579–600.

Stribley, D.P. and Read, D.J. (1980) The biology of mycorrhiza in the Ericaceae. VII. The relationship between mycorrhizal infection and the capacity to utilize simple and complex organic nitrogen sources. *New Phytologist* 86, 365–371.

Taylor, A.F.S. and Alexander, I.J. (1990) Ectomycorrhizal synthesis with *Tylospora fibrillosa*, a member of the Corticiaceae. *Mycological Research* 94, 103–107.

Taylor, C.M.A. (1985) The return of nursing mixtures. *Forestry and British Timber*, May 1985, pp 18–19.

Taylor, C.M.A. and Tabbush, P.M. (1990) Nitrogen deficiency in Sitka spruce plantations. *Forestry Commission Bulletin 89*, HMSO, London.

Trappe, J.M. (1962) Fungus associates of ectotrophic mycorrhizae. *Botanical Review* 28, 538–606.

Twining, S.S. (1984) Flourescein isothiocyanate-labelled casein assay for proteolytic enzymes. *Analytical Biochemistry* 143, 30–34.

Watling, R. (1971) *Identification of the Larger Fungi*. Hulton Educational Publications, Amersham, Bucks.

Weiss, M. and Agerer, R. (1988) Studien an Ectomykorrhizen XII. Drie nicht identifizierte Myckorrhizen an *Picea abies* (L.) karst aus einer Baumschule. *European Journal of Forest Pathology* 18, 26–43.

33 Chitin Degradation by *Hymenoscyphus ericae* and the Influence of *H. ericae* on the Growth of Ectomycorrhizal Fungi

D.T. Mitchell, M. Sweeney and A. Kennedy

Department of Botany, University College Dublin, Dublin 4, Ireland

Introduction

Chitin is a major polymer of fungi and arthropods and may be an important component in heathland ecosystems. Apart from the work of Leake and Read (1990), the ability of *Hymenoscyphus ericae* (Read) Korf and Kernan, the ericoid mycorrhizal fungus, to utilize chitin has received little attention. The biochemistry of its chitinase has not been studied. If *H. ericae* has an active chitinase, this may affect the growth of ectomycorrhizal fungi. This chapter presents preliminary studies on chitin degradation by *H. ericae* and it describes the interactions between ectomycorrhizal fungi and *H. ericae*.

Materials and Methods

A culture of *H. ericae* derived from a single ascospore was employed. The ectomycorrhizal fungi used in the paired culture studies, were Basidiomycetes isolated from excised Sitka spruce mycorrhizas and from basidiocarps collected in Avondhu Forest, Co. Cork, Ireland (Heslin *et al.*, 1992) or those used in inoculation programmes (Grogan and Mitchell, 1990). Before afforestation at Avondhu, the vegetation was dominated by *Molinia caerulea* (L.) Moench and *Calluna vulgaris* L. Hull.

Chitin utilization was determined by growing *H. ericae* at 20°C in 10 ml basal liquid medium in 50-mm Petri dishes and separated from 50 mg purified chitin by means of 47-mm sterile Gelman Supor membrane filter (0.2 μm). The inoculum consisted of a 5-mm disc taken from the growing margin of an agar culture. The liquid medium was similar to that used by Mitchell and Read (1981) excluding yeast extract and ammonium-nitrogen but with 'starter' glucose (0.07 g carbon l^{-1}). At each harvest, the mycelium and chitin were separated, oven-dried and weighed. Exogenous glucose was assayed using the glucose oxidase test (Boehringer) and glucosamine was estimated by the method of Tsuji *et al.* (1969).

Chitinase preparations were obtained from 20-day-old cultures of *H. ericae* grown in 20 ml basal medium containing either chitin (2.2 g carbon l^{-1}) and 'starter' glucose (0.1 g carbon l^{-1}) or ammonium chloride (0.32 g nitrogen l^{-1}) and glucose (2.2 g carbon l^{-1}) and incubated at 20°C. Extracellular, soluble and insoluble chitinase fractions were extracted from three bulked cultures. The culture filtrate passed through a Gelman Supor membrane filter (0.2 μm) constituted the extracellular fraction. Mycelium was homogenized in 0.05 M phosphate buffer (pH 7.2) over ice for 30 s and centrifuged twice at 27000 g for 15 min at 2°C. The supernatants were bulked and the pellet was resuspended in 0.05 M phosphate buffer (pH 7.2). Chitinase activity of each fraction was determined at a range of pH levels by incubation with purified chitin (10 mg) and 0.05 M phosphate buffer in a shaking water bath at 37°C for 22 h and release of glucosamine was assayed colorimetrically. An incubation temperature of 37°C has been used in previous studies on chitinase of fungi (Soderhall and Unestam, 1975).

Interactions of *H. ericae* with ectomycorrhizal fungi were tested in paired culture. Discs of inocula (5 mm diameter taken from the margin of cultures) were placed 3 cm apart on modified Melin Norkrans (MMN) agar medium (Marx, 1969) containing ammonium as the nitrogen source and glucose at 10 g l^{-1}. The plates were incubated at 20°C and diameter of each colony was measured across and towards the opposing culture. These results were compared with mycelial extension of single cultures.

Results

Utilization of chitin and chitinase of *Hymenoscyphus ericae*

Dry mass of the mycelium increased up to day five after inoculation with a concomitant decline in chitin below the filter (Fig. 33.1). Growth on chitin was significantly greater than that in liquid media without a nitrogen source. On day nine, the fungus had penetrated the filter and some mycelial contamination had occurred at the base of the dish. Exogenous glucose declined from 165 μg ml^{-1} to 2 μg ml^{-1} within three days after inoculation, whereas glucosamine increased from 4 to 14 μg ml^{-1} (Fig. 33.1).

Extracellular chitinase was most active at pH 6.0 and some activity was detected in the insoluble fraction (Table 33.1). There was no soluble chitinase activity at pH 7.1 (Table 33.1). Extracellular chitinase appeared to have a broad pH optimum, although the lowest pH of optimum activity was not determined (Fig. 33.2). Chitinase was also demonstrated in ammonium-grown mycelium indicating that the enzyme is not substrate inducible (Table 33.1; Fig. 33.2). When grown on chitin, a dark pigmentation appeared in the medium, which was absent, if ammonium was the nitrogen source.

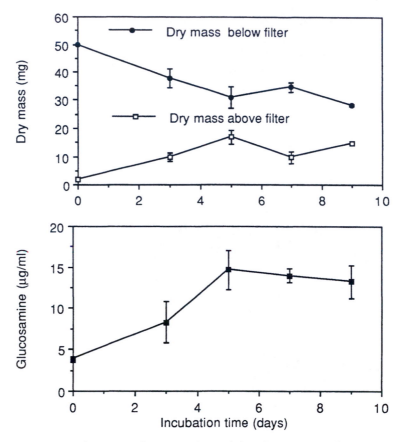

Fig. 33.1. Changes in dry mass of mycelium of *Hymenoscyphus ericae* above the filter, dry mass of chitin below the filter and glucosamine in the liquid medium.

Table 33.1. Chitinase activity of *Hymenoscyphus ericae*.

	pH	Extracellular	Insoluble	Soluble
		μg Glucosamine per culture per hour		
Chitin-grown	6.0	28.6 ± 4.7	6.8 ± 0.3	n.d.
	7.1	2.6 ± 1.5	3.6 ± 2.0	0.0
Ammonium-grown	6.0	42.4 ± 3.1	8.4 ± 1.3	n.d.

Results are means of three replicates ± s.e.m.; n.d. denotes not determined.

Antagonism by *Hymenoscyphus ericae*

An *in vitro* method was used to determine if ectomycorrhizal fungi are antagonistic to *H. ericae*. Growth of *Hebeloma crustuliniforme* S166 and *Laccaria laccata* S238A, two inoculant fungi, was inhibited against *H. ericae* (Table 33.2). Mycelial

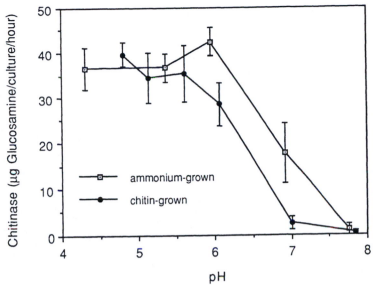

Fig. 33.2. Extracellular chitinase activity of 20-day-old cultures of *Hymenoscyphus ericae* growing in liquid media containing either chitin or ammonium-nitrogen.

extension of *H. crustuliniforme* ceased on contact, whereas *L. laccata* grew onto the margin of *H. ericae*. Other ectomycorrhizal fungi isolated from Avondhu Forest situated on a heathland site, showed no antagonism (Table 33.2). A significant stimulation of growth was observed in *Suillus grevillei* (Table 33.2) and *Paxillus involutus* and *S. grevillei* grew over *H. ericae. Hymenoscyphus ericae*

Table 33.2. Effect of species interactions on colony extension, assessed by significant (Students' *t* test, $P < 0.05$) deviation between either across or towards the opposing fungus compared with diameter of control cultures.

	Test fungus		*Hymenoscyphus ericae*		
	Towards H. ericae	Across	Towards test fungus	Across	Days of incubation
Inoculant fungi					
Hebeloma crustuliniforme (S166)	(i)	(i)	(i)	(0)	17
Laccaria laccata (S238A)	(i)	(0)	(i)	(0)	17
Basidiomycetes from Avondhu Forest					
Paxillus involutus	(0)	(0)	(0)	(0)	10
Sitka spruce 1	(0)	(0)	(0)	(0)	20
Sitka spruce 2	(0)	(0)	(0)	(0)	20
Suillus grevillei	(s)	(s)	(i)	(0)	17

(i), (s) and (0) denote significant inhibition, stimulation and no effect respectively. Results are means of five replicates.

surrounded the Sitka spruce isolates due to the fact that these Basidiomycetes are slow growing. Growth of *H. ericae* over the opposing fungus was not seen in any culture.

Discussion

The ability of *H. ericae* to utilize chitin as a nitrogen source was demonstrated by Leake and Read (1990). In the present study, chitin was both a carbon and nitrogen source. 'Starter' glucose was added but was exhausted soon after growth commenced. Chitinolytic activity was demonstrated by the decline in chitin as well as the appearance of glucosamine in the medium. Growth was depressed, when maximum glucosamine levels were detected in the medium. The inhibitory effect of glucosamine may be due to a fall in phosphate and ATP availability during the formation of the phosphate ester (Herold and Lewis, 1977), Chitinase activity was extracellular, had an acid pH of optimum activity, but was not substrate inducible.

Handley (1963) suggested that in *Calluna* heathlands, toxic compounds from *Calluna* roots may inhibit ectomycorrhizal fungi. The present study suggests that any such antagonistic effects, could be partly attributable to the effects of the chitinase of ericoid mycorrhizal fungi on hyphal growth of competitors. Wall extension of the hyphal tip appears to be finely regulated by chitin synthase and chitinase and the presence of exogenous chitinase may upset this balance (Gooday and Gow, 1990). Paired culture studies showed variations in the interactions between *H. ericae* and several ectomycorrhizal fungi. Although these experiments were not carried out under conditions comparable with those found in the field, the results indicate that there are interactions which are worthy of further examination under more realistic conditions.

References

Gooday, G.W. and Gow, N.A.R. (1990) Enzymology of tip growth in fungi. In: Heath, I.B. (ed.), *Tip Growth in Plant and Fungal Cells*. Academic Press, London, pp. 31–58.

Grogan, H. and Mitchell, D.T. (1990) The mycorrhizal status of some forest sites and the propagation of ectomycorrhizal Sitka spruce seedlings in Ireland. *Aspects of Applied Biology* 24, 123–130.

Handley, W.R.C. (1963) Mycorrhizal associations and *Calluna* heathland afforestation. *Forestry Commission Bulletin* 36, HMSO, London.

Herold, A. and Lewis, D.H. (1977) Mannose and green plants: occurrence, physiology and metabolism, and use as a tool to study the role of orthophosphate. *New Phytologist* 79, 1–40.

Heslin, M.C., Blasius, D., McElhinney, C. and Mitchell, D.T. (1992) Mycorrhizal and associated fungi of Sitka spruce in Irish forest mixed stands. *European Journal of Forest Pathology* 22, 46–57.

Leake, J.R. and Read, D.J. (1990) Chitin as a nitrogen source for mycorrhizal fungi. *Mycological Research* 94, 993–995.

Marx, D.H. (1969) The influence of ectotrophic mycorrhizal fungi on the resistance of pine roots to pathogenic infections. I. Antagonism of ectomycorrhizal fungi to root pathogenic fungi and soil bacteria. *Phytopathology* 59, 153–163.

Mitchell, D.T. and Read, D.J. (1981) Utilization of inorganic and organic phosphates by the mycorrhizal endophytes of *Vaccinium macrocarpon* and *Rhododendron ponticum*. *Transactions of the British Mycological Society* 76, 255–260.

Soderhall, K. and Unestam, T. (1975) Properties of extracellular enzymes from *Aphanomyces astaci* and their relevance in the penetration process of crayfish cuticle. *Physiologia Plantarum* 35, 140–146.

Tsuji, A., Kinoshita, T. and Hoshino, M. (1969) Analytical chemical studies on amino sugars. II. Determination of hexosamines using 3-methyl-2-benzothiazolone hydrazone hydrochloride. *Chemical and Pharmaceutical Bulletin* 17, 1505–1510.

34 Effect of Ferric Iron on the Release of Siderophores by Ericoid Mycorrhizal Fungi

B. Dobernigg and K. Haselwandter

Institut für Mikrobiologie, Universität Innsbruck, Technikerstrasse 25, A-6020 Innsbruck, Austria

Introduction

Plant communities with ericoid mycorrhiza predominate on deep organic soils where, with increasing altitude or latitude, they form the alpine dwarf shrub heath or the arctic tundra (Read, 1984). In such acid soils, ferric iron (Fe^{3+}) can be expected to be available for plant growth due to its greater solubility at low pH (Lankford, 1973). The increase in availability, however, is likely to be counter-balanced by the organic matter which is known for its metal complexing capacity (Mortensen, 1963; Emery, 1977).

Schuler and Haselwandter (1988) demonstrated that ericoid mycorrhizal fungi are capable of producing hydroxamate-type siderophores under pure culture conditions. Additionally, it has been shown that the pH as well as supplementation of the nutrient medium with L-ornithine affect the biosynthesis of siderophores (Federspiel *et al.*, 1991). It is known that the availability of iron determines the extent to which siderophores are synthesized (Neilands and Leong, 1986). Hence, it was appropriate to investigate the effect of various concentrations of ferric iron upon the release of hydroxamate siderophores by ericoid mycorrhizal fungi. The fungi were selected to cover the wide range of different ecological conditions under which ericaceous plants and, hence, the ericoid mycorrhiza become dominant.

Materials and Methods

The following ericoid mycorrhizal fungi were included in the pure culture studies:

1. the ascomycete *Hymenoscyphus ericae*, which is the perfect state of a characteristic endophyte commonly isolated from a wide range of ericaceous plants (Read, 1974, 1983);
2. the hyphomycete *Oidiodendron griseum*, which was reported by Burgeff (1961), Couture *et al.* (1983) and Dalpe (1986) to form typical ericoid mycor-

rhizas in addition to being known as a common soil-borne fungus (Domsch *et al.*, 1980);

3. the ericoid mycorrhizal endophyte of the calcicolous ericaceous plant *Rhodothamnus chamaecistus* which grows on rendzina soils of pH around 6.5 (Haselwandter and Read, 1983).

For comparitive purposes the nutrient medium and incubation conditions described by Szaniszlo *et al.* (1981) as being optimal for hydroxamate siderophore production by ectomycorrhizal fungi were employed. The fungi were sub-cultured four times in Hagem's nutrient solution (Modess, 1941) for four days at 25°C on a gyratory shaker at 200 r.p.m. The fourth sub-culture was used to inoculate a low iron medium (=LIM1 of Szaniszlo *et al.*, 1981) which was deferrated with Chelex 100 and subsequently supplemented with known amounts of ferric iron (FeCl$_3$.6 H$_2$O). The siderophore content of the culture filtrate was determined by a modified version of the *Aureobacterium* (=*Arthrobacter*) *flavescens* JG-9 bioassay with (des-)ferrioxamine B (DFOB) as a standard (Haselwandter *et al.*, 1988). This bioassay is based upon the growth stimulation of this auxotroph by the presence of hydroxamate siderophores. *A. flavescens* JG-9 is not stimulated by catechols, or other naturally occurring chelators such as citric acid, oxalic acid, and 2,3-dihydroxybenzoic acid (Powell *et al.*, 1980).

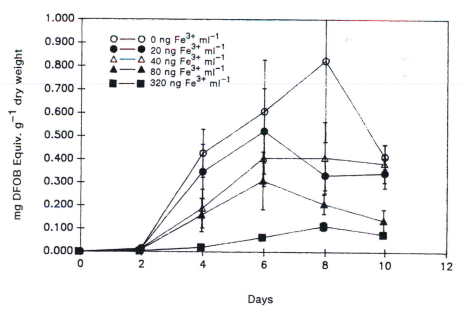

Hymenoscyphus ericae

Fig. 34.1. Time course of siderophore production by *Hymenoscyphus ericae* over a concentration range of 0–320 ng Fe^{3+} ml^{-1} nutrient solution at 25°C and 200 r.p.m. Hydroxamate siderophore concentration expressed as (des-)ferrioxamine B equivalents. Values represent means and standard deviations of nine replications.

Results

The siderophore production by *H. ericae* decreases with increasing iron concentration in the nutrient medium (Fig. 34.1). Within 6 days of incubation the concentration of DFOB equivalents fell to about 50% of the control (0 ng Fe^{3+} ml^{-1}) when the nutrient medium was supplemented with 80 ng Fe^{3+} ml^{-1}. Whereas after 8 days of incubation the effect of ferric iron upon the siderophore concentration in the culture filtrate is clearly visible, this effect is not so marked after 10 days of incubation. This effect is attributable to the siderophore content of the control treatment which was not supplemented with iron. In this case, the siderophore production decreased between 8 and 10 days of incubation by about 30%, probably due to metabolic re-utilization of the siderophores.

Different ericoid fungi differ in their response to iron as indicated by their siderophore biosynthesis. The addition of 80 ng Fe^{3+} ml^{-1} nutrient solution reduces the synthesis of DFOB equivalents to about 30% of the control. Siderophore biosynthesis by *Oidiodendron griseum* is seen to be more sensitive to iron than *H. ericae* (Fig. 34.2). The presence of 160 ng Fe^{3+} ml^{-1} in the nutrient solution led to roughly a 90% reduction in the siderophore production by *O. griseum*.

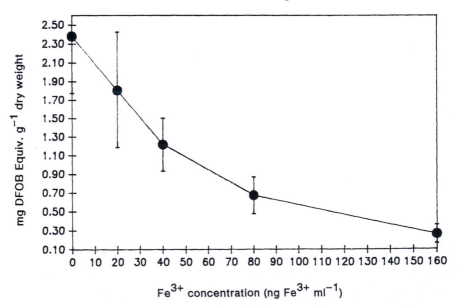

Fig. 34.2. Siderophore production by *Oidiodendron griseum* over a concentration range of 0–160 ng Fe^{3+} ml^{-1} nutrient solution within 10 days of incubation at 25°C and 200 r.p.m. Hydroxamate siderophore concentration expressed as (des-)ferrioxamine B equivalents. Values represent means and standard deviations of nine replications.

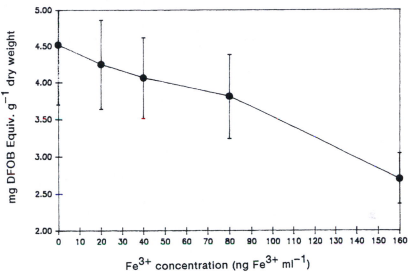

Fig. 34.3. Siderophore production by the ericoid mycorrhizal fungus of the calcicolous ericaceous plant *Rhodothamnus chamaecistus* over a concentration range of 0–160 ng Fe^{3+} ml^{-1} nutrient solution within 10 days of incubation at 25°C and 200 r.p.m. Hydroxamate siderophore concentration expressed as (des-)ferrioxamine B equivalents. Values represent means and standard deviations of nine replications.

In contrast, the endophyte of *Rhodothamnus chamaecistus* tolerated a much higher iron concentration in the nutrient medium. The presence of 80 ng Fe^{3+} ml^{-1} nutrient solution led to a reduction in the siderophore biosynthesis in the range of 10%. At an Fe concentration of 160 ng Fe^{3+} ml^{-1}, the concentration of DFOB equivalents was reduced to *c.* 60% of the control (Fig. 34.3).

Discussion

The importance of siderophores in the iron nutrition of microorganisms has been recognized for some time (Winkelmann *et al.*, 1987). Nevertheless, it is only recently that they have been implicated in the iron nutrition of plants (Reid *et al.*, 1986). A number of species of ectomycorrhizal fungi have been shown to produce hydroxamate siderophores (Szaniszlo *et al.*, 1981). There is also some evidence that *Glomus* species may produce a hydroxamate siderophore; VA mycorrhizal grass showed greater iron uptake than the non-mycorrhizal controls (Cress *et al.*, 1986). Recently, Shaw *et al.* (1990) have demonstrated that mycorrhizal roots of *Calluna* absorbed significantly more iron per unit time than those which were uninfected. It was postulated that this effect may be attributable to the potential of ericoid mycorrhizal fungi to produce siderophores. This hypothesis, however, awaits further experimental testing, in particular as the experiment was carried out at a relatively high iron concentration (120 μmol $FeCl_3.6H_2O$). An iron concentration as low as 1.44 μmol ($= 80$ ng Fe^{3+} ml^{-1}) causes an approximately 50% decrease in the siderophore production by H. ericae (Fig. 34.1) and high iron concentrations are known normally to inhibit the biosynthesis of siderophores (Lankford, 1973; Neilands and Leong, 1986). Nevertheless, the mycorrhizal fungus may play a key role in the iron nutrition of ericaceous plants through its potential to produce hydroxamate siderophores, if for example, the threshold level required for switching-off the hydroxamate siderophore biosynthesis is higher when the mycorrhizal fungus grows in association with its host plant.

The chemical structures of the main siderophores produced by ericoid mycorrhizal fungi have now been determined (Haselwandter et al., 1992). It is interesting to note that both H. ericae and O. griseum produce ferricrocin as their principal siderophore. Ferricrocin is a cyclic hexapeptide containing a tripeptide sequence of an ornithine derivative (N^5-acetyl-N^5-hydroxy-L-ornithine). In contrast, the main siderophore of the calcicolous endophyte of R. chamaecistus is fusigen (= fusarinine C) in which three fusarinine molecules are esterified to build a cyclic triester (Winkelmann, 1986, 1991). It is not clear yet, whether these differences have ecological or taxonomic implications. The feature most likely to be of ecological significance is that pH has a strong influence upon siderophore production by the different isolates. Maximum siderophore production by H. ericae, O. griseum and the endophyte of R. chamaecistus was achieved when the pH in the nutrient medium was adjusted to 4.5, 6.5 or 7.5, respectively (Federspiel 1990; Dobernigg, 1991; Federspiel et al., 1991). This suggests an adaptation of siderophore biosynthesis to the natural environment.

References

Burgeff, H. (1961) Mikrobiologie des Hochmoores. Fischer, Stuttgart.

Couture, M., Fortin, J.A. and Dalpe, Y. (1983) Oidiodendron griseum Robak: an endophyte of ericoid mycorrhiza in Vaccinium spp. New Phytologist 95, 375–380.

Cress, W.A., Johnson, G.V. and Barton, L.L. (1986) The role of endomycorrhizal fungi in iron uptake by Hilaria jamesii. Journal of Plant Nutrition 9, 547–556.

Dalpe, Y. (1986) Axenic synthesis of ericoid mycorrhiza in Vaccinium angustifolium Ait. by Oidiodendron species. New Phytologist 103, 391–396.

Dobernigg, B. (1991) Biosynthese von Hydroxamat-Siderophoren durch ericoide Mykorrhizapilze, insbesondere in Abhägigkeit von der Fe^{+++}-Konzentration. Unpublished MSc thesis, Universität Innsbruck.

Domsch, K.H., Gams, W. and Anderson, T.H. (1980) Compendium of Soil Fungi. Academic Press, New York.

Emery, T. (1977) The storage and transport of iron. In: Sigel, H. (ed.), Metal Ions in Biological Systems. Vol. 7. Iron Model and Natural Compounds. Marcel Dekker, New York, pp. 77–125.

Federspiel, A. (1990) Einfluß von pH-Wert, L-Ornithin und L-Prolin auf die Biosynthese von Hydroxamat-Siderophoren durch ericoide Mykorrhizapilze. Unpublished MSc thesis, Universität of Innsbruck.

Federspiel, A., Schuler, R. and Haselwandter, K. (1991) Effect of pH, L-ornithine and

L-proline on the hydroxamate siderophore production by *Hymenoscyphus ericae*, a typical ericoid mycorrhizal fungus. *Plant and Soil* 130, 259–261.

Haselwandter, K. and Read, D.J. (1983) Die Mykorrhizainfektion von *Rhodothamnus chamaecistus* (L.) Rchb., einer ostalpinen, calcicolen Ericaceae. *Sydowia* 36, 75–77.

Haselwandter, K., Krismer, R., Holzmann, H.P. and Reid, C.P.P (1988) Hydroxamate siderophore content of organic fertilizers. *Journal of Plant Nutrition* 11, 959–967.

Haselwandter, K., Dobernigg, B., Beck, W., Jung, G., Casier, A. and Winkelmann, G. (1992) Isolation and identification of hydroxamate siderophores of ericoid mycorrhizal fungi. *Biology of Metals* 5, 51–56.

Lankford, C.E. (1973) Bacterial assimilation of iron. *Critical Reviews in Microbiology* 2, 273–331.

Modess, O. (1941) Zur Kenntnis der Mykorrhizabildner von Kiefer und Fichte. *Symbolae Botanicae Upsalensis* 5, 1–147.

Mortensen, J.L. (1963) Complexing of metals by soil organic matter. *Soil Science Society of America Proceedings* 27, 179–186.

Neilands, J.B. and Leong, S.A. (1986) Siderophores in relation to plant growth and disease. *Annual Review of Plant Physiology* 37, 187–208.

Powell, P.E., Szaniszlo, P.J., Cline, G.R. and Reid, C.P.P. (1980) Occurrence of hydroxamate siderophores iron chelators in soils. *Nature* 287, 833–834.

Read, D.J. (1974) *Pezizella ericae* sp. nov., the perfect state of a typical mycorrhizal endophyte of Ericaceae. *Transactions of the British Mycological Society* 63, 381–383.

Read, D.J. (1983) The biology of mycorrhiza in the Ericales. *Canadian Journal of Botany* 61, 985–1004.

Read, D.J. (1984) The structure and function of the vegetative mycelium of mycorrhizal roots. In: Jennings, D.H. and Rayner, A.D.M. (eds), *The Ecology and Physiology of the Fungal Mycelium*. Cambridge University Press, Cambridge, pp. 215–240.

Reid, C.P.P., Szaniszlo, P.J. and Crowley, D.E. (1986) Siderophore involvement in plant iron nutrition. In: Swinburne, T.R. (ed.), *Iron, Siderophores and Plant Diseases*. Plenum, New York, pp. 29–42.

Schuler, R. and Haselwandter, K. (1988) Hydroxamate siderophore production by ericoid mycorrhizal fungi. *Journal of Plant Nutrition* 11, 907–913.

Shaw, G., Leake, J.R., Baker, A.J.M. and Read, D.J. (1990) The biology of mycorrhiza in the Ericaceae. XVII. The role of mycorrhizal infection in the regulation of iron uptake by ericaceous plants. *New Phytologist* 115, 251–258.

Szaniszlo, P.J., Powell, P.E., Reid, C.P.P. and Cline, G.R. (1981) Production of hydroxamate siderophore iron chelators by ectomycorrhizal fungi. *Mycologia* 73, 1158–1174.

Winkelmann, G. (1986) Iron complex products (siderophores). In: Rehm, H.-J. and Reed, G. (eds) *Biotechnology*, Vol. 4. Verlagsgesellschaft mbH, Weinheim, pp. 215–243.

Winkelmann, G. (ed.) (1991) *Handbook of Microbial Iron Chelates*. CRC Press, Boca Raton, Florida.

Winkelmann, G., van der Helm, R. and Neilands, J.B. (eds) (1987) *Iron Transport in Microbes, Plants and Animals*. Verlagsgesellschaft mbH, Weinheim.

Part Five
Mycorrhizas in Tropical Ecosystems

35 Short-term Changes in Vesicular-Arbuscular Mycorrhizal Spore Populations in *Terminalia* Plantations in Cameroon

P.A. Mason, M.O. Musoko* and F.T. Last

Institute of Terrestrial Ecology†, Bush Estate, Penicuik EH26 OQB, Midlothian, Scotland, UK

Background

The stable equilibrium between plant and soil in undisturbed tropical moist forest is greatly disrupted when the forest is opened for cultivation of agricultural crops or forest trees (Adejuwon and Ekanade, 1987). Severe methods of site preparation, especially clearfelling, can have both short- and long-term effects on the soil's physical, chemical and microbiological properties (McColl and Powers, 1984). Soil disturbance has been shown to result in a reduction or even complete elimination of VAM fungi from a range of ecosystems (Reeves *et al.*, 1979; Skujins and Allen, 1986; Jasper *et al.*, 1989). Reeves *et al.* (1979) found that VA mycorrhizal infection of colonizers in disturbed soil was 1–2% of root length, compared with 77–99% in roots of colonizers in undisturbed soils. Such a reduction in infectivity could be due to the impact of disturbance on different mycorrhizal propagules (spores, intact hyphae, mycorrhizal roots) within the soil. Ahmad (1989) for example, reported a 30–50% reduction in the number of infective VA mycorrhizal propagules (spores and root fragments) following severe disturbance of a Malaysian forest by logging.

In tropical soils low mineral nutrient availability, especially of phosphorus, is thought to be the reason why many forest trees and perennial crops are obligately mycotrophic (Janos, 1987). If disturbance during the clearance of tropical forests leads to a reduction in the level of VAM inoculum and a subsequent decrease in infection of the planted tree or crop, this may have considerable repercussions for sustainable production of tropical hardwoods and other crops.

In order to examine the impact of forest clearance on the abundance and distribution of VA mycorrhizal fungi, studies have been conducted in secondary moist deciduous forest in Cameroon where the effects of different methods of site clearance on VA mycorrhizal spore populations were assessed for the first

---

* Present address: International Institute of Tropical Agriculture, Oyo Road, PMB 5320, Ibadan, Nigeria.
† A component of the Edinburgh Centre for Tropical Forests.

year after site preparation and replanting with *Terminalia ivorensis* (Musoko, 1991).

Methods

Site description and preparation of research plots

The site was in natural forest in the Mbalmayo Forest Reserve, Cameroon (3°31′N, 11°30′E), characterized by an abundance of Sterculiaceae and Ulmaceae. Soils are highly acidic, ultisols and oxisols. Average annual rainfall is 1522 mm.

Four research plots (100 × 100 m) were set out in May 1987. Three of these were prepared for planting (see below) while one was left undisturbed.

1. *Partial clearance – manual*
All trees and shrubs below 15–20 cm diameter breast height (dbh) were cut down to knee level, using machetes and chain saws, leaving the ground vegetation intact. The resulting slash was left to decompose slowly *in situ*. This method provided sufficient light for young seedlings to be introduced (Eamus *et al.*, 1990) but left physical and chemical soil properties virtually unaffected (Ngeh, 1990).
2. *Partial clearance – mechanical*
A further planting site was cleared using a bulldozer to remove small trees and shrubs. Most trees over 25 cm dbh were retained. Although the bulldozer blade was set above ground level there was nevertheless considerable physical compaction of the soil and damage to the ground vegetation.

The slash was mostly removed and left to decompose at the edge of the plot. This treatment led to subsequent invasion by pioneer trees and weeds.
3. *Complete clearance*
Complete clearance was done by a bulldozer. All slash and logs were removed, with the exception of the two largest tree stumps, and left to decompose at the edge of the plot. Subsequent growth of vegetation was almost entirely of pioneer trees and weeds.

Planting of research plots

All the prepared plots were planted with potted seedlings of *Terminalia ivorensis* in September 1987. Although *T. ivorensis* is indigenous to W. Cameroon it does not occur naturally at Mbalmayo.

Sampling of research plots

Soil samples were collected on three occasions:

1. February 1987 – prior to site preparation in May 1987 (dry season).
2. August 1987 – 3 months after site preparation but before planting with *T. ivorensis* (rainy season).
3. August 1988 – 15 months after clearance. 11 months after planting (rainy season).

As the aim of the study was to examine how the native forest VAM flora, especially that associated with *T. superba*, was affected by site preparation and subsequent planting with *T. ivorensis*, N-S transects were established in each plot based either on naturally occurring *T. superba* trees or in 'open' areas where no trees were present: 6 transects were set up in each plot (3 with a tree at one end and 3 without). Along each 10 m transect, samples were removed to a depth of 20 cm at 2.5 m intervals starting at 2.5 m. This sampling system had to be modified during second and third collections in the cleared plot due to removal of the *T. superba* trees. At these sites for the second collection, all six transects were sampled without reference to the removed *T. superba* trees, while the third set of samples had to be collected from transects related to the planted *T. ivorensis* trees.

Extraction and assessment of spores

Spores were extracted by a centrifugation, sugar flotation method (Jenkins, 1964). Live spores only were counted and different types were distinguished and counted following Perez (1987).

Data analysis

The data presented in this chapter are the plot means. Data were analysed by analysis of variance (ANOVA) using GENSTAT. Numbers of spores were transformed to square roots and percentages converted to angles. Least significant differences (lsd) were determined at $P = 0.05$.

Results

Distribution of VA mycorrhizal fungi in the undisturbed Mbalmayo Forest

In the undisturbed forest a rich VA mycorrhizal flora was found with at least 17 different species, of which 14 have been named (Table 35.1). Twelve named species were in two genera, *Glomus* and *Acaulospora*; there were no species of *Sclerocystis*, *Gigaspora* or *Entrophospora*. In all plots *Glomus etunicatum* made up 55–68% of

Table 35.1. VA mycorrhizal fungal spores found at Mbalmayo Forest Reserve, Cameroon.

Glomus etunicatum Becker and Gerdemann
Glomus macrocarpum (Tul. and Tul.) emend Berch and Fortin
Glomus fasciculatum (Thaxter) Gerdemann and Trappe emend Walker and Koske
Glomus occultum Walker
Glomus geosporum (Nicolson and Gerdemann) Walker
Glomus rubiforme (Gerdemann and Trappe) Almeida and Schenck
Glomus clavisporum (Trappe) Almeida and Schenck
Acaulospora scrobiculata Trappe
Acaulospora spinosa Walker and Trappe
Acaulospora laevis Gerdemann and Trappe
Acaulospora mellea Spain and Schenck
Acaulospora morrowae Spain and Schenck
Scutellospora pellucida (Nicol and Schenk) Walker and Sanders
Scutellospora coralloidea (Trappe, Gerdemann and Ho) Walker and Schenck

the spores: no other fungus gave rise to more than 22% of the total spore number in any one sample.

Effect of site preparation on overall spore numbers

Before site preparation, the Mbalmayo soils possessed a homogeneous spore distribution (Table 35.2) with no significant difference between any of the plots. On average, these soils contained 252 spores $100 \, \text{g}^{-1}$ dry wt.

When spore numbers were assessed 3 months after site preparation the situation had changed considerably (Table 35.3). Although there had been a fall in the number of spores on all four plots during the rainy season, the fall was greatest in the completely cleared plot and least in the undisturbed plot. While on the

Table 35.2. Effect of site treatment on the mean number of VA mycorrhizal spores $100 \, \text{g}^{-1}$ dry wt. soil on four plots at Mbalmayo Forest, Cameroon*.

	Undisturbed forest	Partial clearance (manual)	Partial clearance (mechanical)	Complete clearance
Before clearance (February 1987)	256[a]	262[a]	218[a]	276[a]
3 months after clearance (August 1987)	190	145[a]	160[a]	96[b]
15 months after clearance, 11 months after planting (August 1988)	253	375[a]	475[a]	472[a]

*Tests of significance are omitted where replication within the control plot differed from that of treated plots.
Mean values within a sampling occasion which are followed by different letters are significantly different at $P \leq 0.05$.

Table 35.3. Percentage of two different VAM species aggregates extracted from 100 g dry weight soil samples collected from Mbalmayo Forest, Cameroon in February 1987*, August 1987† and August 1988‡.

		Undisturbed forest	Manually cleared	Mechanically cleared	Completely cleared
G. etunicatum	February 1987	62.4	68.3	55.3	65.1
	August 1987	59.7	56.2	50.7	54.8
	August 1988	64.5	59.2	34.5	29.2
G. occultum/	February 1987	17.3	17.2	19.4	18.7
A. scrobiculata	August 1987	22.0	20.2	29.7	30.8
	August 1988	19.0	24.3	42.0	46.4

*Before clearance; †3 months after clearance; ‡15 months after clearance, 11 months after planting T. ivorensis.

undisturbed plot spore number was 73% of the original number, the completely cleared plot had just 34% of its original spore number.

Fifteen months after clearance, and 11 months after planting *T. ivorensis*, the mean spore number in the three treated plots had increased dramatically (Table 35.2). Spore numbers in the completely cleared plot had risen five-fold. In contrast, in the undisturbed plot, spore numbers had increased by only 30%.

Effect on proportions of individual species

Glomus etunicatum remained the dominant fungus over the 15-month period in the natural undisturbed forest plot (Table 35.3). However, in the plots subjected to the more damaging methods of clearance (i.e. complete clearance and mechanical partial clearance) *G. etunicatum* was replaced by a spore-type containing *G. occultum* and *A. scrobiculata* spores.

Discussion

The presence of at least 17 species of VA mycorrhizal fungi indicates that Mbalmayo Forest is similar to other natural tropical ecosystems (Sieverding, 1989). This species richness is probably a reflection of the rich higher plant flora which is present – more than 200 species (most known to be VA mycorrhizal) have been recorded (Mason *et al.*, 1988). In contrast, land which is farmed with few crop species is characterized by a much lower species richness of VAM fungi (Sieverding, 1989).

In Côte d'Ivoire, Wilson *et al.* (Chapter 36, this volume) found the spores of twice as many species of VA mycorrhizal fungi (41) following forest clearance and planting as were found at Mbalmayo, although the number of species found in the undisturbed forest was similar (16). It is possible that the spores of many of the VA mycorrhizal species in the undisturbed forest occurred in clumps (Walker *et al.*, 1982) in very low numbers and thus became detectable only when disturbance, following site preparation, led to their proliferation.

In both the Cameroon and Côte d'Ivoire, spores of *Glomus* spp. dominated the forest. A lowland rainforest in Singapore was also dominated by a *Glomus* species (Louis and Lim, 1987) while Sharma *et al.* (1986) found that a *Glomus* species accounted for 75–100% of the spores in soil samples from a subtropical evergreen forest in India.

Disturbance due to site preparation led to a considerable reduction in the numbers of VA mycorrhizal spores in the three treated plots compared with that in the undisturbed forest. Moreover, the percentage loss of spores appeared to be related to the severity of the disturbance, with the decreases being greatest in the completely cleared plot (55%) and intermediate in the partially cleared plots (27–45%). Although seasonal effects may have influenced spore numbers, as indicated by the decrease in the control plot, site preparation will have led both to the active removal of spores (via removal of surface soil and erosion)

(Ahmad, 1989) and to a reduction in the viability of some of the residual spores (because of exposure to high temperatures) (Eamus *et al.*, 1990). Such a decrease at the time of planting could have a deleterious impact on the establishment and early growth of an outplanted tree or crop, especially as so many tropical plants are obligately mycotrophic. At Mbalmayo, the considerable reduction in spore numbers may have been one of the factors responsible for the much lower survival of *T. ivorensis* seedlings in the completely cleared plot (Mason *et al.*, 1988).

Eleven months after planting with *T. ivorensis* seedlings, spore numbers had dramatically increased in all three treated plots, especially in those which had received most disturbance. The major cause of these increases was the sharp rise in spore numbers of *G. occultum* and *A. scrobiculata* (Table 35.3). In the completely cleared plot, a six-fold increase in spore numbers of *G. occultum* and *A. scrobiculata* occurred between the second and third samples, leading to this fungal type becoming dominant. In sharp contrast, the manual clearance method maintained the species composition of the undisturbed natural forest, and was least damaging to the indigenous VA mycorrhizal flora. The change in species distribution on the severely disturbed plots appeared to be related to the rapid invasion of the herbaceous weed *Chromolaena odoratum* (Siam weed), with which spores of G. occultum and *A. scrobiculata* were strongly associated (Musoko, 1991).

Such a shift in dominance could be either beneficial or deleterious to the introduced *T. ivorensis* crop. A rapid build-up in spore numbers of a particular VA mycorrhizal species could be highly desirable. However, the dominant species in a particular community may not to be the most effective (Sieverding, 1989). Knowledge of the effectiveness of *G. etunicatum*, *G. occultum* and *A. scrobiculata* as symbionts would enable predictions to be made concerning the symbiotic potential of the mycorrhizal fungi at Mbalmayo particularly following mechanical and complete clearance. Although spore numbers can reflect the relative importance of individual VA mycorrhizal species within communities (Dodd *et al.*, 1990) future studies will also need to take account of the major contribution of infective hyphae, root fragments and vesicles.

Acknowledgements

This project was funded by the CEC, and the UK Overseas Development Administration. We would like to thank Dr C. Walker and Dr J. Dodd for discussions on spore identification.

References

Adejuwon, J.O. and Ekanade, O. (1987) Edaphic component of the environmental degradation resulting from the replacement of tropical rainforest by field and tree crops in S.W. Nigeria. *International Tree Crops Journal* 4, 269–282.

Ahmad, N. (1989) Mycorrhizas in relation to Malaysian forest practice: a study of infection, inoculum and host response. Unpublished PhD thesis, University of Aberdeen. Aberdeen, UK.

Eamus, D., Lawson, G.J., Leakey, R.R.B. and Mason, P.A. (1990) Enrichment planting in the Cameroon moist deciduous forest: microclimate and physiological effects. In: *Proceedings of the XIX World Congress of the International Union of Forest Research Organisations*, Montreal, Canada, pp. 258–270.

Dodd, J.C., Arias, I., Koomen, I. and Hayman, D.S. (1990) The management of populations of vesicular-arbuscular mycorrhizal fungi in acid-infertile soils of a savanna ecosystem. II. The effects of pre-crops on the spore populations of native and introduced VAM-fungi. *Plant and Soil* 122, 241–247.

Janos, D.P. (1987) VA mycorrhizas in humid tropical ecosystems. In: Safir, G.R. (ed.), *Ecophysiology of VA Mycorrhizal Plants*. CRC Press Inc., Boca Raton, Florida, pp. 107–134.

Jasper, D.A., Abbott, L.K. and Robson, A.D. (1989) Hyphae of a vesicular-arbuscular mycorrhizal fungus maintain infectivity in dry soil, except when the soil is disturbed. *New Phytologist* 112, 101–107.

Jenkins, W.R. (1964) A rapid centrifugal-flotation technique for separating nematodes from soil. *Plant Disease Reporter* 48, 692.

Louis, I. and Lim, G. (1987) Spore density and root colonization of vesicular-arbuscular mycorrhizas in tropical soil. *Transactions of the British Mycological Society* 88, 207–212.

Mason, P.A., Leakey, R.R.B., Musoko, M., Ngeh, C.P., Smith, R.I. and Sargent C. (1988) *Endomycorrhizas and Nutrient Cycling in Indigenous Hardwood Plantations in Cameroon: Effects of Different Systems of Site Preparation*. Annual report to UK Overseas Development Administration, ODA/NERC Contract No. F3 CR26 D407.

McColl, J.G. and Powers, R.F. (1984) Consequences of forest management on soil–tree relationships. In: Bowen, G.D. and Nambiar, E.K.S. (eds), *Nutrition of Plantation Forests*, pp. 380–412.

Musoko, M.O. (1991) The ecology of endomycorrhizal associations of trees in Cameroon, with special reference to *Terminalia*. Unpublished PhD thesis, University of Edinburgh.

Ngeh, C.P. (1990) The effects of land clearing methods on a tropical forest ecosystem and the growth of *Terminalia ivorensis* (A. Chev.). Unpublished PhD thesis, University of Edinburgh.

Perez, Y. (1987) *Techniques for recovering and quantifying vesicular-arbuscular mycorrhizal spores from soils*. 7th North American Conference on Mycorrhiza Taxonomy Workshop, IFAS Centre, Gainesville, Florida.

Reeves, F.B., Wagner, D., Moorman, T. and Kiel, J. (1979) The role of endomycorrhizae in revegetation practices in the semi-arid west. I. A comparison of incidence of mycorrhizae in severely disturbed *vs* natural environments. *American Journal of Botany* 66, 6–13.

Sharma, S.K., Sharma, G.D. and Mishra, R.R. (1986) Status of mycorrhizae in sub-tropical forest ecosystem of Meghalaya. *Acta Botanica Indica* 14, 878–892.

Sieverding, E. (1989) Ecology of VAM fungi in tropical agroecosystems. *Agriculture, Ecosystems and Environment* 29, 369–390.

Skujins, J. and Allen, M.F. (1986) Use of mycorrhizae for land rehabilitation. *Mircen Journal* 2, 161–176.

Walker, C., Mize, W. and McNabb, H.S. Jr (1982) Populations of endogonaceous fungi at two locations in Central Iowa. *Canadian Journal of Botany* 60, 2518–2529.

36 Long-term Changes in Vesicular-Arbuscular Mycorrhizal Spore Populations in *Terminalia* Plantations in Côte d'Ivoire

J. Wilson, K. Ingleby, P.A. Mason, K. Ibrahim and G.J. Lawson

Institute of Terrestrial Ecology*, Bush Estate, Penicuik, Midlothian EH26 0QB, UK

Background

Many ecological studies have been done in natural tropical forest ecosystems (e.g. Unesco, 1978; Sutton *et al.*, 1983; Jordan, 1985), but very few in plantations. With extensive deforestation, there is increasing interest in establishing forest plantations using indigenous tree species (Leakey, 1991). However, the silviculture techniques used to establish such plantations are very variable; failures often occur for unknown reasons. There is clearly a need to improve our understanding of the ecology of plantations and of the impacts of different silvicultural techniques. Most of all, it is necessary to understand nutrient cycling in plantations (Jordan 1985; Lugo *et al.*, 1990). Mycorrhizas have an essential role in nutrient cycling in tropical forests (which is often overlooked by foresters), and in determining other aspects of tree health. Stability or change in communities of mycorrhizal fungi may well be important in determining the performance and sustainability of obligately mycotrophic tree plantations. Furthermore, the presence or absence of VA mycorrhizal fungi may determine the species composition of the vegetation present (Miller, 1987). Yet, with the exception of (ectomycorrhizal) exotic pines and eucalypts, studies of mycorrhizas in plantations are practically non-existent apart from recent work in Cameroon (Lawson *et al.* 1990; Leakey 1990; Musoko 1991; Mason *et al.*, Chapter 35, this volume).

In Cameroon, the effects of different methods of clearing native forest on mycorrhizal spore populations were compared. Substantial changes were found in both the numbers and species composition of mycorrhizal spores during a 15-month period following site clearance, during which the sites were planted. The number of mycorrhizal spores decreased over the first 6 months after forest clearance, before replanting. Although there was a recovery in numbers by 1 year after planting, the proportions of different mycorrhizal species were changed in

---

* A component of the Edinburgh Centre for Tropical Forests.

all except the least damaging (manual) method of clearance. The mycorrhizal species which became numerous were those associated with invasive weeds rather than the *Terminalia* trees. These results indicated that changes occurred during the first few months following clearing which may have important effects on the establishment of trees and on production, and that the severity of the effects depended upon the method of clearance. However, they give no indication of the occurrence or severity of long-term effects during the period of a plantation rotation.

In Ghana and Côte d'Ivoire, *T. ivorensis* plantations are susceptible to die-back, the cause of which is unknown (Ofosu-Asiedu and Cannon, 1976). If mycorrhizal populations changed during clearance as found in Cameroon, poor mycorrhizal infection may be a contributory factor. A combination of poor mycorrhizal infection and poor growth of *Terminalia superba* in parts of Côte d'Ivoire has been observed by Blal and Gianinazzi-Pearson (this volume page 372). Distribution maps for this tree (Groulez and Wood, 1984), indicate that poor infection may have occurred when the tree was planted outside its natural range and so possibly lacked appropriate symbionts.

This chapter describes observations on the same *Terminalia* spp. as the Cameroon study and was part of a more extensive appraisal of the Mopri Forest Reserve in Côte d'Ivoire (Lawson *et al.*, 1991). Observations were made on a chronosequence of sites which were cleared and planted up to 23 years previously by contrasting silvicultural methods; seasonal effects were avoided by taking all samples at the same time of year.

Methods

Forest sites and sample collection

The Mopri Forest Reserve is located in southern Côte d'Ivoire, southwest of the town of Tissale (5° 50′ N, 5° 0′ W). The area has an average annual rainfall of 1300 mm (with long and short rainy seasons), a mean annual temperature of 27 °C and a mean annual relative humidity of 80%.

Soil samples for spore counts were taken in late October 1990 (close to the end of the short rains) from a relatively undisturbed area of natural forest and from compartments of mechanically and manually cleared forest of different ages. The mechanically cleared sites had been cleared and planted in 1979, 1985, 1989 and 1990, while the manually cleared sites had been cleared and planted in 1967, 1975, 1989 and 1990. Both 1990 cleared sites had been planted 7 months before sampling.

Methods of manual and mechanical site clearance can vary considerably. At Mopri, mechanical site clearance broadly followed the silvicultural schedule described by Dupuy (1985). The site was felled and totally cleared by bulldozer and the valuable timber extracted; debris and much of the surface organic layer were gathered into windrows and burnt; and weeds were controlled by a combination of manual, mechanical and chemical methods. Manual clearance of the 1967 and 1975 sites followed the 'manual recrû' method (Catinôt, 1965), with partial clearance of the site by hand or chainsaw, leaving many large trees in place. Valuable timber was extracted and the remainder left to decompose. Young trees were planted in the shade of the overstorey, which was progressively removed by a process of girdling and poisoning. The more recent (1989 and 1990) manually cleared sites were probably cleared more extensively than the older sites by a combination of slash and burn combined with charcoaling (Lawson *et al.*, 1991).

It had been intended to take all samples close to *Terminalia ivorensis* trees, but this was not always possible, and some samples (from the 1979 mechanical and the 1990 manually prepared sites) were instead taken near *T. superba* trees. Both species are native to this part of Côte d'Ivoire. In the cleared and planted compartments, 15 replicate samples were taken, in three sets of five. Where possible, each set of five was taken from three adjacent compartments, provided that they had received the same treatment in the same year. If this was not possible, the sets were taken from three different parts of the same compartment. Although *T. ivorensis* is indigenous to this area, it has been heavily logged and very few trees of this species could be found in the undisturbed forest, so the 'control' sample was based on only five trees.

Each soil sample was composed of four subsamples collected 2.5 m to the north, south, east and west of the trunk of the selected tree. Loose material on the soil surface was removed, and the sample was collected from 0–20 cm depth. The subsamples were then mixed and divided to provide 125 g fresh weight of soil for mycorrhizal spore counts, this was kept cool until the counts were done.

Spores were extracted from the soil using the sucrose centrifugation technique of Jenkins (1964), with modifications described in Wilson *et al.* (1990). Live and dead spores (the latter lacking cytoplasmic contents) were counted and different species were distinguished and counted. A sample of each soil was dried at 80°C so that results could be expressed on a dry weight basis.

Data analysis

Analysis of variance (ANOVA) of spore numbers was done after $\log_e$ transformation. Because there was less replication in the undisturbed forest (control) than in the other treatments, and because the time series for the manual and mechanical plantations were different, a one-way ANOVA was conducted with no block structure. Least significant differences (lsd) were determined at $P = 0.05$. With the variation in replicates between treatments, two different lsd's were obtained for each ANOVA; the lsd between cleared and replanted sites (15 replicates each) and the lsd between disturbed and undisturbed sites (15 and five replicates). The latter lsd is the larger and more rigorous; for simplicity it has been applied to all data analysis in this chapter.

Results

Numbers of spores

The total number of live and dead spores at the sites changed considerably with time after clearing and planting (Fig. 36.1). Substantial increases in numbers of spores occurred some time during the first year after site clearance and planting, because the mechanically and manually cleared plots which had been prepared in the previous year (1989) contained two to three times as many spores as occurred in the undisturbed (control) plot. Spore numbers remained high for at least 15 years following site preparation. On the manually cleared sites, with the more extended time series, the numbers of live and dead spores reverted to the preclearance values on the site which had been cleared 23 years previously (in 1967).

Mechanical and manual clearance had similar effects upon spore numbers in the chronosequence although, where comparable samples were taken, there was a tendency for numbers to be higher (although not significantly) with manual than with mechanical clearance.

Numbers of species

A total of 41 mycorrhizal spore types were characterized.

- 31 species of *Glomus*, including *G. occultum* Walker, *G. monosporum* Gerdemann and Trappe, *G. etunicatum* Becker and Gerdemann, *G. constrictum* Trappe, and five species which were previously in the genus *Sclerocystis* Berk and Broome, and have now been placed in the genus *Glomus* following the revisions of Almeida and Schenck (1990). These include *G. clavisporum* (Trappe) Almeida and Schenck and *G. sinuosum* (Gerdemann and Bakshi) Almeida and Schenck.
- 6 species of *Acaulospora*, including *A. scrobiculata* Trappe, *A. mellea* Spain and Schenck, *A spinosa* Walker and Trappe and *A. foveata* Trappe and Janos.
- 2 species of *Gigaspora*.
- 2 species of *Scutellospora* (*S. pellucida* (Nicol and Schenck) Walker and Sanders and *S. gregaria* (Schenck and Nicol) Walker and Sanders).

Five of the types found (four *Glomus* and one *Acaulospora*) are almost certainly new species (C. Walker, personal communication).

Spores of a single unidentified *Glomus* species were dominant or co-dominant at all the sites, including the control. Between 13 and 26% of spores were of this species, except on the mechanically cleared 1989 site which only had 7%. A number of other species represented 10% or more of spore numbers on some sites. On the mechanically cleared sites, all the important types were *Glomus* spp. while on the manually cleared sites and the control undisturbed forest, *Glomus* and *Acaulospora* were both important.

The mean number of species per sample and the total number of species found per site also increased after clearing and planting (Table 36.1). In the manually cleared series of plantations, which extended over a longer time period (23 years),

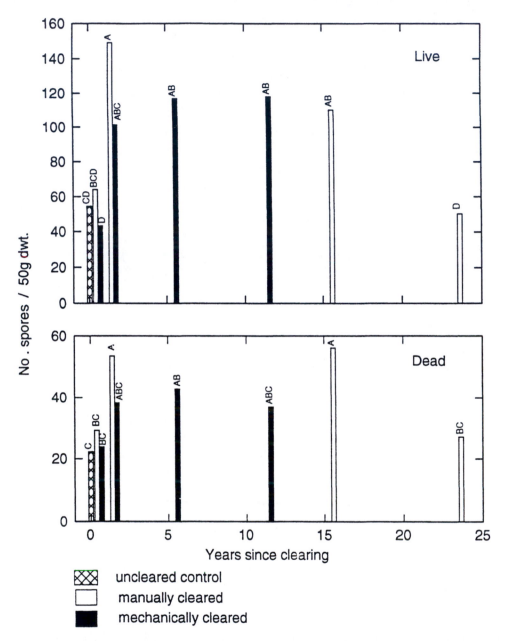

Fig. 36.1. Numbers of live and dead mycorrhizal spores in soil collected near *Terminalia* trees in an undisturbed forest control (year 0) and in manually and mechanically cleared plantations of different ages in the Mopri Forest Reserve of Côte d'Ivoire. (Bars with the same letters are not significantly different at $P = 0.05$.)

Table 36.1. Effects of clearing method, and time since clearing, on mean* and total number of spore types occurring near *Terminalia* trees in an undisturbed forest control and in manually and mechanically cleared plantations of *Terminalia* in the Mopri Forest Reserve of Côte d'Ivoire. Means with the same letter are not significantly different, $P \leq 0.05$.

Number of spore types	Control	Manually cleared and planted during				Mechanically cleared and planted during			
		1990 (0)†	1989 (1)	1975 (15)	1967 (23)	1990 (0)	1989 (1)	1985 (5)	1979 (11)
Mean	8.4 d	10.3 bcd	14.2 a	12.9 ab	9.3 cd	8.4 d	11.7 abc	14.3 a	14.5 a
Total	16	30	35	33	26	26	33	30	30

* Data from five samples (control) and 15 samples (cleared and planted plots);
† Years since clearance.

the mean number of species (like the number of live spores) increased with time, and then declined, so that the mean number 23 years after clearing and planting was not significantly different from that of the undisturbed control. However, while the mean number reverted to the preclearance value, the total number found remained high. The method of clearance did not have a significant effect on the mean number of species occurring in the 1990 and 1989 sites (when both site clearance methods were used), although the data suggest that the increase in species numbers on the mechanically prepared sites lagged behind that on the manually cleared sites.

Discussion

In Côte d'Ivoire, as in Cameroon, major changes in spore populations followed forest clearance and replanting, although there were differences at the two locations. The decrease in spore numbers which occurred 6 months after clearance in Cameroon was not observed in Côte d'Ivoire. The first post-clearance samples were taken 7 months after planting and spore numbers on the cleared sites were the same as in the undisturbed forest control. As in Cameroon, there was a significant increase in spore numbers during the year after planting, so that the 1989 plantations had significantly higher spore numbers than the 1990 plantations. The 1989 manually cleared plots also had significantly more spores than the undisturbed forest control, while on the mechanically cleared site, spore numbers were not significantly higher than the control until the third sample, 6 years after clearance. Spore numbers had reverted to pre-clearance values 23 years after planting on the manually cleared site. This was not a return to the original condition of undisturbed forest because the composition of the mycorrhizal spore population had changed – far more species were present than occurred in the original forest.

The number of spores found in the undisturbed forest in Côte d'Ivoire (109 spores 100 g^{-1} soil) was less than half that found in Cameroon, although the number of species was similar (16 vs. 17), and was high compared with other records from tropical forests. However, species composition in the two forests was very different; *G. etunicatum* spores dominated the forest in Cameroon, but were absent in the undisturbed forest in Côte d'Ivoire. At least three species in the undisturbed forest were the same as in Cameroon – *A. scrobiculata*, *A. mellea* and *A. spinosa*.

The increase in spore numbers observed in Cameroon was mostly due to a massive increase in a spore type containing *G. occultum* and *A. scrobiculata* which appeared to be associated with weeds. In Côte d'Ivoire, the increase in spore numbers was due to increases in spore numbers of several species and also to a large increase in the number of species. It is unlikely that all these were introduced following clearance; it seems more probable that many of them were present in low numbers on all sites, and proliferated after disturbance.

While site preparation had major effects upon spore numbers, the effects of different methods of site preparation were less easily differentiated than they were in Cameroon, where manual clearance reduced spore numbers less than mechanical clearance. This is possibly because the two most recently manually cleared sites in Côte d'Ivoire had been burnt, at least in part. This may have killed spores in the surface layers of soil and made these sites in some respects similar to the mechanically cleared sites. Increase in spore numbers after clearance was more rapid in the manually cleared plots, and species richness increased enormously under both methods of clearance, although the balance of species under manual clearance remained closer to that of the undisturbed forest, suggesting that manual clearance may be preferable ecologically, and perhaps silviculturally, to mechanical clearance.

Acknowledgements

This project was funded by the UK Overseas Development Administration and the Commonwealth Development Corporation, and was contracted to ITE by the International Forest Science Consultancy. Work using imported soil was carried out under DAFS licence number IP/MISC/28/90 issued under the Plant Health (Great Britain) order 1987. The authors would like to thank Dr C. Walker for discussions on spore identification.

References

Almeida, R.T. and Schenck, N.C. (1990) A revision of the genus *Sclerocystis* (Glomaceae, Glomales). *Mycologia* 82 (6), 703–714.

Catinôt, R. (1965) Sylviculture tropicale en forêt dense Africaine. *Bois et forêts des tropiques*, 100, 5–18; 101, 3–16; 102, 3–16; 103, 3–16; 104, 17–30.

Dupuy, B. (1985) *Principales regles de sylviculture pour les plantations a vocation bois d'oeuvre*. Centre Technique Forestier Tropical de Côte d'Ivoire.

Groulez, J. and Wood, P.J. (1984) *Terminalia superba*. Centre Technique Forestier Tropicale, Nogent-sur-Marne, France, and Commonwealth Forestry Institute, Oxford, England.

Jenkins, W.R. (1964) A rapid centrifugal-flotation technique for separating nematodes from soil. *Plant Disease Reporter* 48, 692.

Jordan, C.F. (1985) *Nutrient cycling in tropical forest ecosystems – principles and their application in management and conservation*. John Wiley and Sons, New York.

Lawson, G.J., Mason, P.A., Ngeh, C.P., Musoko, M., Eamus, D. and Leakey, R.R.B. (1990) Endomycorrhizal and nutrient cycling in indigenous hardwood plantations in Cameroon – effects of different systems of site preparation. Final Report to UK Overseas Development Administration.

Lawson, G.J., Wilson, J., Ingleby, K. and Mason, P.A. (1991) Mycorrhizal and chemical changes after land clearance in the Mopri Forest Reserve, Côte d'Ivoire. Report to UK Overseas Development Administration.

Leakey, R.R.B. (1990) The domestication of tropical forest trees by cloning: a strategy for increased production and for conservation. In: Werner, D. and Muller, P. (eds), *Proceedings of OECD Conference 'Fast growing trees and nitrogen fixing trees'* Gustav Fischer Verlag, Stuttgart, New York, pp. 22–31.

Leakey, R.R.B. (1991) Towards a strategy for clonal forestry: some guidelines based on experience with tropical trees. In: Jackson, J. (ed.), *Proceedings of the Tree Breeding and Improvement Symposium*. Royal Forestry Society, London, UK.

Lugo, A.E., Cuevas, E. and Sanchez, M.J. (1990) Nutrients and mass in litter and top soil of ten tropical tree plantations. *Plant and Soil* 125, 263–280.

Miller, R.M. (1987) The ecology of vesicular-arbuscular mycorrhizae in grass-and shrublands. In: Safir, G.R. (ed.), *Ecophysiology of VA mycorrhizal plants*, CRC Press Inc., Boca Raton, Florida, pp. 135–170.

Musoko, M.O. (1991) The ecology of endomycorrhizal associations of trees in Cameroon, with special reference to *Terminalia*. Unpublished PhD thesis, University of Edinburgh.

Ofosu-Asiedu, A. and Cannon, P. (1976) *Terminalia ivorensis* decline in Ghana. *Pest Articles and News Summaries* 22, 239–242.

Sutton, S.L., Whitmore, T.C. and Chadwick, A.C. (1983) *Tropical Rain Forest: Ecology and Management*. A special publication of the British Ecological Society, Blackwell Scientific Publications, Oxford.

Unesco/UNEP/FAO (1978) *Tropical forest ecosystems*. Natural resources research XIV, Unesco, Paris.

Wilson, J., Munro, R.C., Ingleby, K., Dick, J. McP., Mason, P.A., Jefwa, J., Muthoka, P.N., Newton, A.C. and Leakey, R.R.B. (1990) *Agroforestry and Mycorrhizal Research for Semi-arid Lands of East Africa*. UK Overseas Development Administration/Natural Environment Research Council contract no. R4319, Institute of Terrestrial Ecology, Edinburgh.

37 Heterogeneity and Scale in Tropical Vesicular-Arbuscular Mycorrhiza Formation

D.P. Janos

Department of Biology, University of Miami, P.O. Box 249118, Coral Gables, FL 33124 USA

Introduction

To ascertain the role of mycorrhizas in plant ecology, two general questions must be answered: (1) do mycorrhizas affect plant fitness (i.e., growth and survival, or reproduction), and (2) how probable is the level of infection required to produce fitness effects? Mycorrhizas can benefit plant growth, and improve seedling survival in a competitive milieu. For example, vesicular-arbuscular mycorrhizas (VAM) can be indispensable for survival of rainforest plants on typically phosphorus-poor tropical soils (Janos, 1980a, 1985). However, mycorrhizas have only as much time to form as seedlings can persist without them (Janos, 1980b).

The probability of mycorrhiza formation while seedlings survive ranges from inevitable to improbable. High root densities in the top five centimetres of tropical wet forest soils (Janos, 1984) and the ubiquity of VAM on rainforest plant species (Janos, 1980a) suggest that VAM fungi are always present in intact tropical forest. On the other hand, VAM formation is unlikely after relatively large-scale, succession-initiating disturbances such as volcanism (Gemma and Koske, 1990) and landslides. A 'hyphal network' conceptual model for intact vegetation (e.g. Read *et al.*, 1985) implies that seedlings inevitably form sufficient mycorrhizas to affect fitness. Alternatively, a 'habitat mosaic' model views mycorrhiza formation as patchy in space and time such that some seedlings of ecologically obligate mycotrophic species (Janos, 1984) fail to survive.

Which model most accurately depicts the probability of mycorrhiza formation by a given plant species depends upon the spatial and temporal scale of inoculum heterogeneity. Because seedlings of different species in phosphorus-poor soil attain different maximum sizes before mycorrhiza formation (Janos, 1980a), their root systems explore different volumes of soil. Thus, soil which appears to be part of a 'habitat mosaic' for a host species that explores a very limited volume of soil without mycorrhizas, would be part of an unbroken 'hyphal network' to a different species with a large pre-mycorrhizal root system. In this chapter, I examine factors that contribute to heterogeneity in VAM formation, and I discuss the applicability of these two models to lowland wet tropical habitats.

Inoculum Availability

Three kinds of propagules can produce VAM: spores, hyphae emerging from dead or dying root fragments, and 'runner' hyphae associated with live plants (Friese and Allen, 1991). Of these, spores are the most likely to be dispersed into large, bare areas from which mycorrhizal plants are absent. In intact, aseasonal vegetation spread of hyphae from root to root might be the primary means of VAM transmission.

Root Inocula

Hyphal spread of VAM infection under intact vegetation depends upon: the abundance of VAM, the relative importance in initiating new infections of hyphae emerging from roots attached to shoots versus those from detached roots, root turnover, and the longevity of infections in roots and root fragments. Runner hyphae attached to live roots can produce new VAM infections on the same or on a different plant (Friese and Allen, 1991). The latter activity by runner hyphae

Table 37.1. The percentage of seedlings of *Psidium guajava* L. exposed to different VAM inocula which had height increments >0.22 cm wk^{-1}. The total number of seedlings is indicated in parentheses. Treatments are: VAM effect = inoculated with field-collected cacao roots or receiving fumigated roots and a microbial filtrate (Janos, 1980a); Shoot effect = preceded by decapitated or intact *Virola koschnyi* Warb. saplings with or without VAM; Predecessor effect = preceded by decapitated VAM *Cassia fruticosa* Mill. (mycotrophic plant), decapitated inoculated *Phytolacca rivinoides* Kunth. & Bouche (non-mycorrhizal plant), or by uninoculated, decapitated plants of both species. nd = no data (experiment terminated).

Treatment	Weeks after treatment		
	3	5	11
VAM effect:			
Inoculated	62.5% (24)	75.0 (24)	95.7 (23)
Not inoculated	19.1 (21)	23.8 (21)	11.8 (17)
Shoot effect:			
Shoot removed	62.5 (16)	68.8 (16)	n.d.
Shoot not removed	6.3 (16)	50.0 (16)	n.d.
Not inoculated	3.1 (32)	6.3 (32)	n.d.
Predecessor effect:			
Mycotrophic plant	87.5 (16)	93.8 (16)	100.0 (16)
Non-mycorrhizal plant	37.5 (16)	6.3 (16)	81.3 (16)
Not inoculated	20.0 (15)	0.00 (15)	0.0 (15)

forms hyphal bridges among plants (Read *et al.*, 1985). Excised VAM roots and root fragments collected for use as inoculum also can produce new mycorrhizas without intervening sporulation (Powell, 1976). Here, I describe the transmission of VAM from roots attached to shoots or from detached roots, and assess the longevity of excised root inoculum in the presence of non-mycorrhizal plants.

In two bioassays of VAM transmission, I used the response to mycorrhizal infection of *Psidium guajava* L. seedlings to estimate when they had formed sufficient VAM to affect growth. Under conditions of low phosphorus availability (approximately $6 \mu g\, g^{-1}$ available P) in an acid, infertile lowland wet tropical soil, seedlings of *P. guajava* increase in height very slowly without mycorrhizas (Janos, 1980a). Table 37.1 ('VAM effect') shows that few non-inoculated, but most inoculated, seedlings exceeded an arbitrary height increment of $0.22\, cm\, wk^{-1}$.

Inoculum from living versus dying roots

In order to examine the transfer of VAM from roots attached to shoots or not attached, I transplanted *P. guajava* seedlings to the vicinities of *Virola koschnyi* Warb. saplings in pots that I had inoculated or not inoculated approximately 26 weeks previously. Inoculated *V. koschnyi* responded to VAM with increased growth (Janos, 1980a). One week after transplanting *P. guajava*, I cut off the shoots of half the inoculated and control *V. koschnyi*. None resprouted. Table 37.1 ('Shoot effect') shows that bioassay *P. guajava* seedlings in the presence of mycorrhizal *V. koschnyi* root systems from which shoots had been removed, achieved growth rates which indicated that they had mycorrhizas two weeks before those planted with intact *V. koschnyi*.

This experiment suggests that hyphae associated with dying root systems comprise a more active VAM inoculum than that derived from intact plants. This result, if confirmed, poses a problem for extractive bioassays of the VAM inoculum potential of soils. Soil collection severs roots from shoots, therefore may cause more rapid mycorrhiza formation than would occur *in situ*.

Inoculum persistence in the presence of non-mycorrhizal plants

In a similar experiment, I examined the transfer of VAM to *P. guajava* from the root systems of mycotrophic and non-mycotrophic plants. I inoculated mycotrophic *Cassia fruticosa* Mill. and non-mycotrophic *Phytolacca rivinoides* Kunth and Bouche and grew them for 16 weeks in pots. The mycotrophic species responded to VAM while the non-mycotrophic species neither formed VAM nor responded to inoculation (Janos, 1980a). I cut the shoots off all inoculated and non-inoculated control individuals of both species 1 week after transplanting *P. guajava* seedlings into the pots. Neither *C. fruticosa* nor *P. rivinoides* resprouted after shoot removal. Table 37.1 ('Predecessor Effect') reveals that within 3 weeks of being transplanted to pots that held inoculated *C. fruticosa*, most *P. guajava* seedlings had rates of height increment typical of plants with mycorrhizas. *P. guajava* seedlings in pots that had not been inoculated, and those preceded by non-mycorrhizal plants did not grow rapidly. After 11 weeks, however, *P. guajava* seedlings in pots that had contained inoculated *P. rivinoides*, showed growth rates typical of VAM plants. Sieving revealed the presence of a few spores of *Glomus mosseae* (Nicolson and Gerdemann) Gerdemann and Trappe (c. five spores 400 ml^{-1} soil) in these pots.

This experiment demonstrates that in a fumigated lowland tropical wet forest soil, excised VAM root inoculum loses most of its infectivity in 4 months under cover of a non-mycorrhizal plant species. A few spores, however, are produced and persist in a viable state for approximately 6 months. These may be upper limits for inoculum persistence, however, because soil fumigation probably killed hypha and spore predators. This experiment does not resolve whether the decline in infectivity is strictly passive and would have occurred in the absence of plants, or if non-mycorrhizal *P. rivinoides* is actively antagonistic to VAM fungi (Janos, 1985).

Spores

When viable hyphae are not present, residual or introduced spores alone comprise inocula. Spore numbers in lowland wet tropical soils can be small. This may be because few spores are produced in aseasonal climates, or because predation on spores by microbes, protozoa, and invertebrates is high when there is no adverse season to limit their populations. Published spore counts for lowland wet tropical soils range from a low of approximately 2.5 spores 100 g^{-1} soil under rainforest (top 15 cm: calculated from Redhead, 1977 assuming a bulk density of 0.8 g cm^{-3}) to 1100 spores 100 g^{-1} air-dried soil in a rubber plantation (unspecified depth: Waidyanatha, 1980). Published data are too disparate and sparse precisely to calculate a modal number of spores for lowland wet tropical soils, but it is likely to be between 50 and 500 spores 100 g^{-1} dry soil (top 10 cm). I am not aware of published counts of glomalean fungus sporocarps in tropical forests. I have found aggregated *Sclerocystis* sporocarps just three times, and did not find *Glomus* sporocarps in the litter, or on the soil surface, of more than 300 0.25-m^2 plots in wet forest in Costa Rica and Panama.

Catherine Sahley and I examined faecal samples that were collected each month for 1 year from six species of rodents of the genera *Proechimys* and *Oryzomys* in an Amazonian Peruvian rainforest. We found that 69% of samples contained spores or sporocarps of glomalean fungi, and that the presence of these was correlated with season. We estimated that on average 34 *Sclerocystis coremioides* Berk. and Broome sporocarps and 8531 *Glomus* spores are deposited in faeces m^{-2} yr^{-1}. This estimate is equivalent to 107 *Glomus* spores 100 g^{-1} soil, which accords well with the modal number of spores suggested earlier. Four of the five *Glomus* species that we found in faeces appeared to be sporocarpic.

Provided that the rodents we studied ingest sporocarps as they are encountered, our data indicate that sporulation is seasonal in an evergreen tropical forest. Moreover, it appears that sporocarps are seasonally common, but are rapidly found and ingested by rodents, which explains my failure to find sporocarps in the forest. I have also found spores and sporocarps during preliminary examination of *Oryzomys* faeces from Costa Rica.

Inoculum Heterogeneity in the Field

Although the experiments and observations described above suggest factors that may contribute to spatiotemporal heterogeneity in VAM inocula in lowland wet tropical soils, heterogeneity needs to be demonstrated in the field. I used the analytical approach taken with *Psidium guajava*, i.e., determination of the length of time before individuals of an obligately mycotrophic species manifest growth rates typical of being fully mycorrhizal, to examine spatial heterogeneity in mycorrhiza formation in two field experiments.

In the first experiment, I transplanted moribund, 6-month-old saplings of *Pithecellobium longifolium* (H. and B.) Standl. without VAM and vigorous VAM saplings of the same age into an overgrazed, sedge-filled pasture in Costa Rica. Sedges (Cyperaceae) frequently are non-mycorrhizal (Tester *et al.*, 1987), and I hypothesized that they would reduce VAM inoculum in a manner similar to *Phytolacca rivinoides*. Pre-inoculated saplings continued to grow well in the pasture, but most individuals lacking VAM remained moribund for 7 months (Table 2 in Janos, 1988), suggesting that there was very slow mycorrhiza formation in this site. After 7 months, many individuals that initially lacked VAM attained growth rates comparable to those of mycorrhizal plants. Rows of non-inoculated plants were separated from rows of pre-inoculated plants by only 1.5 m and the pasture also contained scattered adult VAM host trees. Both pre-inoculated transplants and adult trees may have been sources of VAM inocula.

In a second, similar experiment, I transplanted 7-month-old, *Inga edulis* var. *minutula* Schery saplings with VAM and saplings lacking VAM into lowland, wet tropical, seral vegetation of two ages in Costa Rica. Sites had been clear-cut, but not burned, 3 and 15 months before. The younger seral vegetation contained abundant *Phytolacca rivinoides*, but also contained mycorrhizal plants, many of which had sprouted from cut stumps. The older vegetation lacked *P. rivinoides*, which is relatively short-lived. Differences in relative growth rate between pre-inoculated plants and those initially lacking mycorrhizas persisted for 5 months in the young seral vegetation site, and for 3 months in the old (Fig. 2 in Janos, 1988). These results imply that the inferred delay of mycorrhiza formation was greatest in the site with abundant non-mycorrhizal plants.

Discussion and Conclusions

At a large scale (within habitats), the applicability of the hyphal network versus the habitat mosaic model of VAM formation depends upon disturbance regime, provided that succession-initiating disturbance eliminates VAM propagules. At an intermediate scale (near individual adult plants), the presence of non-mycorrhizal species as dominants (e.g. sedge-filled pasture), or as a vegetation component (e.g. 3-month-old seral vegetation), causes heterogeneity in mycorrhiza formation. At the small scale of the volume of soil explored by a seedling root system before mycorrhiza formation, mycorrhiza formation may be delayed, even within intact forest (e.g., 15-month-old seral vegetation).

Rates of root death probably influence heterogeneity in mycorrhiza formation within intact forest. Gradual, slow root death in response to seasonal environmental changes stimulates sporulation by VAM fungi, but catastrophic root death (e.g., as a consequence of tree fall, or of predation on roots and shoots by animals) may stimulate spread of runner hyphae. My 'shoot effect' experiment suggests that severing roots from shoots might cause VAM fungi to change from producing predominantly non-infective 'absorptive hyphal networks' (Friese and Allen, 1991) to producing runner hyphae capable of initiating new VAM. Under warm, wet tropical soil conditions, however, hyphae associated with severed roots and root fragments probably do not persist long, because high respiration rates rapidly exhaust energy stores (Perry *et al.*, 1987) or because of predation. Spores are likely to survive in a dormant state longer than detached VAM. Even if the presence of spores compensates for reduced root inocula, spore dispersal may be extremely patchy if rodents are the primary dispersers of spores, because rodents do not defaecate indiscriminately.

Deciding which model best fits different habitats is important for predicting the likelihood of seedling establishment, and for predicting competitive interactions among established seedlings (Janos, 1985). The hyphal mat model suggests that hyphal bridges are abundant. These might transport photosynthate from overstorey plants, thereby allowing shaded seedlings to be partially heterotrophic for carbon (Francis and Read, 1984). Julie Whitbeck (Stanford University, unpublished manuscript) found, however, that the total length of VAM on *Inga* saplings in rainforest understorey in Costa Rica was correlated with shoot illumination. This suggests that neither spore-derived infection networks nor runner hyphae colonize roots from which they fail to gain carbon compounds. Although the energetic cost of mycorrhizal infection to shaded seedlings might be reduced by connection to overstorey plants even if net carbon flow to shoots of shaded individuals does not occur, the need for profuse absorptive hyphal networks, which do not form bridges (Freise and Allen, 1991), may render such cost reduction insignificant. The dependence on sudden root death for enhanced runner hypha formation, patchy distribution of spores dispersed by rodents, and limited mycorrhiza formation on shaded seedlings support the habitat mosaic model for intact Neotropical rainforests.

Acknowledgements

I thank Blase Maffia for discussion of this paper. This is contribution No. 374 from the Program in Tropical Biology, Ecology, and Behavior of the Department of Biology, University of Miami.

References

Francis, R. and Read, D.J. (1984) Direct transfer of carbon between plants connected by vesicular-arbuscular mycorrhizal mycelium. *Nature* 307, 53–56.
Friese, C.F. and Allen, M.F. (1991) The spread of VA mycorrhizal fungal hyphae in the soil: inoculum types and external hyphal architecture. *Mycologia* 83, 409–418.

Gemma, J.N. and Koske, R.E. (1990) Mycorrhizae in recent volcanic substrates in Hawaii. *American Journal of Botany* 77, 1193–1200.

Janos, D.P. (1980a) Vesicular-arbuscular mycorrhizae affect lowland tropical rain forest plant growth. *Ecology* 61, 151–162.

Janos, D.P. (1980b) Mycorrhizae influence tropical succession. *Biotropica* 12 (Supplement), 56–64.

Janos, D.P. (1984) Methods for vesicular-arbuscular mycorrhiza research in the lowland wet tropics. In: Medina, E., Mooney, H.A. and Vasquez-Yanes, C. (eds), *Physiological Ecology of Plants of the Wet Tropics*. de Junk, The Hague, pp. 173–187.

Janos, D.P. (1985) Mycorrhizal fungi: agents or symptoms of tropical community composition? In: Molina, R. (ed.), *Proceedings of the Sixth North American Conference on Mycorrhizae*. Forest Research Laboratory, Corvallis, Oregon, pp. 98–103.

Janos, D.P. (1988) Mycorrhiza applications in tropical forestry: are temperate-zone approaches appropriate? In: Ng, F.S.P. (ed.), *Trees and Mycorrhiza*, Forest Research Institute, Malaysia, Kuala Lumpur, pp. 133–188.

Perry, D.A., Molina, R. and Amaranthus, M.P. (1987) Mycorrhizae, the mycorrhizosphere, and forestation: current knowledge and research needs. *Canadian Journal of Forest Research* 17, 929–940.

Powell, C.Ll. (1976) Development of mycorrhizal infections from Endogone spores and infected root segments. *Transactions of the British Mycological Society* 66, 439–445.

Read, D.J., Francis, R. and Finlay, R.D. (1985) Mycorrhizal mycelia and nutrient cycling in plant communities. In: Fitter, A.H., Atkinson, D., Read, D.J. and Usher, M.B. (eds), *Ecological Interactions in Soil*. Blackwell Scientific, Oxford, pp. 193–217.

Redhead, J.F. (1977) Endotrophic mycorrhizas in Nigeria: species of the Endogonaceae and their distribution. *Transactions of the British Mycological Society* 69, 275–280.

Tester, M., Smith, S.E. and Smith, F.A. (1987) The phenomenon of 'nonmycorrhizal' plants. *Canadian Journal of Botany* 65, 419–431.

Waidyanatha, U.P.de S. (1980) Mycorrhizae of *Hevea* and leguminous ground covers in rubber plantations. In: Mikola, P. (ed.), *Tropical Mycorrhiza Research*. Clarendon Press, Oxford, pp. 238–241.

38 Mycorrhizal Studies in Dipterocarp Forests in Indonesia

W.T.M. Smits

Tropenbos-Kalimantan, P.O. Box 319, Balikpapan, Indonesia

Introduction

Indonesia, a country consisting of more than 13 000 islands arranged along the Equator, possesses 143 million hectares of forest land. Most of this forest is dominated by trees belonging to the family of the Dipterocarpaceae. Out of this total forest area some 65 million hectares have been allocated as production forest. Most timber trees belong to the genus *Shorea* of the Dipterocarpaceae. These trees make up about 80% of the timber exported from Southeast Asia and total Dipterocarp timber export amounts to some 30% of the total tropical hardwood timber trade. Within the forest Dipterocarpaceae can make up almost 80% of the total canopy. It is therefore evident that both economically and ecologically, trees belonging to this family are of great importance.

Mycorrhizal studies in the tropics are still rare. For Africa the situation has changed recently (Högberg, 1982) and for Latin America, Lodge (1987) has brought some more information. In Southeast Asia relatively little is known about the role of mycorrhizas in the mixed Dipterocarp forests. A number of articles and student theses have appeared (Singh, 1966; Bakshi, 1974; Shamsuddin, 1979; Hong, 1979; Alwis and Abeynayake, 1980; Iskandar, 1982; Becker, 1983; Smits, 1983a,b; Jülich, 1985; Nuhamara *et al.*, 1985; Chalermpongse, 1987; Hadi, 1987; Lee, 1987; Louis and Scott, 1987; Smits *et al.*, 1987). Four decades before Singh (1966) made reference of Dipterocarps having ectomycorrhiza, Dutch foresters in Indonesia (Roosendael and Thorenaar, 1924; Voogd, 1933), noticed them. Since then most studies have been concentrating on inventories of ectomycorrhizal types on roots or the appearance of ectomycorrhizal sporocarps near Dipterocarps.

Dipterocarp Mycorrhizal Studies in Indonesia

At present there are a number of people working on Dipterocarp ectomycorrhiza in Indonesia, mainly in Yogyakarta, Bogor and Samarinda. Most of these studies are done in nurseries or in small experimental plantations of Dipterocarps outside the area of their natural occurrence. Not all results may therefore be of equal importance for an understanding of the functioning of Dipterocarp ectomycorrhizas under natural conditions. In this chapter I want to present some Dipterocarp mycorrhizal work that is presently going on in East Kalimantan.

The work is being executed within the framework of a cooperative project between the Agency for Forestry Research and Development of the Indonesian Ministry of Forestry, the state forestry enterprises P.T. Inhutani I and P.T. Inhutani II, and the Tropenbos programme from the Netherlands. The aim of this project is a very practical one, to contribute to the wise utilization and conservation of the tropical rain forests. One of the goals for the East Kalimantan site is the production of good quality Dipterocarp planting stock. Most of the work concentrates on techniques for vegetative propagation and hedge orchards for Dipterocarps, but part involves practical work on Dipterocarp ectomycorrhiza. Basic scientific research in Indonesia is seen as the responsibility of Universities, while the Agency for Forestry Research and Development has to concentrate on applied research to support forestry practice in Indonesia. Facilities, equipment and financial means to do more detailed mycorrhizal work are therefore not yet available.

Practical Aspects

Management of mixed Dipterocarp forest after selective harvesting of timber has been problematic. One of the problems has been the provision of planting stock. The need for suitable ectomycorrhiza has proved to be a key factor for successful production and survival of this planting stock.

Practical techniques to produce ectomycorrhizal Dipterocarp planting stock have been developed and applied on large scale in Indonesia (Smits, 1986; Smits *et al.*, 1990). An estimated 30 million young Dipterocarps have now been produced in Indonesia alone (Smits, unpublished). For the production of mycorrhizal planting stock, two methods are used. The first one is based upon the use of Dipterocarp wildlings from natural forest. These seedlings already have been infected with unknown ectomycorrhizal fungi in the forest. A disadvantage of this technique is that the seedlings are only available for limited periods of time due to the irregular flowering and fruiting of Dipterocarps, in so-called mast flowering years that normally occur once every 3–4 years after a pronounced dry period. Seeds that are produced during such rare events cannot be stored for prolonged periods of time except for two *Dipterocarpus* species from the deciduous mixed Dipterocarp forest. After one year, due to natural mortality, the number of available seedlings decreases considerably and their size and consequent survival in nurseries becomes very unfavourable. It becomes economically very unattractive to use these plants for reforestation purposes. Therefore it was necessary to develop an alternative method for Dipterocarp planting stock production. This method is based upon vegetative propagation. Numerous practical problems had to be resolved but now this method can be used, and is being applied, for large scale operations involving the production of Dipterocarp planting stock. The stem cuttings produced are non-mycorrhizal and need to be inoculated at the time of transplanting to plastic containers. This can be done with several practical methods but the one now commonly used is to apply a small amount of soil inoculum in the transplanting hole of the rooted cutting, in close contact with the roots. This soil inoculum can be collected underneath a mother tree of the species that is being transplanted but this has several disadvantages. First of all the method is relatively expensive and laborious and demands that the labour collecting the inoculum is capable of recognizing the correct trees. Secondly the soil collected does not necessarily contain the right fungus. The risk of carrying root pathogens to the nursery, often mentioned as another disadvantage, has not been found to occur. Thirdly the inoculum potential of this soil inoculum rapidly decreases and in practice less than 50% of the inoculated plants form ectomycorrhizas when the inoculum is stored for 8 days or more, varying somewhat with the storage conditions. This means that collecting fresh soil inoculum has to be done frequently and in small quantities, making it even more costly.

Another method uses soil inoculum obtained from plants specially grown for this purpose. First a large number of wildlings of the species to be propagated are grown in relatively poor soil, comparable to the soil condition near the mother tree. Then the wildings showing best growth are selected. These plants tend to have the same type of ectomycorrhizas. After 8 weeks roots and ectomycorrhizal hyphae have colonized all of the soil and the seedlings and their adhering soil are removed from the plastic containers. The soil is carefully removed from the roots and the seedlings are transplanted to new plastic containers filled with fresh soil, without ectomycorrhizal spores, and grown for another 8 weeks before this procedure is repeated. When the vigour of these plants becomes less, some of the inoculated cuttings can be used for inoculum production. The soil that is removed from the seedling root system, containing many little root and ectomycorrhizal pieces and hyphae, is thoroughly mixed and used to inoculate the transplanted cuttings. One seedling thus yields fresh inoculum for 100–150 cuttings. More than 90% of the inoculated cuttings develop the same type of ectomycorrhizas within 2–3 weeks. The cost of this procedure is very low and viable inoculum is available at all times. The few plants that do not develop ectomycorrhizas within 2–3 weeks tend to get infected later from the mycorrhizal hyphae growing from one container to another on the plant bed. Normally these plants will be somewhat smaller after 6 months, at which time the infected cuttings are ready for planting in the field. These smaller cuttings are then kept an additional 4–6 weeks in the nursery to reach a plantable size.

With the above system the identity of the ectomycorrhizal fungus is not known, but growth of at least young Dipterocarp trees is good and has enabled production of Dipterocarp planting stock.

Practical research for supporting this production system involves studies on optimal storage conditions of soil inoculum, and on heating of soil to exclude unwanted ectomycorrhizal inoculum from the medium to be used for transplanting the cuttings, on the influence of pH, light, temperature, soil compaction and fertilizer upon development and performance of the ectomycorrhizas (Noor and Smits, 1987). When too much (more than 50%) peat was used in the medium for transplanting the cuttings many *Thelephora terrestris* ectomycorrhizas developed (Smits and Noor, 1987). Growth of *Hopea nervosa* cuttings infected with *Thelephora terrestris* was slower than cuttings having a white ectomycorrhiza probably formed by *Scleroderma* cf. *lycoperdoides*. After outplanting in light shade the plants with the *Thelephora terrestris* mycorrhizas became yellowish and stunted and lost their mycorrhizas. These practical results show the importance of the correct choice of Dipterocarp ectomycorrhizal fungi in the nursery. Fertilizer application after the establishment of abundant ectomycorrhiza on cutting root systems has no negative effect. Exposure to full sunlight results in heating of the soil, kills the ectomycorrhizas and is followed by stunted growth of the plants. Heating soil with sunlight is used to prepare medium free of ectomycorrhizal inoculum. Heating slightly moist soil spread in layers 5–10 cm thick and covered with agricultural plastic results in a temperature above 50°C. Soil treated in this way for more than 1 hour never infected non-mycorrhizal Dipterocarp material.

The basic approach of this practical type of research is very simple. Empirical observations are used to select the best procedures. Identity of mycorrhizal types is seldom known and selection of other mycorrhizal fungi may result in better performance of the young Dipterocarps. Although simple, and perhaps not optimal, this modest production system has enabled foresters in Indonesia to produce large numbers of inoculated young Dipterocarps that are widely used in reforestation. Basic research into many aspects of Dipterocarp mycorrhizas can now be used to improve on the present production method and to contribute to a better understanding of the role of mycorrhizas in species-rich tropical rain forest. Some of the ongoing research is presented below.

Basic Studies Concerning Dipterocarp Mycorrhiza

At the Tropenbos-Kalimantan site, at the Wanariset station in East Kalimantan, permanent inventory plots have been established in natural undisturbed forest. A number of variables, including the position of trees, fruit bodies and litter depth, have been recorded over a 5-year period and entered in a database. Other files can be imported for analyses. One of these contains data on the extent of root systems of different tree species of different diameters. This type of data is very hard to obtain, especially in the very species-rich rain forest in East Kalimantan. In the course of this work it proved possible to recognize many tree families and even genera from the morphological appearance of their roots, e.g. Guttiferae, Myristicaceae and Sapotaceae. The tree root system diameters are correlated with tree and tree crown diameters. To estimate the root system diameter of a tree the stem diameter is used to predict a value. Another file contains information on the mycorrhizal status of tree species. Tree roots are sampled, cleared and stained and investigated for presence of mycorrhiza. Some results of this work are presented in Table 38.1. In general most of these confirm the mycorrhizal status already reported in other publications. All Dipterocarps investigated in East Kalimantan possessed ectomycorrhiza of the classical type with a Hartig net of only one cell layer deep. Roots were always ectomycorrhizal and this proved one of the most practical ways to identify Dipterocarp roots. The records of ectomycorrhiza in the

Table 38.1. Mycorrhizal status of tree species at Wanariset, East Kalimantan, VAM = VA mycorrhizal, ECM = ectomycorrhizal, Non = non-mycorrhizal.

Family	Species	Mycorrhiza
Annonaceae	*Polyalthia sumatrana*	VAM
Bombacaceae	*Durio griffithii*	ECM?
Burseraceae	*Canarium pilosum* Benn. spp. *pilosum*	VAM
	Dacryodes rostrata	VAM
	Dacryodes sp.	VAM
Celastraceae	*Bhesa* sp.	ECM
Compositae	*Vernonia arborea*	VAM
Dipterocarpaceae	*Anisoptera marginata*	ECM
	Dipterocarpus confertus	ECM
	Dipterocarpus cornutus	ECM
	Dipterocarpus elongatus	ECM
	Dipterocarpus gracilis	ECM
	Dipterocarpus grandiflorus	ECM
	Dipterocarpus hasseltii	ECM
	Dipterocarpus humeratus	ECM
	Dipterocarpus tempehes	ECM
	Dryobalanops aromatica	ECM
	Dryobalanops keithii	ECM
	Dryobalanops lanceolata	ECM
	Hopea dryobalanoides	ECM
	Hopea mengerawan	ECM
	Hopea nervosa	ECM
	Shorea assamica	ECM
	Shorea faguetiana	ECM
	Shorea johorensis	ECM
	Shorea laevis	ECM
	Shorea lamellata	ECM
	Shorea leprosula	ECM
	Shorea ovalis	ECM
	Shorea parvifolia	ECM
	Shorea pauciflora	ECM
	Shorea pinanga	ECM
	Shorea polyandra	ECM
	Shorea smithiana	ECM
	Shorea stenoptera	ECM
	Vatica chartacea	ECM
	Vatica rassac	ECM
	Vatica sp. 1	ECM
	Vatica umbonata	ECM
Ebenaceae	*Diospyros* sp.	VAM
Euphorbiaceae	*Aporusa frutescens*	Non
	Aporusa sp.	VAM
	Baccaurea pyriformis	VAM
	Baccaurea racemosa	VAM
	Baccaurea sp.	VAM
	Breynia sp.	VAM
	Chaetocarpus castanocarpus	Non

Table 38.1. *cont.*

Family	Species	Mycorrhiza
	Fahrenheitia pendula	VAM
	Macaranga gigantea	VAM
	Macaranga lowii	Non
	Macaranga motleyana	VAM
	Mallotus penangensis	VAM
	Pimelodrendron griffithianum	?
	Trigonostemon laevigatus	VAM
Fagaceae	*Lithocarpus* sp.	ECM
Flacourtiaceae	*Hydnocarpus* sp.	VAM
Guttiferae	*Calophyllum* sp.	VAM
	Garcinia parvifolia (Miq.) Miq.	VAM
Lauraceae	*Alseodaphne* sp.	VAM
	Eusideroxylon zwageri	VAM
	Litsea sp.	VAM
Lecythidaceae	*Barringtonia macrostachya*	VAM
	Barringtonia sp.	VAM
Leguminosae	*Archidendron* sp.	VAM
	Crudia sp.	ECM
	Pithecellobium cf. *splendens*	VAM
Melastomataceae	*Pternandra* sp.	VAM
Meliaceae	*Aglaia* sp.	VAM
Moraceae	*Artocarpus dadah*	Non
	Artocarpus nitida spp. *griffithii*	VAM
	Ficus sp.	VAM
Myristicaceae	*Horsfieldia grandis* (Hook. f.) Warb.	VAM
	Horsfieldia reticulata Warb.	VAM
	Knema cf. *furfuracea*	VAM
	Knema latericia Elm ssp. *albifolia*	VAM
	Knema latifolia Warb.	VAM
	Knema latifolia Warb.	VAM/ECM
	Knema laurina (Bl.) Warb. var. *laurina*	VAM
	Myristica iners Bl.	VAM
	Myristica maxima Warb.	VAM
	Myristica villosa Warb.	Non
Myrtaceae	*Eugenia* sp.	VAM
	Eugenia sp. 2	VAM
	Rhodamnia cinerea	VAM
Polygalaceae	*Xanthophyllum* sp.	VAM
Rubiaceae	*Gardenia* sp.	VAM
	Ixora sp.	VAM
	Porterandia anisophyllea	VAM
	Urophyllum sp. 1	Non
	Urophyllum sp. 2	Non
Rutaceae	*Euodia* sp.	VAM
Sapotaceae	*Ganua pallida*	VAM
	Madhuca cf. *sericea (Miq.)*	VAM
	Palaqium sp.	VAM
	Palaquium sp.	VAM

Table 38.1. *cont.*

Family	Species	Mycorrhiza
	Payena ludica	VAM
Sterculiaceae	*Sterculia rubiginosa*	VAM
Theaceae	*Schima wallichii*	VAM
	cf. *Pyremaria sp.*	VAM
Thymelaeaceae	*Aquilaria malaccensis*	?
Ulmaceae	*Gironniera sp.*	ECM
	Trema orientalis	VAM

Table 38.2. Numbers of potential phytobionts per mycobiont for the ectomycorrhizal fungi associated with eight Dipterocarp species.

	Number of potential associated phytobionts		
	1	2	3
Number of mycobionts	24	13	3
Percentage of mycobionts (%)	60	33	7

Celastraceae and one *Durio* species (Bombacaceae) are new although the latter needs to be confirmed. Mycorrhizas were consistently absent in *Urophyllum* spp. (Rubiaceae).

From analysis of the database a number of findings were obtained. For example most mycobionts were found associated with only one or two host *Dipterocarp* spp. (Table 38.2). Members of the Boletales appeared to show a marked preference for areas where organic matter accumulated.

There was an increasing relative importance of ectomycorrhizal trees in the higher diameter classes (Fig. 38.1). This was not only because of the dominance of Dipterocarps, whose relative importance among the ectomycorrhizal trees actually decreased toward the higher diameter classes (Fig. 38.2). It will be interesting to see whether this is a general trend that can be related to the better nutrient efficiency of the ectomycorrhizal association of these tree species on nutrient poor soils. Dipterocarps made up 35% of the total basal area in the plot. In view of their better developed crowns their root system will therefore also be relatively more important, probably amounting to more than 50% of the roots in the soil. The mixed Dipterocarp forests are therefore ectomycorrhizal forests.

A number of long lasting experiments has been started to investigate the performance of different Dipterocarp species inoculated with different ectomycorrhizal fungi and planted on various geomorphological units under different light conditions. These plots make up some 200 ha and already, after $1\frac{1}{2}$ years, some strong differences in field performance are apparent. *Shorea leprosula* inoculated with *Scleroderma lycoperdoides* planted on well draining soil in 50% light intensity is performing best, showing an average height growth $>1.5\,\mathrm{m\,yr^{-1}}$ and diameter increment $>2\,\mathrm{cm\,yr^{-1}}$. Individual trees of this species, 5 years old,

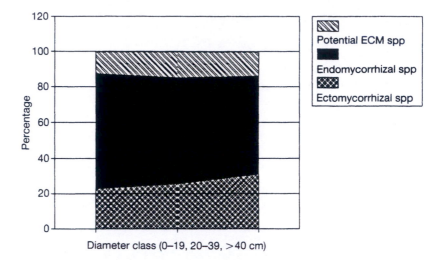

Fig. 38.1. Percentages of ectomycorrhizal, potential ectomycorrhizal and VA mycorrhizal tree species in three diameter classes.

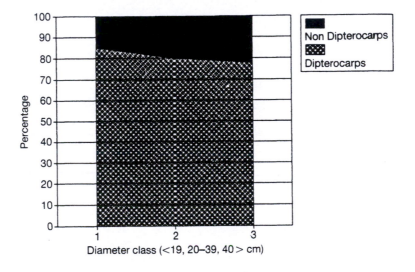

Fig. 38.2. Percentage of Dipterocarps among the ectomycorrhizal tree species in three diameter classes.

planted under the same conditions are growing at >3 cm diameter yr^{-1}, which is a number of times faster than previously assumed possible for Dipterocarps. The same host/fungus combination, but growing on flat wet terrain, is performing very poorly with less than 50 cm height growth per year. More of these trials are under way.

Other investigations conducted in the nursery and in the greenhouse involve the effect of different mycorrhizal fungi upon leaf morphology of Dipterocarps, the performance of seedlings of different mother trees of one Dipterocarp species as well as clones, inoculated with the same ectomycorrhizal fungus, and the morphology of ectomycorrhiza of one ectomycorrhizal fungus with different Dipterocarp species.

Conclusions

Many interesting aspects of Dipterocarp ectomycorrhiza remain to be investigated. Results so far show that Dipterocarps are obligately ectomycorrhizal and that large variation in host growth occurs with different ectomycorrhizal fungi. The data from the research in the natural Dipterocarp forest in East Kalimantan show that suprisingly few ectomycorrhizal fungi are shared between large Dipterocarp host species.

References

Alwis, de D.P. and Abeynayake, K. (1980) A survey of mycorrhizae in some forests of Sri Lanka. In: Mikola, P. (ed.), *Tropical Mycorrhiza Research*. Clarendon Press, Oxford. pp. 146–153.

Bakshi, B.K. (1974) *Mycorrhizae and its Role in Forestry*. P.L. 480 Project Report. Dehra Dun, 89 pp.

Becker, P. (1983) Ectomycorrhizae on *Shorea* (Dipterocarpaceae) seedlings in a lowland Malaysian rain forest. *Malaysian Forester* 46, 146–170.

Chalermpongse, A. (1987) Mycorrhizal survey of dry deciduous and semi evergreen Dipterocarp forest ecosystems in Thailand. In: Kostermans, A.J.G.H. (ed.), *Proceedings of the Third Round Table Conference on Dipterocarps*. pp. 81–103.

Hadi, S. (1987) The association of fungi with Dipterocarps. In: Kostermans, A.J.G.H. (ed.), *Proceedings of the Third Round Table Conference on Dipterocarps*. pp. 73–79.

Högberg, P. (1982) Mycorrhizal associations in some woodland and forest trees and shrubs in Tanzania. *New Phytologist* 97, 591–599.

Hong, L.T. (1979) A note on dipterocarp mycorrhizal fungi. *Malaysian Forester* 42, 280–283.

Iskandar, E. (1982) Ektomikoriza dari Dipterocarpaceae. Thesis Fakultas Kehutanan, Universitas Mulawarman, Samarinda, Indonesia, 64 pp.

Jülich, W. (1985) Fungi associated with Dipterocarpaceae in southeast Asia. I. The genera Riessia and Riessiella. *International Journal of Mycology and Lichenology* 2(1), 123–140.

Lee, S.S. (1987) The ectomycorrhizas of *Shorea leprosula*. In: Ng, F.S.P. (ed.), *'Trees and Mycorrhiza' Proceedings of the Asian Seminar 13–17 April 1987 Kuala Lumpur*. pp. 189–206.

Lodge, D.J. (1987) Resurvey of mycorrhizal associations in the El Verde rain forest Puerto Rico. In: Sylvia, D.M., Hung L.L. and Graham, J.H. (eds), *Mycorrhizae in the Next Decade*. University of Florida, Gainesville, p. 127.

Louis, I. and Scott, E. (1987) In vitro synthesis of mycorrhizae in root organ cultures of a tropical dipterocarp species. *Transactions of the British Mycological Society* 88(4), 565–568.

Noor, M. and Smits, W.T.M. (1987) Pengaruh intensitas cahaya dan suhu tanah terhadap ektomikoriza dan pertumbuhan anakan Shorea cf. polyandra. In: Soemarna, K. *et al.*

(eds), *Prosiding Simposium Hasil Penelitian Silvikultur Dipterocarpaceae.* pp. 11–31.

Nuhamara, S.T., Hadi, S., De La Cruz, R.E. and Hendromono A.Q. (1985) *Growth of some Dipterocarp Seedlings in Different Soil Types and Inoculated with Different Ectomycorrhizal Fungi.* BIOTROP publication. 17 pp.

Roosendael van, J. and Thorenaar, A. (1924) De natuurlijke verjonging van Ngerawan (*Hopea mengarawan* Miq.) in Zuid-Sumatera. *Tectona* Dl. XVI, 519–567.

Shamsuddin, M.N. (1979) Mycorrhizas of tropical forest trees. In: Furtado, J.I. (ed.), *Abstracts V International Symposium of Tropical Ecology,* Kuala Lumpur, Malaysia, 16–21 April 1979. Univ. Malaya, Kuala Lumpur. p. 173.

Singh, K.G. (1966). Ectotrophic mycorrhiza in equatorial rain forest. *Malaysian Forester* 29(1), 13–19.

Smits, W.T.M. (1983a) Dipterocarps and mycorrhiza, an ecological adaptation and a factor in forest regeneration. *Flora Malesiana Bulletin* 36, 3926–3937.

Smits, W.T.M. (1983b) Kepentingan mycorrhiza untuk Dipterocarpaceae. *Duta Rimba* No. 65–66/IX/1983, 16–17.

Smits, W.T.M. (1986) Pedoman sistem cabutan bibit Dipterocarpaceae. In: Priasukmana, S. dan Tangketaasik, J. (eds), *Edisi khusus* No 2. 37 pp.

Smits, W.T.M. and Noor, M. (1987) Pemakaian pot-tray dengan gambut untuk penyapihan stek *Hopea* sp. In: Soemarna, K. *et al.* (eds), *Prosiding Simposium Hasil Penelitian Silvikultur Dipterocarpaceae,* p. 31–44.

Smits, W.T.M., Oldeman, R.A.A and Limonard, T. (1987) Mycorrhizae and Dipterocarpaceae in East-Kalimantan rain forests. *WOTRO-report for the year 1986,* pp. 67–78.

Smits, W.T.M., Leppe, D., Yasman, I and Noor, M. (1990) Ecological approaches to commercial Dipterocarp forestry. Keynote address, *International Conference on Forest Biology and Conservation in Borneo.* Kota Kinabalu, July 30–August 3, 16 pp.

Voogd, de, C.N.A. (1933) Cultuurproeven met *Shorea platyclados* v.Sl. in Redjang en Lebong. *Tectona* XXVI(9) 703–713.

39 Controlled Mycorrhization of Eucalypts

F. Lapeyrie[1], J. Garbaye[1], V. de Oliveira[2] and M. Bellei[2]

[1]INRA, Centre de Recherches Forestières de Nancy, 54280 Champenoux, France: [2]Universidade Federal de Santa Catarina, 88049 Florianopolis SC, Brazil

Introduction

Many species of the genus *Eucalyptus* grow quickly and produce large quantities of wood when grown in well-managed plantations, both within and outside their natural range. This ability is of great value in Mediterranean, tropical or subtropical areas, as demand for wood becomes greater. Furthermore many *Eucalyptus* species can grow on sites of very low nutrient status, especially those deficient in nitrogen and phosphorus.

Controlled mycorrhizal inoculation of *Eucalyptus* seedlings with selected ectomycorrhizal fungi, before plantation, could become a very successful practice to improve productivity. Research on *Eucalyptus* mycorrhizas has a long history beginning in Australia and then spreading in other countries, as the demand for *Eucalyptus* plantations increased. Eucalypt appears to be a good model to study plant–fungus symbiotic interactions, and therefore it is one of the few examples where applied and basic mycorrhizal research are carried out with the same tree.

Some of this work will be reviewed here focusing on field results, plant fungus specificity and VA/ectomycorrhiza co-infection on the same root system.

Field Results in *Eucalyptus* Industrial Plantations

In Australia

Large growth response (+40–80%) of indigenous Eucalypts to ectomycorrhizal inoculation with local fungi has been recorded under a wet mediterranean climate in South Western Australia (Malajczuk, personal communication).

In Congo

Hybrid *Eucalyptus* are planted extensively in the coastal region of the Congo near Pointe-Noire (Delwaulle *et al.*, 1982). In spite of the very poor soils, productivity is good (about $30\ m^3\ ha^{-1}\ yr^{-1}$ of wood) because of a favourable climate (sub-equatorial with a 3–4 month dry season) and the use of genetically improved plant material with high growth potential. As a consequence, the main limiting factor for a further increase in productivity is mineral nutrition. Experiments are presently being carried out to improve this by fertilization and/or optimization of the mycorrhizal symbioses (Garbaye et al., 1988).

Pisolithus tinctorius (strain MARX-270), when compared with the native mycorrhiza, stimulated the growth of E. urophylla × E. kirtoniana, first at the nursery stage, then 20, 27, and 50 months after plantation. The most important effect was recorded at 27 months, with 61% extra wood production per hectare, but the extra wood production at 50 months was only 29%. Comparing these results with mycorrhizal infection at 25 and 50 months suggests that although this Pisolithus tinctorius strain is very efficient in growth promotion it is a poor competitor, and it was replaced by a eucalypt-specific Scleroderma species after outplanting in the field. Therefore the aim of further field experiments will be to look for new fungal strains with high growth-promoting ability but that are better adapted to the local conditions or to Eucalyptus as a host plant, in order to have a long-lasting benefit from controlled inoculation.

In the Philippines

Impressive field results have recently been published by De La Cruz and collaborators in the Philippines (De La Cruz and Lorilla, 1989; De La Cruz, 1990): in 3-year-old experimental E. deglupta plantations on industrial sites, inoculated with Pisolithus tinctorius, the extra wood-production can reach 154%. The inoculated trees were not significantly taller but their diameter was 51% larger. Fertilizing intensive plantations is more and more being considered, and selected mycorrhizal fungi can be either seen as 'fertilizer savers' or as efficient 'fertilizer users'. Indeed, studies were conducted to determine the fertilizer replacement ability of mycorrhiza under field conditions, comparing the wood production of non-inoculated and inoculated plots with increasing levels of a complete fertilizer. Mycorrhiza inoculation replaced about 86% of the fertilizers required by uninoculated E. deglupta to reach maximum production. If fertilizers are readily available at a reasonable cost, then, the same experimental results show that even at high fertility levels the inoculated plots make a much better use of the fertilizers added.

Infectivity of Pisolithus tinctorius Isolates toward Eucalyptus Roots

Aggressiveness comparison in vitro

As mentioned earlier, an otherwise efficient fungal strain can be used successfully in plantations only if it is able to remain associated with the host plant till the end of the rotation. The experiment set up in Congo shows that even on short rotation this might be a critical point. In vitro studies have demonstrated that isolates of

Pisolithus tinctorius, a broad host-range species, vary in their ability to colonize and infect roots of tree species (Malajczuk *et al.*, 1982, 1990). A strain originally isolated from under *Eucalyptus* in Australia infected *Eucalyptus* much more rapidly and aggressively than the strain used in field experiments in Congo, isolated from under *Pinus* in North America. Such specific aggressiveness, assessed either using the paper-sandwich technique (Chilvers *et al.*, 1986) or by the simple placement of pregerminated sterile seedlings on aerial hyphae in a Petri dish, could be related to plant–fungus co-evolution in their native ecosystem.

Ultrastructural comparison of pre-infection stages

A comparison of the early pre-infection stages between *Eucalyptus urophylla* and strains of *Pisolithus tinctorius*, using transmission electron microscopy, shows some important differences at the root interface within both the host and the fungal tissue (Lapeyrie *et al.*, 1989; Lei *et al.*, 1990).

Two days after inoculation with the *Eucalyptus* strain, hyphae were in contact with cortical cells and polysaccharide fibrillar material, reacting to the PATAg test, could be observed between the two organisms. These fibrils reacted positively to the Swift test indicating the presence of cystine-rich proteins. Four days after inoculation, the fungal sheath was established and the typical Hartig net was developed. Acidic phosphatase activity was in evidence at the surface of the fungal plasmalemma as well as in the vacuoles. The fungal plasmalemma activity was detected when the hyphae were close to the root while it was absent in the hyphae away from the root.

With the pine strain, most of the hyphae were still at some distance from the root surface after 4 days. Whenever the hyphae came in contact with the root, an internal deposition of dense granular or fibrillar polysaccharide materials was observed on the root cell wall. No fibrillar bridge could be detected between root and fungal cells and acidic phosphatase activity was detected only in fungal vacuoles. However, this pine strain was known to be a good colonizer of pine and a typical fibrillar bridge was present in a pine root–fungus interface four days after inoculation (Lei *et al.*, 1989). The appearance of such fibrils and the induction of an acidic phosphatase activity along fungal plasmalemma, seem to depend on the interaction between the two partners involved and it might be an indicator of different levels of compatibility.

At present we do not have any knowledge regarding the nature of the signal from each organism which initiates this succession of early events of mycorrhizal infection and which determines this selective aggressiveness. However, some models are now available for further study. Other plant–microorganism recognition interactions have been extensively studied in relation to plant pathology and *Rhizobium*–legume symbiosis (Halverson and Stacey, 1986). Comparatively, the recognition mechanisms between ectomycorrhizal fungi and host plants are poorly understood. To explain the differences in ectomycorrhiza formation and specificity, the action of antifungal compounds originating from the plant have been suggested (Malajczuk *et al.*, 1982; Duchesne *et al.*, 1987), as well as the role of elicitors from mycorrhizal fungi (Coleman and Anderson, 1985), or fungal lectins (Guillot *et al.*, 1983).

Aggressiveness comparison in nursery

In view of these results strains of diverse geographical origin (Brazil, USA, Australia), collected either under pine or *Eucalyptus* were used to inoculate *E. viminalis* and *E. dunii* in a Brazilian nursery. Three months after inoculation some strains could be recognized as good colonizers, others as bad colonizers, but no obvious relationship links aggressiveness in nursery with geographical or ecological origin.

Selection of aggressive strains in nursery as well as in plantation ecosystems has always been the key for successful introduction of ectomycorrhizal technology into forestry. Recent work shows that aggessiveness and competition between strains could be largely under bacterial control: rhizobacteria specifically stimulate or inhibit ectomycorrhizal infection (Duponnois and Garbaye, 1991).

Dual VA–Ectomycorrhizal Infection of *Eucalyptus* Roots

Most vascular plants form either ectomycorrhizas or VA mycorrhizas. In contrast, the genus *Eucalyptus* forms both VA mycorrhizas and ectomycorrhizas. *Eucalyptus* VA mycorrhizas were first described by Asai (1934) and then by Maeda (1954), but the first synthesis in controlled conditions (Malajczuk *et al.*, 1981), as well as the first ulstrastructural studies (Boudarga and Dexheimer, 1989) are fairly recent. In contrast, the anatomy, ecology and physiology of *Eucalyptus* ectomycorrhizas have been extensively studied (Chilvers and Pryor, 1965; Chilvers, 1968; Ling Lee *et al.*, 1975; Horan *et al.*, 1988; Hilbert and Martin, 1988).

Evidence of succession

In pot experiments, with natural soil, the VA mycorrhizas are more prevalent on young seedlings whereas ectomycorrhizas take over as the plant ages (Lapeyrie and Chilvers, 1985). However, both types of symbionts can be present simultaneously in the same root apex: the VA mycorrhizal fungus colonizes the inner part of the cortex while the ectomycorrhizal fungus is restricted to the outer cell layer. In Brazilian *Eucalyptus* plantations, the VA ectomycorrhiza succession can be followed during the first year.

Some evidence suggests that the succession between VA mycorrhizas and ectomycorrhizas during host plant ageing could be related to competition for infection sites (Chilvers *et al.*, 1987). The possible existence, however, of a direct antagonism between symbionts within the plant needs further investigation. The fungi could also compete for a limiting substrate.

Ultrastructure comparison

A technique has been recently described for the *in vitro* synthesis of VA mycorrhizas and ectomycorrhizas on the same root apex (Boudarga *et al.*, 1990). Interactions between three symbiotic organisms could henceforth be studied in detail.

A preliminary ultrastructural study shows that both types of mycorrhiza can be found in an active state, even when they are involved in a dual association. In such dual mycorrhiza the ultrastructure of each individual symbiotic association was indistinguishable from that found in single VA or ectomycorrhiza. The VA endophyte was functional within the ectomycorrhiza as indicated by the occurrence of a continuous host cell plasma membrane around the arbuscule. However the ectomycorrhizas may not become totally operational until the VA mycorrhizal arbuscules are degenerating (Boudarga, 1989).

Nursery experiments

Glasshouse or *in vitro* studies should be complemented by nursery experiments. Would early VA mycorrhizal colonization of the root system favour young plant development and subsequent ectomycorrhizal infection? Amorim and Muchovej (1991) give valuable information.

Seedlings were inoculated with a mixed VA–ecto inoculum. After 2 months the infection by either symbiont was not dependent on the presence or absence of the co-symbiont. Ectomycorrhizal inoculation stimulated plant gowth compared with uninoculated control and stimulated growth of seedlings co-inoculated with VA strains compared with the treatment with VA alone. In contrast, VA mycorrhizal inoculation had no effect on plant gowth compared with non-inoculated plants and more surprisingly it had a depressive effect on plant growth when added to ectomycorrhizal plants (Muchovej and Amorim, 1990; Amorim and Muchovej, 1992). This unexpected result is difficult to explain as the interactions between symbionts and the host plant are still poorly understood. However it could be related to the transmission electron microscope observations previously mentioned. Complementary studies on different soils and in different nurseries with different fungal strains and Eucalypt species are required before rejecting potential use of VA mycorrhizas for nursery inoculation.

Conclusion

There is a great potential in ectomycorrhizal fungal inoculation for industrial *Eucalyptus* plantation in tropical areas. However, intensive *Eucalyptus* plantations are often thought dramatically to reduce soil fertility due to high dry matter production, total tree utilization, including sometimes the leaves for fuel, and short rotations. The heartwood does not start to form until the tree is more than 15 years old and the sapwood which is therefore exported contains, for example, 33 times more phosphorus than the heartwood.

Extra-production due to mycorrhizal inoculation will result in higher nutrient cost. How much of this cost could be balanced by fungal activity in the soil, mineral alteration (Lapeyrie *et al.*, 1987, 1991), or nutrient leaching is not known. Understanding mycorrhizal fungi–soil interface processes could be essential for intensive and sustainable *Eucalyptus* plantation management on tropical soils.

References

Amorim, E.F.C. and Muchovej, R.M.C. (1992) Development and effect of endo- and or
  ectomycorrhizal fungi on seedlings of *Eucalyptus grandis*. *Plant and Soil* (in press)
Asai, T. (1934) Uber das Vorkommen und die Bedeutung der Wurzelpilze in den Land-
  pflanzen. *Japanese Journal of Botany* 7, 107–150.
Boudarga, K. (1989) Etude des mycorhizes de l'*Eucalyptus camaldulensis*, application prati-
  que à la mycorhizatioin de vitro-plants. Unpublished Thesis, Nancy I University,
  pp. 209.
Boudarga, K. and Dexheimer, J. (1989) Sur la mycorhization controlée de semis
  d'*Eucalyptus camaldulensis* Dehnardt par *Gigaspora margarita* Becker and Hall.
  Annales des Sciences Forestières 46 2, 131–139.
Boudarga, K., Lapeyrie, F. and Dexheimer, J. (1990) A technique for dual VA-
  endomycorrhizal/ectomycorrhizal infection of *Eucalyptus in vitro*. *New Phytologist*
  114, 73–76.
Chilvers, G.A. (1968) Low power electron microscopy of the root cap region of eucalypt
  mycorrhizas. *New Phytologist* 67, 663–665.
Chilvers, G.A. and Pryor, L.D. (1965) The structure of eucalypt mycorrhizas. *Australian
  Journal of Botany* 13, 245–259.
Chilvers, G.A., Douglass, P.A. and Lapeyrie F.F. (1986) A paper sandwich technique for
  rapid synthesis of ectomycorrhizas. *New Phytologist* 103, 397–402.
Chilvers, G.A., Lapeyrie, F.F. and Horan, D.P. (1987) Ectomycorrhizal versus endomycor-
  rhizal fungi within the same root system. *New Phytologist* 107, 441–448.
Coleman, M.E. and Anderson, A.J. (1985) The role of elicitors in ectomycorrhizal forma-
  tion. *Proceedings of the 6th NACOM*, Oregon State University, pp. 361.
De La Cruz, R.E. (1990) Current status of nursery and field applications of ectomycorrhiza
  in the Philippines. *Abstracts of the 8th NACOM*, Jackson Wyoming, 75.
De La Cruz, R.E. and Lorilla, E.B. (1989) Ectomycorrhizal tablets for *Eucalyptus* species.
  International Conference on Fast Growing and Nitrogen Fixing Trees. Marburg RFA,
  Gustav Fischer, Verlag, Stuttgart and New York.
Delwaulle, J.C., Garbaye, J. and Okombi, G. (1982) Stimulation de la croissance initiale
  de *Pinus caribaea* Marelet dans une plantation du Congo par contrôle de la mycorrhiza-
  tion. *Bois et Forêts des Tropiques* 196, 25–32.
Duchesne, L.C., Peterson, R.L. and Ellis, B.E. (1987) The accumulation of plant-produced
  antimicrobial compounds in response to ectomycorrhizal fungi: a review. *Phytoprotec-
  tion* 68, 17–27.
Duponnois, R. and Garbaye, J. (1991) Mycorrhization helper bacteria associated with
  Douglas fir-*Laccaria laccata* symbiosis : effects in aseptic and in glasshouse conditions.
  Annales des Sciences Forestières 48, 239–251.
Garbaye, J., Delwaulle, J.C. and Diangana, D. (1988) Growth response of eucalypts to
  mycorrhizal inoculation in the Congo. *Forest Ecology and Management* 24, 151–157.
Guillot, J., Genaud, L., Gueugnot J. and Damez, M. (1983) Purification and properties
  of two hemagglutinins of the mushroom *Laccaria amethystina*. *Biochemistry* 22,
  5365–5369.
Halverson, L.J. and Stacey, G. (1986) Signal exchange in plant-microbe interactions.
  Microbiological Reviews 50, 193–225.
Hilbert, J.L. and Martin, F. (1988) Regulation of gene expression in ectomycorrhizas. I.
  Protein changes and the presence of ectomycorrhiza-specific polypeptides in the
  Pisolithus – Eucalyptus symbiosis. *New Phytologist* 110, 339–346.
Horan, D.P., Chilvers, G.A. and Lapeyrie F.F. (1988) Time-sequence of the infection pro-
  cess in Eucalypt ectomycorrhizas. *New Phytologist* 109, 451–458.

Lapeyrie, F.F. and Chilvers, G.A. (1985) An endomycorrhiza-ectomycorrhiza succession associated with enhanced growth of *Eucalyptus dumosa* seedlings planted in calcareous soil. *New Phytologist* 100, 93–104.

Lapeyrie, F.F., Chilvers, G.A. and Bhem, C.A. (1987) Oxalic acid synthesis by the mycorrhizal fungus *Paxillus involutus*. *New Phytologist* 106, 139–146.

Lapeyrie, F., Lei, J., Malajczuk, N. and Dexheimer, J. (1989) Ultrastructural and biochemical changes at the preinfection stage of mycorrhizal formation by two isolates of *Pisolithus tinctorius*. *Annales des Sciences Forestières* 46 s, 754–757.

Lapeyrie, F., Ranger, J. and Vairelles, D. (1991) Phosphates solubilizing activity of ectomycorrhizal fungi in vitro. *Canadian Journal of Botany* 69, 342–346.

Lei, J., Ding, H., Lapeyrie, F., Piche, Y., Malajczuk, N. and Dexheimer, J. (1989) Ectomycorrhizal formation on the roots of *Eucalyptus globulus* and *Pinus caribea* with two isolates of *Pisolithus tinctorius*: Structural and cytochemical observations. *Fourth International Colloquium on Endocytobiology and Symbiosis, Lyon* INRA, Paris, pp. 123–126.

Lei, J., Lapeyrie, F., Malajczuk, N. and Dexheimer, J. (1990) Infectivity of pine and eucalypt isolates of *Pisolithus tinctorius* (Pers.) Coker and Couch on roots of *Eucalyptus urophylla* S.T. Blake *in vitro*. 2-Ultrastructural and biochemical changes at the early stage of mycorrhizal formation. *New Phytologist* 116, 115–122.

Ling Lee, M., Chilvers, G.A. and Ashford, A.E. (1975) Polyphosphate granules in three different kinds of tree mycorrhiza. *New Phytologist* 75, 551–554.

Maeda, M. (1954) The meaning of mycorrhiza in regard to systematic botany. *Kumamoto Journal of Science Series B* 3, 57–84.

Malajczuk, N., Linderman, R.G., Kough, J. and Trappe, J.M. (1981) Presence of vesicular-arbuscular mycorrhizae in *Eucalyptus* sp. and *Acacia* sp., and their absence in *Banksia* sp. after inoculation with *Glomus fasciculatus*. *New Phytologist* 87, 567–572.

Malajczuk, N., Molina, R. and Trappe, J.M. (1982) Ectomycorrhiza formation in *Eucalyptus*. I-Pure culture synthesis, host specificity and mycorrhizal compatibility with *Pinus radiata*. *New Phytologist* 91, 467–482.

Malajczuk, N., Lapeyrie, F. and Garbaye, J. (1990) Infectivity of pine and eucalypt isolates of *Pisolithus tinctorius* (Pers.) Coker and Couch on roots of *Eucalyptus urophylla* S.T. Blake *in vitro*. I-Mycorrhizal formation in model systems. *New Phytologist* 114, 627–631.

Muchovej, R.M.C. and Amorim, E.F.C. (1990) Development and effect of endo- and or ectomycorrhizal fungi on seedlings of *Eucalyptus grandis*. *Abstracts of the 8th NACOM*, Jackson Wyoming, 251.

Part Six
Physiological Ecology
of Mycorrhizas

40 A Functional Comparison of Ecto- and Endomycorrhizas

P.B. Tinker[1], M.D. Jones[2] and D.M. Durall[2]

[1]*Natural Environment Research Council, Terrestrial and Freshwater Sciences Directorate, Polaris House, North Star Avenue, Swindon SN2 1EU, UK:* [2]*Okanagan College, 1000 KLO Road, Biology Department, Kelowna, British Columbia, V1Y 4X8 Canada*

Introduction

The two largest and most important classes of mycorrhizas are the ecto (EC) and vesicular-arbuscular (VA) mycorrhizas (Harley and Smith, 1983). The mechanisms of the growth enhancement of higher plants following infection with EC and VA fungi are therefore of very high importance.

Research on the two groups has proceeded along rather different lines. Attention in the EC mycorrhizas has focused on descriptions of the morphological structure and nutrient uptake characteristics of the sheath, external hyphae and hyphal strands, and the identification and succession of the fungi. Mechanisms of growth enhancement proposed have included the rapid uptake and storage of phosphorus, hydrolysis of organic phosphates, 'solubilization' of poorly soluble inorganic phosphates, hydrolysis of nitrogen-containing compounds, and siderophore excretions.

Much of this work had no clear parallel in the VA mycorrhizas, where attention centred on the physiology of growth enhancement of the host (Tinker and Gildon, 1983; Smith and Gianinazzi-Pearson, 1988). It was shown that phosphorus could be absorbed by external hyphae at a distance from the root, that root inflows could only be explained by hyphal uptake (from the same chemical pool as uptake by the root alone), and that hyphae translocated nutrients at reasonable rates. Thus attention gradually shifted to the external hyphal network of VA fungi (Sanders *et al.*, 1977; Sylvia, 1990).

In some work with VA mycorrhizas, growth reduction was seen to follow infection (Tinker, 1978). The usual explanation has been the carbon drain by the fungus, regarded as a 'cost' of infection (Stribley *et al.*, 1980; Whipps, 1990).

The resulting theory remains as the best current explanation of growth effects of VA infections (Reid, 1990), and we refer to it as the 'standard model' (Table 40.1). The possibility of chemical displacement effects, pH change and siderophore excretion remain. However, these root surface processes appear to be more important for elements such as iron, or to represent special cases. The results of Bolan

Table 40.1. Effects and processes involved in the 'standard model' of growth enhancement by VA mycorrhizas.

1. Growth increase occurs by improved supply of elements of low mobility in growth medium, predominantly phosphate.
2. This arises by increased uptake rate per unit amount of root length (inflow).
3. This is caused by proliferation of a considerable length of external hyphae.
4. Hyphae absorb, translocate and transfer phosphorus to the host, from soil outside the root depletion zone.
5. Uptake is normally from the isotopically labile pool of nutrient, from which the root also absorbs.
6. There is a feedback effect by absorbed phosphorus on the percentage of infected root.
7. Infection of phosphate-deficient plants is accompanied by a rapid but temporary increase in internal phosphorus concentration.
8. Much of the phosphorus in the fungal partners is in the form of polyphosphate.
9. The fungus is maintained by carbon supplies from its host, and infection results in a larger proportion of total fixed carbon being allocated below ground.
10. The uptake efficiency should be larger for this mechanism than for uptake by the uninfected root system.

et al. (1984) are difficult to explain on this model, but these results have not been confirmed or followed up since their publication.

The Natural Environment Research Council set up a small group in Oxford in 1988 to make a comparative investigation of the mechanism underlying growth promotion in VA and EC systems. This terminated in 1990 and some results have been published or are in press (Jones *et al.*, 1990; Tinker *et al.*, 1990; Tinker *et al.*, 1992; Jones *et al.*, 1991). This is the first assessment of the general conclusions.

Species of Host and Fungi

Whereas the 'standard model' has been tested for VA mycorrhizas with a reasonable number of species, equivalent results are still very sparse for EC systems. Ideally, the comparison between VA and EC mycorrhizas would be made with a host species that accepted infection with both types of fungi. In our work, we have tested willow, poplar and eucalyptus. Willow was an excellent experimental plant, but our clones failed to become infected with VA fungi. Poplar was impractical to use in the growth chambers. Eucalyptus can be infected by both EC and VA fungi (Chilvers *et al.*, 1987) and we report results with both types. However, slow and irregular growth of this host caused difficulties. EC mycorrhizal infection was typical in both hosts.

The comparisons given here are all for young plants. A 2–3-month experiment with VA mycorrhizas gives a reasonable test of an annual crop plant with a 100–120-day life. However, our work is only the first step towards understanding the processes in mature forest trees infected with ectomycorrhizal fungi, where both the relative nutrient accumulation rate (Miller, 1981) and the fungal carbon demand (Vogt *et al.*, 1982) are likely to be very different from those in seedlings.

The 'Standard Model' Applied To EC ectomycorrhizas – EC–VA Comparison

Phosphorus supply as growth-enhancing process

A growth increase following VA infection normally results from an increased uptake of phosphorus. This is less clear for EC mycorrhizas, because few experiments give unambiguous answers on this point (Harley and Smith, 1983). In 1990 the simple determination of the response curve of a plant to added phosphorus with and without EC mycorrhiza infection was first reported (Chilvers *et al.*, 1987; Jones *et al.*, 1990). This basic experiment shows that phosphorus supply was the sole mechanism whereby growth enhancement occurred in these particular situations (other mechanisms are not excluded in other soils and circumstances).

Uptake from labile pool

As far as we know, no experiment has tested whether phosphorus taken up by plants with EC mycorrhizas comes solely from the isotopically labile pool. This point has been repeatedly tested for VA mycorrhiza (Tinker *et al.*, 1992). In view of frequent suggestions that ectomycorrhizal fungi cause the hydrolysis of soil organic phosphates (Harley and Smith, 1983), this seems strange, though it may be due to practical problems with experiments in EC systems. Our *Salix* systems might well allow the appropriate experiment to be made.

Uptake, translocation and transfer of phosphorus by external mycorrhizal hyphae

There is no reason to believe that the uptake processes of the VA and EC hyphae differ for phosphate ions. Whereas the translocation processes may differ in details, because of the different internal morphology of VA and EC fungi, there is no doubt that translocation can occur in both types. Major differences in the structures believed to effect transfer (arbuscules and Hartig net respectively) could suggest major differences in mechanism. However, in both cases the essential steps are analogous.

Phosphorus concentration and hydration changes

After infection of a plant with VA fungus there is normally a sharp increase in the percentage of phosphorus in the shoot (Snellgrove *et al.*, 1986). Subsequently this declines to a value much closer to that of the uninfected controls. This agrees with the observation of Sanders *et al.* (1977) that the hyphal inflow was exceptionally large in young VA infections. A similar effect has also been detected in *Salix* infected with *Thelephora terrestris* (Jones *et al.*, 1991).

It is interesting that this surge in phosphorus concentration appeared to be associated with increased wet:dry ratios in plants with VA mycorrhizas (Snellgrove *et al.*, 1986). This effect was also found in the work with *Salix*, where the shoot wet:dry ratio, in the four harvests, were: mycorrhizal – 3.9, 4.1, 3.6, 3.0; non-mycorrhizal – 3.2, 3.5, 3.5, 3.0. The differences due to infection were significant at the first two harvests.

Polyphosphate in storage or transport

Polyphosphate has been implicated in hyphal transport (Tinker and Gildon, 1983) for VA, and in storage for EC systems (Harley and Smith, 1983). However, the storage hypothesis requires some simple calculations on potential quantities relative to plant demand before it can be credible. Polyphosphate is present in both systems, and there is no real evidence that its function differs in VA and EC mycorrhiza.

Phosphorus inflows

The density of roots of trees tends to be low compared with that of annual crop plants. Despite this, inflow values for plants with ectomycorrhiza (Table 40.2)

Table 40.2. Some reported values of phosphorus inflow, and its enhancement by mycorrhizal infection.

	Root inflows, $\times 10^{-12}$ mol m^{-1} s^{-1}		
	Mycorrhizal (M)	Non-mycorrhizal (NM)	Ratio M:NM
Vesicular-arbuscular mycorrhiza			
Sanders and Tinker, 1973 (onion)	12.0	3.7	3.2
Sanders *et al.*, 1977 (final harvest) (onion)	5.0	2.0	2.5
Smith, 1982 (avg 30–38 days low P) (clover)	0.3	0.1	3.0
Smith *et al.*, 1986 (onion)	7.0	2.0	1.0–4.0
Jakobsen, 1986 (peas)	7.8	3.0	2.6
Jones *et al.* (unpublished) (eucalyptus)	0.8	0.3	2.5
Ectomycorrhiza			
Jones *et al.*, 1991 (willow)	2.3	1.0	2.3
Jones *et al.* (unpublished) (eucalyptus)	1.2	0.3	4.0

appear rather low in relation to those with VA infections. However, the ratios between inflows with and without mycorrhizas are remarkably similar for VA and EC. The experiments with EC were done in conditions where it was known that this mycorrhizal growth enhancement resulted from improved phosphorus supply (Jones *et al.*, 1990). The conclusions that were made for VA infections (Sanders and Tinker, 1973) therefore apply also to EC infections.

Effect of phosphorus supply on percentage root infection

The extent of VA infection is normally measured by an intercept technique, and is expressed as a percentage of total root length occupied. Other methods have normally been used for ectomycorrhizal roots but Jones *et al.* (1990) applied the intercept method to ectomycorrhiza-infected *Salix* root systems, and stated their results in values that are comparable to VA data.

It is known that infection with VA and EC fungi is reduced by added phosphorus. Use of the intercept technique now allows a quantitative comparison. The effects are similar for VA and EC systems (e.g. Sanders and Tinker, 1973; Amijee *et al.*, 1989; Jones *et al.*, 1990), though the sensitivity to soil phosphorus may differ greatly.

Extent of external hyphal network

The amount of mycelium is best expressed as length of hyphae per unit length of *infected* root, which is a meaningful value in relation to the function of the mycorrhizal system. Sylvia (1990) quotes data of 2–59 200 m hyphae m^{-1} infected root, but these extremes seem atypical. More usual values for VA are around 70–300 m hyphae m^{-1} of infected root (Smith and Gianinazzi-Pearson, 1988), with some values above 1000 m hyphae m^{-1}.

Quantitative data for hyphal development and length in EC mycorrhiza are extremely sparse (Reid, 1990). However, Jones *et al.* (1990) found that the total length of hyphae on root of *Salix viminalis* infected with *Laccaria proxima* or

Table 40.3. Percentage infection of mycorrhizal *Salix viminalis* roots at six levels of soil phosphorus and the length of fungal hyphae associated with the roots (after Jones et al., 1990).

	Soil phosphorus concentration (mg kg^{-1})					
	4	6	10	21	60	90
Length (m) of mycorrhizal hyphae m^{-1} of mycorrhizal root						
L. proxima	289	208	313	193	n.c.	n.a.
T. terrestris	319	185	106	103	n.c.	n.a.
Percentage infection						
Laccaria proxima	44.0	44.3	43.3	30.2	0.59	0
Thelephora terrestris	50.6	41.7	49.8	43.0	0.06	0

n.c. = not calculated; n.a. = not applicable.

Thelephora terrestris ranged from 100 to 300 m hyphae m^{-1} infected root, corresponding well with the range for VA mycorrhiza (Table 40.3).

The amount of external hyphae may be correlated with the growth enhancement caused by different VA fungi (see Sylvia, 1990). Recent work (Jones *et al.*, in preparation) found a close correlation of growth response and *living* hyphal length for both VA and EC infections of eucalyptus.

Carbon allocation and demand

The additional percentage of carbon fixed that is directed below ground is normally some 6–12% greater when the host is infected by VA fungi (Whipps, 1990). This 'cost' may be a considerable penalty on a plant. Normally, it is compensated by increased photosynthetic rates or physiological changes (Whipps, 1990), but if there were no compensating advantages of infection, it could cause major yield depressions.

Reid *et al.* (1983) reported a comparable study, with *Pinus taeda* and *Pisolithus tinctorius*. The allocation to the root was increased from 16% in non-mycorrhizal to 39% in ectomycorrhizal plants at 2 months, and from 26% in non-mycorrhizal to 38% in infected plants at 10 months, both being above the range found for VA infections. Work with young *Salix viminalis* and *Thelephora terrestris* (Durall *et al.*, in preparation) has obtained results which range from 6% to 11% additional allocation to the root, increasing regularly over the experimental period from 80 to 120 days. These values agree well with those already found in VA mycorrhizas, but the progressive increase may indicate considerably larger final values. We conclude that carbon 'costs' are similar for both infections in relatively young plants, though values may rise for older EC infections.

Prosphate acquisition efficiency

Against the 'cost' of the extra carbon allocated belowground in mycorrhizal plants can be set the 'benefit' of the extra phosphorus supply. The phosphate acquisition efficiency $\Delta P/\Delta C_b$ can therefore be calculated as a measure of cost:benefit ratio (Koide and Elliott, 1989). Jones *et al.* (1991) found interesting changes with time in this parameter.

We have made one experiment with *Eucalyptus coccifera* infected separately with EC and VA fungi, from which such calculations could be made. Full results

Table 40.4. Phosphorus uptake efficiency for *Eucalyptus coccifera* over time period 28–89 days from seedling emergence.

Treatment	ΔP (μmol P)	ΔC_b (mmol C)	$\Delta P/\Delta C_b$
+ *Thelephora terrestris*	67.4	6.0	11.2
+ *Glomus* 'E3'	34.8	3.5	10.0
NM − P	11.3	1.6	7.2
NM + P	69.4	11.4	6.1

will be reported elsewhere; these were somewhat variable, but were similar for VA and EC mycorrhiza on average (Table 40.4).

Conclusions

We have shown that when the concepts in the 'standard model' and appropriate experimental measurements are applied to some EC systems, the results can be surprisingly similar to those obtained earlier with VA systems. EC fungi may possess properties advantageous to the host over and above those of VA. Nevertheless, the core property of enhancing phosphate uptake in the two systems seems to function similarly. The very obvious differences in form may have led too readily to the assumption that function differs also, and hence to a very different research approach to these two classes of mycorrhiza.

In view of the antiquity of the VA system (Nicholson, 1975), it is tempting to speculate that the 'standard model' was the original basis of mycorrhizal associations, and that some fungi have subsequently developed processes which aid in the exploitation of specific habitats and soil types.

References

Amijee, F., Stribley, D.P. and Tinker, P.B. (1989) Development of endomycorrhizal root systems. VII. A detailed study of effects of soil phosphorus on colonization. *New Phytologist* 111, 435–446.

Bolan, N.S., Robson, A.D., Barrow, N.J. and Aylmore, I.A. (1984) Specific activity of phosphorus in mycorrhizal and non-mycorrhizal plants in relation to the availability of phosphorus in plants. *Soil Biology and Biochemistry* 16, 299–304.

Chilvers, G.A., Lapeyrie, F.F. and Horan, D.P. (1987) Ectomycorrhizal *vs* endomycorrhizal fungi within the same root system. *New Phytologist* 107, 441–448.

Harley, J.L. and Smith, S.E. (1983) *Mycorrhizal Symbiosis*. Academic Press, London, pp. 483.

Jakobsen, I. (1986) Phosphorus inflow into roots of mycorrhizal and non-mycorrhizal peas under field conditions. In: Gianinazzi-Pearson, V. and Gianinazzi, S. (eds), *Physiological and Genetic Aspects of Mycorrhizae* INRA, Paris, pp. 317–322.

Jones, M.D., Durall, D.M. and Tinker, P.B. (1990) Phosphorus relationships and production of extramatrical hyphae by two types of willow ecto-mycorrhizas at different soil phosphorus levels. *New Phytologist* 115, 259–267.

Jones, M.D., Durall, D.M. and Tinker, P.B. (1991) Fluxes of carbon and phosphorus between symbionts in willow ectomycorrhizas and their changes with time. *New Phytologist* 119, 99–106.

Koide, R. and Elliott, G. (1989) Cost, benefit and efficiency of the vesicular-arbuscular mycorrhizal symbiosis. *Functional Ecology* 3, 252–255.

Miller, H.G. (1981) Forest fertilization: some guiding concepts. *Forestry* 54, 172–175.

Nicholson, T.H. (1975) Evolution of vesicular-arbuscular mycorrhizas. In: Sanders, F.E., Mosse, B. and Tinker, P.B. (eds), *Endomycorrhizas*. Academic Press, London, pp. 25–34.

Reid, C.P.P. (1990) Mycorrhizas. In: Lynch, J.M. (ed.), *The Rhizosphere* Wiley, Chichester, pp. 281–316.

Reid, C.P.P., Kidd, F.A. and Ekwebelam, S.A. (1983) Nitrogen nutrition, photosynthesis and carbon allocation in ectomycorrhizal pine. *Plant and Soil* 71, 415–432.

Sanders, F.E. and Tinker, P.B. (1973) Phosphate flow into mycorrhizal roots. *Pesticide Science* 4, 385–395.

Sanders, F.E., Tinker, P.B., Black, R.L.B. and Palmerley, S.M. (1977) The development of endo-mycorrhizal root systems. I – Spread of infection and growth-promoting effects with four species of vesicular-arbuscular endophyte. *New Phytologist* 78, 257–268.

Smith, F.A., Smith S.E., St John, B.J. and Nicholas, D.J.D. (1986) Inflows of N and P into roots of mycorrhizal and non-mycorrhizal onions. In: Gianinazzi-Pearson, V. and Gianinazzi, S. (eds), *Physiological and Genetic Aspects of Mycorrhizae*. INRA, Paris, pp. 371–378.

Smith, S.E. (1982) Inflow of phosphate into mycorrhizal and non-mycorrhizal plants of *Trifolium subterraneum* at different levels of soil phosphate. *New Phytologist* 90, 293–303.

Smith, S.E. and Gianinazzi-Pearson, V. (1988) Physiological interactions between symbionts in vesicular-arbuscular mycorrhizal plants. *Annual Review of Plant Physiology and Plant Molecular Biology* 39, 221–244.

Snellgrove, R.C., Stribley, D.P., Tinker, P.B. and Lawlor, D.W. (1986) The effect of vesicular-arbuscular mycorrhizal infection on photosynthesis and carbon distribution in leek plants. In: Gianinazzi-Pearson, V. and Gianinazzi, S. (eds), *Mycorrhizae: Physiology and Genetics*. INRA, Paris, pp. 421–424.

Stribley, D.P., Tinker, P.B. and Rayner, J.H. (1980) Relation of internal phosphorus concentration and plant weight in plants infected by vesicular-arbuscular mycorrhizas. *New Phytologist* 86, 261–266.

Sylvia, D.M. (1990) Distribution, structure and function of external hyphae of vesicular-arbuscular mycorrhizal fungi. In: Box, J.E. and Hammond, L.C. (eds), *Rhizosphere Dynamics*. AAAS Selected Symposium, 113, 144–167.

Tinker, P.B. (1978) Effects of vesicular-arbuscular mycorrhizas on plant nutrition and plant growth. *Physiologie Végétale* 16, 743–751.

Tinker, P.B. and Gildon, A. (1983) Mycorrhizal fungi and ion uptake. In: Robb, D. and Pierpoint, W.S. (eds), *Metals and Micronutrients: Uptake and Utilisation by Plants*. Academic Press, New York, pp. 21–32.

Tinker, P.B., Jones, M.D. and Durall, D.M. (1990) Phosphorus and carbon relationships in willow ectomycorrhizae. *Symbiosis*, 9, 43–50.

Tinker, P.B., Jones, M.D. and Durrall, D.M. (1992) Principles of use of radioisotopes in mycorrhizal studies. *Methods in Microbiology* 23, 295–307.

Vogt, K.K., Grier, C.C., Meier, C.E. and Edmonds, R.L. (1982) Mycorrhizal role in net primary production and nutrient cycling in *Abies amabilis* ecosystems in Western Washington. *Ecology* 63, 370–380.

Whipps, J.M. (1990) Carbon economy. In: Lynch, J.M. (ed.), *The Rhizosphere*. Wiley, Chichester, pp. 59–98.

41 Spatial Distribution of Nitrogen Assimilation Pathways in Ectomycorrhizas

F. Martin[1], M. Chalot[2], A. Brun[2], S. Lorillou[1,2], B. Botton[2] and B. Dell[3]

[1]Laboratoire de Microbiologie Forestière, INRA-Nancy, Champenoux, France:
[2]Laboratoire de Physiologie Végétale et Forestière, Université de Nancy I,
Vandoeuvre-Nancy, France: [3]School of Biological Sciences, Murdoch University,
Murdoch, Western Australia

Introduction

Recent studies have highlighted the ecological role of ectomycorrhizas in N cycling in forest ecosystems. As a consequence, increasing attention has been directed toward the N metabolism of ectomycorrhizal fungi and ectomycorrhizas. A number of studies have shown that enzyme compartmentation in symbiotic tissues is a key element in understanding the high degree of metabolic integration during mycorrhiza development and function (Martin *et al.*, 1987; Martin and Botton, 1993). In this contribution we will discuss more recent developments in knowledge of the spatial distribution and regulation of N assimilation pathways in symbiotic tissues.

Pathways for N Assimilation in Ectomycorrhizal Fungi

From the composition of free amino acid pools, the ^{15}N-labelling patterns and effects of inhibitors of N assimilating enzymes, N assimilation in *Cenococcum geophilum*, *Laccaria laccata*, and *Hebeloma* sp. appears to proceed via the glutamate dehydrogenase (GDH) pathway (Martin and Botton, 1993). The label was however predominantly incorporated into the amino-N of glutamine. These data are consistent with a pivotal role for glutamine synthetase (GS) and indicate that the primary N assimilation is brought about by successive activity of GDH and GS.

Compartmentation of the Enzymes in Ectomycorrhiza

Ammonium assimilation in higher plant roots is now firmly established as occurring via the glutamate synthase cycle, involving the sequential action of GS and glutamate synthase (GOGAT) (Oaks and Hirel, 1985). There is little evidence for the participation of GDH in N assimilation in higher plants. In contrast, N assimilation in several ectomycorrhizal fungi occurs predominantly through the GDH pathway. The possibility that in ectomycorrhizas, fungal and root cells have different N assimilation pathways raises the intriguing problem of their coexistence and interactions. During the last decade, the spatial distribution of N assimilation pathways in various ectomycorrhizal associations has been studied. It is not our intention here to give a full description of these data (for a review see Martin and Botton, 1993). Rather, we wish to highlight more novel aspects arising from comparative studies of the spatial distribution of the pathways.

Type 1: Beech ectomycorrhizas

In beech ectomycorrhizas, incorporation of $^{15}NH_4^+$ successively into amino-N of glutamine, glutamate, alanine and, after additional incubation, other amino acids is consistent with the GS/GOGAT pathway. The activity and the amount of the fungal NADP-GDH polypeptide were strongly suppressed in *Fagus sylvatica* associations with *Cenococcum geophilum*, *Hebeloma crustuliniforme*, *Lactarius* sp. and *Paxillus involutus* confirming the minor role plays by the GDH pathway. Similarly, the fungal aspartate aminotransferase (AAT) which was very active in free-living mycelium was not detected in the symbiotic tissues whereas the two root AAT isoenzymes remained active. Dissection of the mycorrhizal tissues (vascular cylinder, cortical region and mycelial sheath), immunochemical assays and immunocytochemistry were consistent with the repression at the protein biosynthesis level of the fungal AAT. Repression of the fungal NADP-GDH and AAT was also demonstrated in several eucalypt associations (Botton and Dell, unpublished results).

Type 2: Spruce ectomycorrhizas

The NADP-GDH was studied in extracts of spruce roots, mycelium of *Hebeloma* sp. and associated ectomycorrhizas. Evidence from enzyme activities in mycelial extracts, electrophoresis patterns and immunological assays consistently showed that *Hebeloma* NADP-GDH was active in spruce ectomycorrhizas. Additional ^{15}N studies demonstrated that N assimilation proceeds essentially via the GDH pathway. However, evidence from cytochemical immunogold labelling indicated that the density of GDH polypeptide decreased from the peripheral cells of the fungal sheath to the most internal hyphae of the Hartig net. The main N assimilation pathways in beech and spruce mycorrhizas are therefore different (GDH vs. GS/GOGAT), but in both cases the fungal GDH polypeptide is repressed in the vicinity of root cells.

Thus, the development and metabolic activity of ectomycorrhiza alters the biosynthesis and distribution of N-assimilating enzymes, and the nature of these changes depends on the plant and fungal associate.

Effect of the N Source on NADP-GDH in *Laccaria laccata*

The mechanisms underlying the regulation of the N-assimilating enzymes in mycorrhizas are not yet understood, but there is no reason to postulate the existence of mechanisms peculiar to mycorrhizas, although work may conceivably show that some exist. It has been observed in filamentous fungi that the amount and type of N source available to the cells during growth considerably affect the activities of both GDH and GS. Certain N sources, such as glutamate and glutamine, are preferentially utilized by these fungi, and synthesis of the enzymes needed to utilize NO_3^- and NH_4^+ requires the lifting of nitrogen catabolite repression, a regulatory system acting at the level of gene transcription (Fu and Marzluf, 1990). The N control circuit of *Neurospora crassa* and *Aspergillus nidulans* consists of regulatory genes which mediate N repression and govern the expression of the structural genes which specify biosynthetic enzymes. Glutamine is the key metabolic repressor in this nitrogen circuit. It presumably interacts with the product of the major regulatory gene *nit*-2, a DNA-binding protein. Similar regulatory systems may operate in ectomycorrhizal fungi and control the biosynthesis of N-assimilating enzymes in symbiotic tissues.

We therefore designed studies to investigate the action of glutamine on the biosynthesis of GDH in the basidiomycete *Laccaria laccata*. Mycelium of *L. laccata* was grown on media containing NH_4^+ or glutamine as the sole source of N and NADP-GDH activity, quantity, and biosynthesis rate were measured. Mycelium grown on glutamine possessed low GDH activity compared with mycelium grown on NH_4^+. Protein gel blot analysis and immunoprecipitation of *in vivo* pulse-labelled NADP-GDH demonstrated that the enzyme biosynthesis was repressed, correlating with decreasing enzyme activity in glutamine-grown mycelium. This repression seems to be under translational controls as shown by immunoprecipitation of the *in vitro* translation products of poly(A)-containing mRNA from repressed and derepressed mycelium.

Regulation of N-Assimilating Enzymes in Mycorrhiza: A Proposal

It is proposed that cellular pools of glutamine in symbiotic tissues govern enzyme activity and biosynthesis of N-assimilating enzymes as in free-living mycelium. Fungal genes coding for NADP-GDH and AAT would be repressed by glutamine that accumulated in storage layers of the fungal sheath, or by those supplied by root cells. These regulatory mechanisms would give rise to a spatial distribution of N assimilation pathways within the mycelium. Absorbed N would be assimilated in extramatrical hyphae and glutamine would be immediately translocated to the sheath and Hartig net, preventing any accumulation of glutamine at the site of N assimilation. Translocated glutamine would then accumulate in the sheath triggering the N catabolite repression of NADP-GDH and AAT. Glutamine catabolism would take place in the Hartig net and host cells.

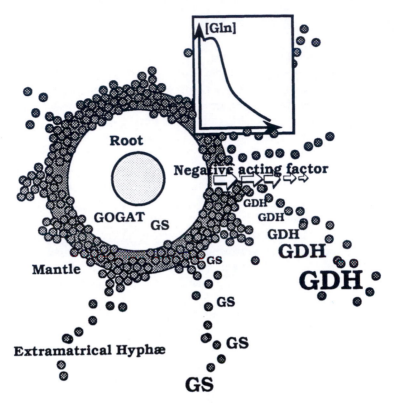

Fig. 41.1 A model for the repression of N assimilating enzymes in ectomycorrhizas. Glutamine is synthesized in extramatrical hyphae and translocated to the ectomycorrhizal sheath where it accumulates. Following the accumulation of glutamine in the ectomycorrhizal sheath, the concentration of a negative acting factor increases leading to the repression of the fungal genes encoding glutamine synthetase (GS) and NADP-glutamate dehydrogenase (GDH). Ultimately, the level of the enzymes is decreased in the Hartig net and the ectomycorrhizal sheath.

Concluding Remarks

The range of N assimilation patterns that occurs in spruce and beech ectomycorrhizas reflects the diversity encountered among ectomycorrhizas of other species, as demonstrated by recent data obtained on eucalypt species (Botton and Dell, unpublished results). This diversity correlates with anatomical features of the symbioses (e.g. the sheath:Hartig net ratio). Clearly, there is still much to learn about N assimilation in ectomycorrhizas and its regulation, particularly in relation to enzyme compartmentation and how it is affected by the different N sources. Elucidation of the N regulatory systems in the specialized symbiotic environment is essential to our understanding of N metabolism in mycorrhizas.

References

Fu, Y.H. and Marzluf, G.A. (1990) *nit-2*, the major nitrogen regulatory gene of *Neurospora crassa*, encodes a protein with a putative zinc finger DNA-binding domain. *Molecular and Cellular Biology* 10, 1056–1065.

Martin, F. and Botton, B. (1993) Nitrogen metabolism of ectomycorrhizal fungi and ectomycorrhizas. In: Tommerup, I.C., Ingram, D.S. and Williams, P.H. (eds), *Mycorrhiza: A Synthesis, Advances in Plant Pathology 8*. Academic Press, London.

Martin, F., Ramstedt, M. and Söderhäll, K. (1987) Carbon and nitrogen metabolism in ectomycorrhizal fungi and ectomycorrhizas. *Biochimie* 69, 569–581.

Oaks, A. and Hirel, B. (1985) Nitrogen assimilation in roots of higher plants. *Annual Review of Plant Physiology* 36, 345–365.

42 Ectomycorrhizas – Organs for Uptake and Filtering of Cations

I. Kottke

Universität Tübingen, Institut für Botanik, Spezielle Botanik und Mykologie, Auf der Morgenstelle, 1, 7400 Tübingen, Germany

Introduction

With evolution of ectomycorrhizas in trees the pathway of cation uptake has switched from root hairs to transfer of ions via fungal hyphae (Fig. 42.1a). Ectomycorrhizas may have evolved from saprophytic fungi which used root exudates and were themselves a source of cations for the plant (Fig. 42.1b). The ability of fungi to accumulate metals may have assisted to drive evolution in this direction. Mycorrhizas represent a special adaptation to stress in the habitat (Allen, 1985). The stresses we consider here are low cation availability and high concentrations of toxic metals. Both situations are most typically found together in acidic soils. It may therefore be that the same mechanisms are involved in uptake and in detoxification of metals.

While the physiological processes involved in the source–sink relations of cation transport from fungal to root cells is still mostly unknown, we can show structural pecularities in ectomycorrhizas which appear to have evolved to favour cation transfer and accumulation.

Formation of Cell Wall Papillae

The formation of wall papillae on the outermost cortical or epidermal cell walls may be considered as a first, but not very efficient adaptation to enlarge the absorptive surface of the root (Fig. 42.1c). This sort of papilla formation has been described from *Pisonia grandis* mycorrhizas (Ashford and Allaway 1982, 1985), *Alnus crispa* mycorrhizas (Massicotte *et al.*, 1986) and from the artificial association *Suillus grevillei–Picea abies* (Kottke and Oberwinkler 1988), but is a more widespread phenomenon (Kottke and Oberwinkler, 1986 and unpublished data). The apoplast and plasmamembrane of the root cells are considerably enlarged. Ashford and Allaway (1985) calculated a 1.7-fold increase of length of wall profile by the wall ingrowths. These structures can therefore be considered functionally to

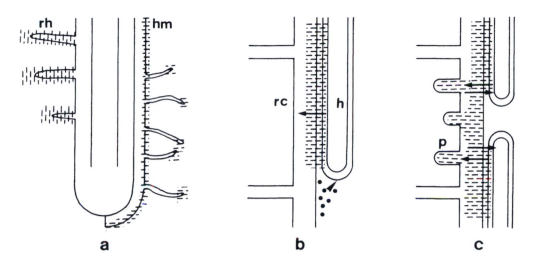

Fig. 42.1. (a) Scheme showing solute transfer in non-mycorrhizal roots (left side; rh root hairs) and mycorrhizas (right side; hm hyphal mantle);
(b) hypha of a mycorrhizal fungus (h) absorbing root exudates and eventually releasing cations to root cell (rc);
(c) outer tangential wall of root cortex enlarged by wall protuberances (P) induced by a mycorrhizal fungus, thus enlarging region for solute transfer.

be typical transfer cells (Gunning, 1977).

The presence of large numbers of negatively charged sites in the wall papillae is suggested by strong metachromasy after staining with toluidin blue O (Ashford and Allaway, 1985). The exchange capacity of mycorrhizal fungal walls has been found to occupy a lower range than in saprophytic fungi and to be equivalent to about half that of trees (McKnight *et al.*, 1990). Transfer of cations from the fungal to the root cell wall may thus be enabled by higher exchange capacity of the latter.

Hartig Net (HN) Architecture

The efficiency of the apoplast in ion transport has been greatly increased by the evolution of the Hartig net (HN). While at first glance, the structure of the Hartig net seems to favour the fungus only, a more detailed investigation reveals that its architecture is also an advantage for the root (Kottke and Oberwinkler 1990a). Fully developed but still young mycorrhizas display a dense Hartig net region, where hyphal branches are in intimate contact with each other and also with the cortical cell walls, thus forming an apoplasm in common with both symbiotic partners (Fig. 42.2b). The cellular structure at this stage indicates high physiological activity and bidirectional solute transfer (Kottke and Oberwinkler, 1987).

We have interpreted this stage as a transfer-cell like organization, because the lack of regular septa means that the hyphal branches form a common compart-

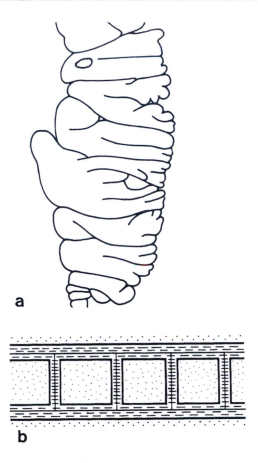

Fig. 42.2. (a) Outlining of hyphal walls in the Hartig net reveals transfer cell like organization; (b) a square cut of the Hartig net region reveals the common fungal-cell apoplasm.

ment in which the surface is enlarged in a similar way to that seen in transfer cells (Fig. 42.2a). One distinction is that, in the HN the 'protuberances' are not wall ingrowths, but derive from hyphae in intimate contact (Fig. 42.3a). Hyphal growth in the Hartig net is characterized by a negative autotropism. Apical dominance and regular septation are suppressed (Fig. 42.2a). The functional meaning of this growth mode is now further considered.

Determination of surface area of the Hartig net by means of MOP-Videoplan, showed that the calculated surface to volume ratio was found to be less in the case of the observed Hartig net architecture than the theoretical case where the fungus grew inside intercellular spaces by its normal growth mode of individual, separated hyphae (Kottke and Oberwinkler, 1990a). This means that the absorptive surface of the hyphae would be larger in the latter, normal growth mode of hyphal colonies. The Hartig net structure thus is not the optimum for the nutrient absorption of the fungus, but may be induced by the root in favour of its own solute supply.

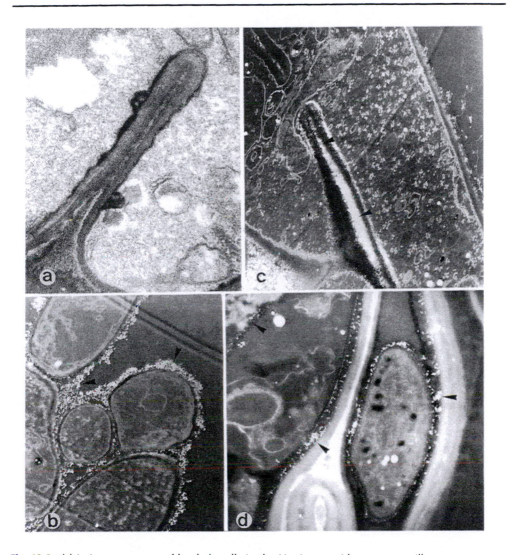

Fig. 42.3. (a) Intimate contact of hyphal walls in the Hartig net with narrow capillary space inbetween;
(b) cerium (arrows) fixed to fungal mucilage of mantle hyphae after short time exposure to cerium chloride; ESI taken at 12 000-fold magnification;
(c) cerium passing apoplasmically via hyphal mucilage into the mantle and Hartig net; ESI taken at 20 000-fold magnification;
(d) cerium accumulated in the interhyphal space; ESI taken at 30 000-fold magnification.

One such advantage for nutrient transport would be provided by the channel-like arrangement of the hyphae (Fig. 42.2a), in which the conducting elements are radially disposed in the direction of the endodermis. The other advantage is the enlarged, common apoplasm which consists of the root cell walls plus hyphal walls (Fig. 42.2b) including the small capillary interhyphal spaces (Fig. 42.3a). By obtaining this apoplastic unity the membrane surface of the hyphae can most probably be used as a very large transport surface to facilitate transfer of ions from the fungus. Such a situation is then comparable to that seen in excretory transfer cells (Gunning, 1977). The apoplastic enlargement of the surface by the Hartig net is much bigger than that obtained by the wall papilla formation. Ashford and Allaway (1985) calculated a 10–13-fold increase of the epidermal cell surface in mycorrhizas of *Fagus silvatica* or *Eucalyptus fastigiata*.

Visualization of Apoplastic Pathways of Elements

Element energy-loss spectrometry and imaging (EELS/ESI) on a Zeiss TEM902, was used to visualize pathways and deposition of elements in ectomycorrhizas (Kottke, 1991). Mycorrhizas from Petri dish cultures were exposed in solutions of 3 mmol cerium chloride, lanthanum nitrate or 0.5 mmol aluminium sulphate for 10 min up to 24 h. After the shortest time exposure (10 min) the elements were observed to be fixed on the mucilaginous surface of the hyphae and by this route to be penetrating the outer and inner mantle up to the Hartig net (Fig. 42.3b, d). A significant infiltration of the small channels between adjacent hyphae was observed (Fig. 42.3d). With longer exposure time cerium and lanthanum was found in high concentrations in the hyphal walls of the inner Hartig net and also in root cell walls. Cations seemingly pass by fast diffusion through the common apoplast of fungus and root cell. The passage of ions seems to be facilitated by the binding capacities of the outer mucilaginous layers of the hyphal walls. The narrow interhyphal spaces in the Hartig net probably provide capillary channels in ion transport.

Visualization of Apoplastic Deposition of Elements

Obviously the apoplastic space can provide sites for precipitation of cations and may be interpreted as a filter that prevents toxic elements from entering the cell. This possibility has been discussed in the literature frequently and metals have been found to accumulate on dead cell walls after exposure to solutions. However, a more detailed localization is enabled now with the electron energy loss imaging techniques. Aluminium has been found to be precipitated in considerable quantities in the mucilage on the surface of the hyphae, but to be less evident in the hyphal or root cell walls (Kottke and Oberwinkler, 1990b). Cadmium, copper and aluminium were found in high concentrations in the pigment layer of mantle hyphae from mycorrhizas of polluted plots in Poland (Turnau *et al.*, in prep.).

Symplastic Pathway and Deposition of Elements in Fungal Vacuoles

With the same analytical method the symplastic pathway of cations can be visualized also. Elements can be shown to pass the septal porus or to be deposited in the fungal vacuole after a few minutes (Kottke, 1991). The distinctive composition of osmophilic depositions in the fungal vacuoles has been detected by element energy-loss spectroscopy (Turnau *et al.*, in prep.). Some droplets were revealed to contain large amounts of P and may therefore be polyphosphate granules. Aluminium was frequently associated with these droplets (Kottke and Oberwinkler, 1990b). Other osmophilic, vacuolar precipitations contained large amounts of N and S and may be protein bodies of some kind. Cadmium and copper were bound to these droplets (Turnau *et al.*, in prep.). Potentially toxic elements can obviously be precipitated in the fungal vacuole.

Conclusion

The importance of apoplastic transport of elements in ectomycorrhizas can be determined by examination of structural features and can be visualized by means of electron energy-loss imaging. The symplastic pathway can also be examined by this method. Apoplastic transport of ions may occur primarily when the soil macropores inhabited by the mycorrhizas are filled with soil solution. The symplastic transport may become important when soil dries. Then, though the macropores are filled with air, hyphae extending from mycorrhizas can invade the smaller still water filled soil pores. These binding facilities of the hyphae are important for deposition and filtering of elements such as aluminium, cadmium or copper. The exclusion mechanism is shown to be very effective as only small amounts of toxic elements were found in the root tissue.

References

Allen, M.F. (1985) Phytohormone action: an integrative approach to understanding diverse mycorrhizal responses. In: Molina, R. (ed.), *Proc. 6th NACOM*, Forest Research Laboratory, Corvallis, Oregon, pp. 158–160.

Ashford, A.E. and Allaway, W.G. (1982) A sheathing mycorrhiza on *Pisonia grandis* R. Br. (Nyctaginacea) with development of transfer cells rather than a Hartig net. *New Phytologist* 90, 511–519.

Ashford, A.E. and Allaway, W.G. (1985) Transfer cells and Hartig net in the epidermis of the sheathing mycorrhiza of *Pisonia grandis* R. Br. from Seychelles. *New Phytologist* 100, 595–612.

Gunning, B.E. (1977) Transfer cells and their role in transport of solutes in plants. *Science Progress, Oxford* 64, 539–568.

Kottke, I. (1991) Electron energy loss spectroscopy and imaging techniques for subcellular localization of elements in mycorrhiza. *Methods in Microbiology* 23, 369–382.

Kottke, I. and Oberwinkler, F. (1986) Root–fungus interactions observed on initial stages of mantle formation and Hartig net establishment in mycorrhizas of *Amanita muscaria* (L. ex Fr.) Hooker on *Picea abies* (L.) Karst. in pure culture. *Canadian Journal of Botany* 64, 2348–2354.

Kottke, I. and Oberwinkler, F. (1987) Cellular structure and function of the Hartig net: coenocytic and transfer cell-like organization. *Nordic Journal of Botany* 7, 85–95.

Kottke, I. and Oberwinkler, F. (1988) Comparative studies on the mycorrhization of *Larix decidua* and *Picea abies* by *Suillus grevillei*. *Trees* 2, 115–128.

Kottke, I. and Oberwinkler, F. (1990a) Amplification of root–fungus interface by Hartig net architecture. In: Dreyer, E. *et al.* (eds), Forest Tree Physiology, Int. Symp. Nancy 1988, *Annales Sciences Forestières* 46 (suppl.), 737s–740s.

Kottke, I. and Oberwinkler, F. (1990b) Aufnahme und Verbleib von Elementen in Ektomykorrhizen in Abhängigkeit von den Bodenfaktoren. I. Lokalisation von Aluminium, Calcium und Phosphor in Mykorrhizen und Wurzelspitzen aus Steril kulturen mittels der Elektronen-Energieverlustanalyse am TEM 902, ZEISS. Projekt Europäisches Forschungszentrum für Massnahmen zur Luftreinhaltung (*KfK-PEF*) 61, 65–73, Kernforschungszentrum Karlsruhe.

Massicotte, H.B., Peterson, R.L., Ackerley, C.A. and Piche, Y. (1986) Structure and ontogeny of *Alnus crispa–Alpova diplophloeus* ectomycorrhizae. *Canadian Journal of Botany* 64, 177–192.

McKnight, K.B., McKnight, K.H. and Harper, K.T. (1990) Cation exchange capacities and mineral element concentrations of macrofungal stipe tissue. *Mycologia* 82, 91–98.

43

The Effects of Ectomycorrhizal Status on Plant–Water Relations and Sensitivity of Leaf Gas Exchange to Soil Drought in Douglas Fir (*Pseudotsuga menziesii*) Seedlings

J.-M. Guehl[1], J. Garbaye[2] and A. Wartinger[1]

[1]*Laboratoire de Bioclimatologie-Ecophysiologie Forestières, INRA Centre de Nancy, F-54280 Champenoux, France; [2]Laboratoire de Microbiologie Forestière, INRA Centre de Nancy, F-54280 Champenoux, France*

Introduction

While it has been shown that VA fungal associates can modulate the drought resistance of the host plants through mechanisms as diverse as enhanced water uptake, osmotic adjustment, altered elasticity of the cell walls or altered symplastic water content (Augé and Stodola, 1990), information on the role of ectomycorrhizal associations in plant water relations remains scarce (Duddridge *et al.*, 1980; Boyd *et al.*, 1986; Dosskey *et al.*, 1991).

The anatomy of ectomycorrhizas differs from that of VA endomycorrhizas, their most specific feature being the fungal sheath which completely covers the absorbing short roots: most of the flux of water between the soil and the plant has to pass through the fungus. Such a difference in structure is likely to induce differences in drought resistance mechanisms.

The present study was aimed at comparing plant water relations and sensitivity of leaf gas exchange to soil drought in Douglas fir seedlings associated with two ectomycorrhizal fungi.

Materials and Methods

Fungal and plant material

Cold-pretreated germinating seeds of Douglas fir (*Pseudotsuga menziesii* (Mirb.) Franco) were sown in 95 ml containers filled with sphagnum peat–vermiculite (1:1, v:v), with and without the addition of fungal inoculum (1:10, v:v, mixed into the potting substrate). The inoculum was *Laccaria laccata* mycelium aseptically grown in vermiculite peat (9:1, v:v) contained in glass jars. The seedlings were grown for one summer in a glasshouse, watered twice a week per excess with a nutrient solution (14.8 mg Nl^{-1} from nitrate and 2 mg Pl^{-1}). At the end of the growing season, 30 plants of each treatment were transplanted in 1 litre containers filled with methyl bromide-sterilized soil from a forest nursery (sandy loam, pH 5.6, C:N 12.5, 8.9% organic matter, $0.52 g kg^{-1}$ available P determined by the method of Duchaufour and Bonneau, 1969). At the time of transplanting, non-inoculated seedlings were infected (50–70% of mycorrhizal short roots) with *Thelephora terrestris* (Tt) from air-borne basidiospores in the glasshouse, and the inoculated seedlings were mycorrhizal (80–100%) with *L. laccata* (Ll).

Experimental design

After shoot extension growth was complete, 12 plants per treatment were selected for the experiment after eliminating the largest and smallest individuals. They were transferred into a climate room with: photoperiod, 16 h; photosynthetic photon flux density (400–700 nm), $400 \mu mol m^{-2} s^{-1}$ provided by fluorescent tubes; air temperature, $22 \pm 0.2°C$ (day) and $16.0 \pm 0.2°C$ (night); relative air humidity, 60% (day) and 90% (night); ambient CO_2 concentration, $420 \pm 30 \mu mol mol^{-1}$. The plants were kept under optimal watering conditions for 4 weeks. Thereafter water supply was withheld. At the end of the experiment (day 39 after withholding of water) the aerial parts of the plants were processed for needle area (ΔT area metre; ΔT Devices, Cambridge, UK) and dry-weight (oven-drying at 60°C for 48 h) determinations. The mycorrhizal status and total dry-weight of roots were resulted.

Leaf gas exchange measurements

Gas exchange measurements were made periodically (3 h after start of illumination period) during the soil drying cycle in the climate room with a portable gas exchange measurement system (Li-Cor 6200, Li-Cor, Lincoln, NE, USA). Only current year shoots were used. The ambient CO_2 concentration in the climate room was kept constant ($C_a = 420 \pm 15 \mu mol mol^{-1}$) during the time of measurements. Gas exchange parameters (CO_2 assimilation rate, A; leaf conductance for water vapour, g; and internal CO_2 concentrations in the sub-stomatal cavities, C_i) were calculated with the classical equations taking into account simultaneous CO_2 and H_2O diffusion through the stomatal pores. Internal CO_2 concentrations were calculated in order to assess whether differences for A between treatments and A changes in response to drought could be accounted for by stomatal factors or whether other factors could be involved (Jones, 1985). Previous measurements made on conifers (unpublished data) did not show any patch pattern in stomatal closure, so that reliable C_i calculations can be made from leaf gas exchange data.

Soil and leaf water status

Soil water content ($g g^{-1}$) was estimated gravimetrically on the occasions of gas exchange measurements, soil water potential (Ψ_{soil}) was determined from the water retention curve previously determined by means of a pressure plate apparatus (Soil Moisture Equipment corp., Santa Barbara, CA, USA).

Leaf water potential (Ψ_w) was measured on single needles with a Scholander pressure chamber (1) just before illumination commenced (predawn water potential, Ψ_{wp}) and (2) just after the gas exchange measurements (minimum water potential, Ψ_{wm}). Needle osmotic potential was assessed on days 1, 22 and 39 of the drying cycle. Three needles were harvested from each plant simultaneously to the water potential measurements, put in 1 ml syringes, and immediately frozen in liquid nitrogen. Samples were then kept deep frozen. Before the sap was expressed from the syringes, the samples were thawed for 30–60 min at room temperature. The osmotic potential of the sap (7 μl) was then measured with a calibrated vapour pressure osmometer (Wescor 5500, Logan, Utah, USA).

Hydraulic efficiency determinations

Plant hydraulic efficiency (C) was assessed as the ratio of stomatal conductance to the difference between soil water potential and minimum leaf water potential ($C = g/(\Psi_{soil} - \Psi_{wm})$) which is analogous to an apparent hydraulic conductance of the whole soil–plant system when g is multiplied by the prevailing leaf-to-air vapour pressure gradient (ΔW). Plant hydraulic efficiency could be used here as a relevant parameter, since ΔW was maintained constant for all measurements.

Mineral analyses

Dried leaf samples were ground and mineralized either in sulphuric acid and hydrogen peroxide for the determination of total N (sodium salicylate colorimetric method, autoanalyser Technicon) or in perchloric acid for the determination of other elements (plasma torch).

Table 43.1. Mycorrhizal status (proportion of short roots either dead, alive but non-mycorrhizal, mycorrhizal with *T. terrestris* or mycorrhizal with *L. laccata*) and mineral concentrations (mean values ± s.e.) in the needles of the plants at the end of the experiment (day 39) in the two treatments. No statistically significant differences were found between treatments ($P \leq 0.05$, Scheffe-*F* test).

	Thelephora terrestris	*Laccaria laccata*
Mycorrhizal status (% of short roots)		
dead	18.1	12.9
alive non-myc	12.6	19.0
myc Tt	62.1	17.9
myc Ll	7.2	50.2
myc total	69.3	68.1
Mineral concentrations (g 100g^{-1})		
nitrogen	1.70 ± 0.07	1.89 ± 0.06
phosphorus	0.086 ± 0.003	0.099 ± 0.005
potassium	0.89 ± 0.09	1.11 ± 0.07

Results

Plant biomass, mycorrhizal status and nutrient status

There was no significant difference between the two treatments in the biomass of the different plant compartments (data not shown) and in the root:shoot ratio (0.70). At the end of the experiment about 70% of the short roots were mycorrhizal in both treatments (Table 43.1). Among these roots Tt was predominant as the fungal associate in the non-inoculated treatment, while Ll was predominant in the inoculated treatment. Slightly higher mean values of nitrogen, phosphorus and potassium concentrations were found in the needles of the Ll treatment as compared with the Tt treatment (Table 43.1), but these differences were not statistically significant ($P \leq 0.05$, one-factor ANOVA followed by the Scheffe F-test).

Plant water relations and hydraulic efficiency

The time course of soil water depletion was the same in the two treatments (Fig. 43.1). Soil water content (SWC) (g g^{-1}) at field capacity was about 30% and dropped to 7.5% at the end of the drying cycle.

In both treatments, predawn leaf water potential remained unaffected until soil water content reached 15%, corresponding to a value of about one-third for the ratio of actual:maximum extractable water. The Tt plants exhibited a lower sensitivity of Ψ_{wp} to decreasing soil water content than the Ll plants, since at the

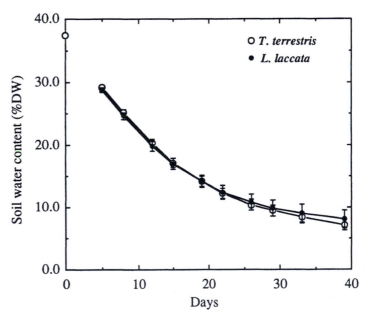

Fig. 43.1. Time-course of soil water content in the two treatments (days after water withholding). Bars denote ± s.e., $n = 12$.

end of the drying cycle the Ψ_{wp} values were -1.3 MPa and -2.15 MPa for the Tt and the Ll plants, respectively (Fig. 43.2). The Ll plants also exhibited a higher sensitivity of Ψ_{wm} to decreasing SWC, but the difference between the two treatments was less pronounced than for Ψ_{wp}.

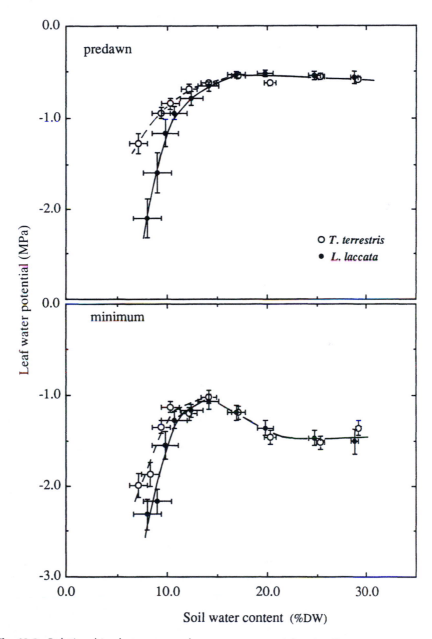

Fig. 43.2. Relationships between predawn water potential and soil water content; and between minimum leaf water potential and soil water content in the two treatments. Bars denote ± s.e., $n = 12$.

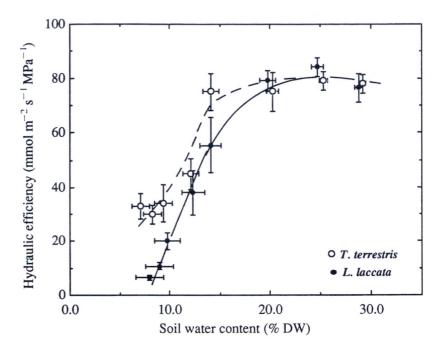

Fig. 43.3. Relationship between hydraulic efficiency and soil water content in the two treatments. Bars denote ± s.e., *n* = 12.

In the Ll plants a steep decrease in hydraulic efficiency was noticed from SWC values as high as 20% (Fig. 43.3) whereas in the Tt plants the decrease occurred only below 15% SWC and was less dramatic.

In Fig. 43.4, the values of needle osmotic potential obtained for the three measurement days have been plotted against the corresponding values of needle water potential. There was a general tendency for the Tt plants to exhibit lower osmotic potentials (Ψ_π) and thus higher turgor potentials ($\Psi_\tau = \Psi_w - \Psi_\pi$) than the Ll plants (difference of about 0.20–0.25 MPa), but this difference was statistically significant for Ψ_w superior to −1.0 MPa only.

Leaf gas exchange

In both treatments, leaf conductance to water vapour (g) presented a sharp decrease starting from a SWC of about 25% (Fig. 43.5a). In the Tt treatment it remained slightly higher than for Ll throughout the soil water depletion cycle. Carbon dioxide assimilation rate (A) was affected by the drought in the soil only below SWC = 20% and A was clearly higher in Tt than in Ll during the phase of decreasing A.

The ($A, C_i/C_a$) relationships characterizing the decrease in A during the soil water depletion cycle are shown in Fig. 43.6. Above a threshold value of A of about 50% of the maximum value, C_i/C_a decreased from 0.96 to 0.63 (Ll) and to 0.56

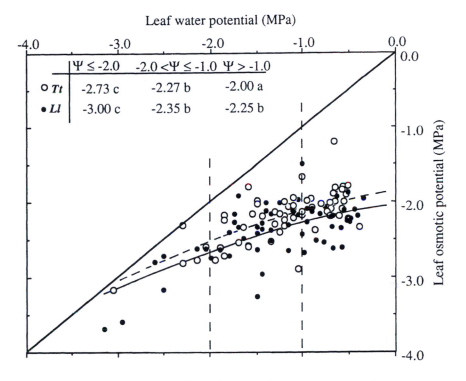

Fig. 43.4. Leaf osmotic potential as a function of leaf water potential in the two treatments. Data are individual values of one plant and correspond to days 1, 22 and 39 after water withholding. The inset gives the mean values of osmotic potential for different leaf water potential intervals. Numbers not sharing a common letter differ significantly ($P \le 0.05$) by Scheffe's F test.

(Tt) with decreasing A; below that threshold value C_i/C_a reincreased with decreasing A.

Discussion

In contrast with previous results (Guehl and Garbaye, 1990) with the same plant–fungal systems but in different nutritional conditions, A and g were the same for plants in both fungal treatments at the beginning of the drying cycle (Fig. 43.5), and P concentration in the needles did not differ at the end of the cycle (Table 43.1). Therefore, nutritional factors are not involved in the differential behaviour of the plants in drying soil conditions.

Osmotic adjustment was more efficient in plants associated with *L. laccata* than in those associated with *T. terrestris* (Fig. 43.4), resulting in turgor potentials about 0.2 MPa higher in the former, for a same leaf water potential. However, though soil water content did not differ between the two treatments at the end of the cycle, predawn leaf water potential was markedly lower (about 0.8 MPa) for the plants

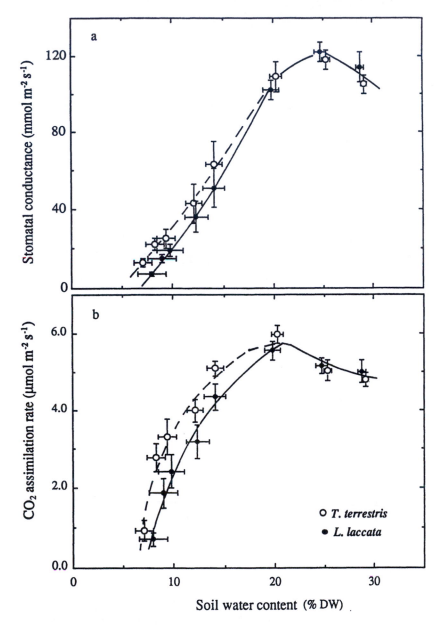

Fig. 43.5. Relationships between leaf conductance for water vapour and soil water content and between CO_2 assimilation rate and soil water content in the two treatments. Bars denote ± s.e., $n = 12$.

associated with *L. laccata* (Fig. 43.2). Thus, at the end of the cycle, turgor potential in these plants was about 0.6 MPa (0.8–0.2) lower than in those with *T. terrestris*. This is consistent with the fact that carbon dioxide assimilation and stomatal conductance also dropped faster with *L. laccata* (Fig. 43.5). This differential sensitivity of A can at least partly be attributed to non-stomatal processes as suggested by

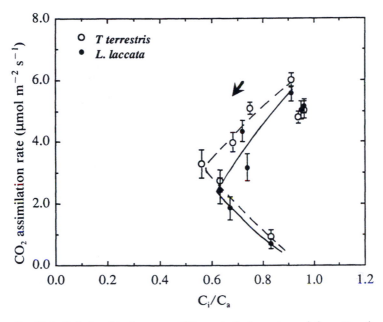

Fig. 43.6. Relationship between CO_2 assimilation rate and the ratio of internal to ambient CO_2 concentration in the two treatments. The arrow indicates increasing soil drought. Bars denote ± s.e., $n = 12$.

(1) increasing C_i/C_a with decreasing A at the low A values and (2) the discrepancy in the (A, C_i/C_a) pathway between the two treatments (Fig. 43.6).

The poor stability of leaf water potential of the Ll plants in response to soil drying is not associated with lower transpirational losses (Fig. 43.5), but is due to a lower hydraulic efficiency (Fig. 43.3). This efficiency involves water flow limitations in both the soil–plant interface (largely through the fungal symbiont) and in the conducting system of the plant. These results are consistent with those of Coleman *et al.* (1990) who, directly measuring root hydraulic conductivity on detached mycorrhizal root systems of Douglas fir, also found lower values for *L. laccata* than for *T. terrestris*.

In conclusion, the ectomycorrhizal fungus *L. laccata*, in spite of increased osmotic adjustment, is less efficient than *T. terrestris* for conferring drought resistance to Douglas fir. Many experimental results in forest nurseries and plantations in wet temperate climate with the same tree species and the same two fungi have shown that *L. laccata* was the more efficient at stimulating plant growth (Le Tacon *et al.*, 1988). The present results suggest that drought resistance is not the main factor involved in these responses.

References

Augé, R.M. and Stodola A.J.W. (1990) An apparent increase in symplasmic water contributes to greater turgor in mycorrhizal roots of droughted *Rosa* plants. *New Phytologist* 115, 285–295.

Boyd, R., Furbank, R.T. and Read D.J. (1986) Ectomycorrhiza and the water relations of trees. In: Gianinazzi-Pearson, Y. and Gianinazzi, S. (eds), *Physiology and Genetics, Proceedings 1st European Symposium on Mycorrhizae*, 1–5 July 1985, Dijon, INRA, Paris, pp. 689–693.

Coleman, M.D., Bledsoe, C. and Smit, B.A. (1990) Root hydraulic conductivity and xylem sap levels of zeatin riboside and abscisic acid in ectomycorrhizal Douglas fir seedlings. *New Phytologist* 115, 275–284.

Dosskey, M.G., Boersma, L. and Linderman, R.G. (1991) Role of the photosynthate demand of ectomycorrhizas in the response of Douglas fir seedlings to drying soil. *New Phytologist* 117, 327–334.

Duchaufour, P. and Bonneau, M. (1969) Une méthode nouvelle de dosage du phosphore assimilable dans les sols forestiers. *Bulletin de l'Association Française pour l'Etude du Sol* 4.

Duddridge, J.A., Malibari, A. and Read, D.J. (1980) Structure and function of mycorrhizal rhizomorphs with special reference to their role in water transport. *Nature* 287, 834–836.

Guehl, J.M. and Garbaye, J. (1990) The effects of ectomycorrhizal status on carbon dioxide assimilation capacity, water-use efficiency and response to transplanting in seedlings of *Pseudotsuga menziesii* (Mirb) Franco. *Annales Sciences Forestières* 21, 551–563.

Jones, H.G. (1985) Partitioning stomatal and non-stomatal limitations to photosynthesis. *Plant Cell and Environment* 8, 95–104.

Le Tacon, F., Garbaye, J., Bouchard, D., Chevalier, G., Olivier, J.M., Guimberteau, J., Poitou, N. and Frochot, H. (1988) Field results from ectomycorrhizal inoculation in France. In: Lalonde, M. and Piché, Y. (eds), *Proceedings Canadian Workshop, Mycorrhizae in Forestry*, Université Laval, Québec, pp. 51–74.

44 Protein Activities as Potential Markers of Functional Endomycorrhizas in Plants

S. Gianinazzi, V. Gianinazzi-Pearson, B. Tisserant and M.C. Lemoine

Laboratoire de Phytoparasitologie, INRA-CNRS, SGAP, INRA, BV 1540, 21034 Dijon cédex, France

Introduction

Mycorrhizal effectiveness is usually estimated in terms of plant growth or nutrient uptake relative to that seen in non-mycorrhizal plants, and there has been little research into the use of endomycorrhiza-specific protein activities to identify functional endomycorrhizal associations *in situ*. One of the best known physiological functions of proteins is their role as enzymes. Proteins with different enzyme activities have been studied in endomycorrhizas in attempts to identify fungal species (Sen and Hepper, 1986), to understand nutritional phenomena (Gianinazzi-Pearson and Smith, 1993), as a means of evaluating amounts of living mycelium (Kough *et al.*, 1987; Sylvia, 1987) and in analyses of cellular interactions between symbionts (Gianinazzi *et al.*, 1983).

Since the primary effect of endomycorrhiza in many situations is to improve phosphate uptake by the host plant through transport from the soil by the fungal mycelium, the most obvious protein expression to study in relation to mycorrhiza function is that represented by enzymes associated with phosphate metabolism. Previous biochemical and cytochemical studies have shown that (1) the activity of some enzymes involved in phosphate metabolism is modified in endomycorrhizal plants, and (2) endomycorrhizal fungi produce different types of phosphatases which appear to be activated by the host plant and to be related to crucial steps for the symbiotic expression of plant–fungal interactions (Table 44.1).

From what we know about these enzymes, all appear to be potential indicators of the functional state of the symbiosis. However, their relative activities in terms of functional mycorrhizal symbioses have been little studied. In the present chapter, we discuss the use of two fungal phosphatases: a wall-bound acid phosphatase of ericoid endomycorrhizal fungi and a vacuole localized alkaline phosphatase of vesicular-arbuscular (VA) endomycorrhizal fungi.

Table 44.1. Phospatases studied in endomycorrhizal symbioses.

Undetermined origin	
Surface phosphatases	Gianinazzi-Pearson and Gianinazzi, 1981
(acid, phytase)	Gianinazzi-Pearson and Gianinazzi, 1986
	Dodd et al., 1987
Host origin	
Neutral phosphatases	Gianinazzi et al., 1983
	Jeanmaire et al., 1985
ATPase/H⁺ ATPase	Marx et al., 1982
	Gianinazzi-Pearson et al., 1991
Fungal origin	
Polyphosphatases/polyphosphate kinases	Capaccio and Callow, 1982
Acid phosphatases	Straker and Mitchell, 1986
	Lemoine et al., 1992
Alkaline phosphatases	Gianinazzi-Pearson and Gianinazzi, 1983, 1986
	Tisserant et al., 1992
ATPase	Marx et al., 1982
	Gianinazzi-Pearson et al., 1991

Wall Bound Acid Phosphatase of Ericoid Endomycorrhizal Fungi

Ericoid endomycorrhizal fungi produce acid phosphatases which can mobilize phosphate from complexed or condensed phosphate esters (Straker and Mitchell, 1986). The major enzyme is localized in the fungal wall and its activity can be histochemically visualized over the hyphal surface of mycelium in pure culture (Straker et al., 1989). This fungal acid phosphatase activity can be inhibited by high levels of soluble phosphate (Lemoine et al., 1992), that is under conditions where endomycorrhiza would not be effective in improving host phosphate nutrition. Furthermore, this enzyme activity appears to be modulated by the host plant in the endomycorrhizal association. It is enhanced as hyphae approach the host root surface and is repressed as soon as hyphae penetrate and develop within living host cells (Gianinazzi-Pearson et al., 1986). All these observations suggest that fungal acid phosphatase activity could represent a potential marker of functional symbiosis in ericoid endomycorrhiza. The histochemically visualized fungal wall-bound phosphatase activity cannot, however, be distinguished from other phosphatases which may be present and, in particular, from those of the host plant. In an alternative immunocytochemical approach, using polyclonal antibodies specific to the wall bound acid phosphatase of *Hymenoscyphus ericae*, we have found that no substantial differences exist in the amount of acid phosphatase protein when either the level of phosphate changes in the culture medium, or whether hyphae develop outside or inside roots (Lemoine et al., 1992). This has led us to conclude that neither the activity nor the synthesis of the fungal acid phosphatase could be used to assay the functional state of ericoid endomycorrhiza by simple histochemical or immunocytochemical techniques.

Vacuole-localized Alkaline Phosphatase of VA Endomycorrhizal Fungi

An enzyme activity that we have studied in detail for diagnosing a functional VA endomycorrhizal symbiosis is fungal alkaline phosphatase activity. This activity which is localized in the fungal vacuole (Gianinazzi *et al.*, 1979; Dexheimer *et al.*, 1982) and can be detected by polyacrylamide gel electrophoresis (PAGE) in extracts of roots infected by different VA endomycorrhizal fungi (Fig. 44.1) (Gianinazzi-Pearson *et al.*, 1978; Gianinazzi-Pearson and Gianinazzi, 1983; Ollivier *et al.*, 1983), is linked to the mycorrhizal growth effect. For comparable infection levels, the extractable phosphatase activity diminishes with decreases in mycorrhizal effects, either following phosphate additions to soil (Gianinazzi-Pearson and

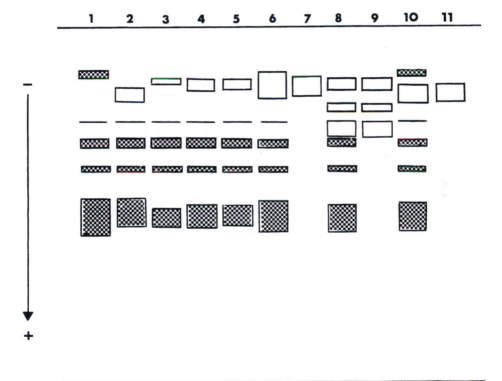

Fig. 44.1. Diagrammatic representation of electrophoretic patterns of 10% PAGE-separated alkaline phosphatases (☐) of VA endomycorrhizal fungi belonging to the genus *Glomus*, extracted from infected onion roots (2–6, 8, 10) or mycelium isolated from enzymically-digested roots (7, 9, 11). 1, Non-mycorrhizal root; 2, *G. vesiculifer* (LPA14); 3, *Glomus* spp. (LPA 15); 4, *G. intraradix* (LPA 8); 5, *Glomus* spp. (LPA 9); 6 and 7, *G. mosseae* (LPA 5); 8 and 9, *Glomus* E₃ (LPA 6); 10 and 11, *G. monosporum* (LPA 13). Other bands represent aspecific plant phosphatases.

Table 44.2. Alkaline phosphatase activities, endomycorrhizal infection and growth effects of VA endomycorrhizal fungi on soyabean and wheat growing in P-sufficient (amended with 0.25 g KH_2PO_4 kg^{-1}) and P-deficient (without fertilizer) soil (modified from Gianinazzi-Pearson and Gianinazzi, 1983).

	Infection (% root length)	Mycorrhizal effect on shoot fresh weight $\left(\dfrac{M-NM}{NM}\times 100\right)$	Relative fungal alkaline phosphatase activity (%)
Soyabean/*Glomus mosseae*			
P-sufficient soil	63.8	6.5	36.5
P-deficient soil	68.8	139.7	100.0
Soyabean/E_3			
P-sufficient soil	82.8	14.4	38.5
P-deficient soil	92.1	126.7	100.0
Wheat/*Glomus mosseae*			
P-sufficient soil	25.7	−13.0	32.9
P-deficient soil	22.3	15.2	100.0

M, mycorrhizal; NM, non-mycorrhizal.

Gianinazzi, 1983) (Table 44.2) or with changes in the host plant variety (Ollivier *et al.*, 1983). Furthermore, the VA fungal alkaline phosphatase activity is particularly enhanced when the mycorrhizal effect on plant growth becomes evident in young plants (Fig. 44.2) (Gianinazzi-Pearson and Gianinazzi, 1978). We have recently developed a simple histochemical test to estimate microscopically VA endomycorrhizal infections in terms of fungal alkaline phosphatase activity (Tisserant *et al.*, 1992). This enzyme activity appears to be enhanced by the presence of the host tissues since it is limited to the tip of germ tubes of VA fungi growing *in vitro* without the plant, while it can be detected throughout active fungal structures when the fungus develops inside the root cortex. Values for infection levels estimated in terms of fungal alkaline phosphatase activity are always lower than those obtained using the non-vital trypan blue staining technique which reveals both living and dead fungal structures. They are also consistently lower than infection levels detected by vital succinate dehydrogenase staining, which could be an indication that only a part of the living intraradical mycelium is functionally active in phosphate turnover at any given time. The proportion of both succinate dehydrogenase and alkaline phosphatase active mycelium reaches a maximum when plant growth stimulation occurs and decreases as the infection ages, in a similar way to the extractable enzyme (Fig. 44.2). Comparisons of succinate dehydrogenase and alkaline phosphatase activities in VA fungal mycelium could provide simple biochemical tests for investigating where VA endomycorrhiza are functioning in a given root system at a given time, and for more precisely estimating the effects of different environmental conditions on the functional state of the symbiosis.

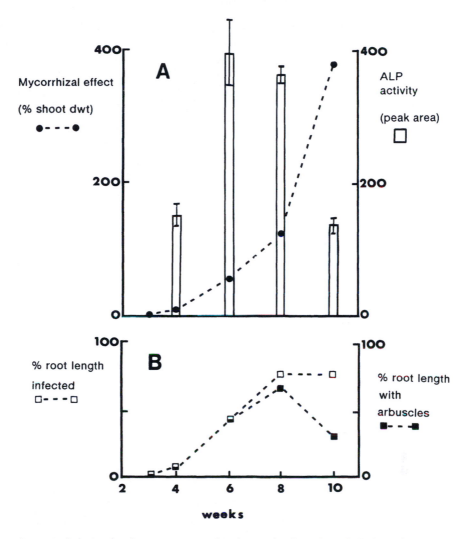

Fig. 44.2. Relationship between mycorrhizal growth effect, fungal alkaline phosphatase (ALP) activity detected by PAGE (A) and intensity of infection (B) of *Glomus mosseae* in *Allium cepa* roots.

Conclusion

The results obtained using different approaches have clearly shown that studies of protein expression in endomycorrhizal fungi can provide simple biochemical tests for evaluating the functional state of endomycorrhizas. The data accumulated in studies of the vacuolar-localized alkaline phosphatase of VA endomycorrhizal fungi are promising. Application of molecular biological techniques (PCR, mRNA, cDNA analyses) should make it possible to improve understanding of protein expression in terms of endomycorrhiza functioning. Eventually this should lead to a better definition of effective biochemical markers of the functional state of the symbiosis *in planta*, as well as providing essential tools for ecophysiological studies and for the rational use of endomycorrhizas in agroecosystems.

References

Capaccio, L.C.M. and Callow, J.A. (1982) The enzymes of polyphosphatase metabolism in vesicular-arbuscular mycorrhizas. *New Phytologist* 91, 81–91.

Dexheimer, J., Gianinazzi-Pearson, V. and Gianinazzi, S. (1982) Acquisitions récentes sur la physiologie des mycorhizes VA au niveau cellulaire. *Les Colloques de l'INRA* 13, 61–73.

Dodd, J.C., Burton, C.C., Burns, R.G. and Jeffries, P. (1987) Phosphatase activity associated with the roots and the rhizosphere of plants infected with vesicular-arbuscular mycorrhizal fungi. *New Phytologist* 107, 163–172.

Gianinazzi, S., Gianinazzi-Pearson, V. and Dexheimer, J. (1979) Enzymatic studies on the metabolism of vesicular-arbuscular mycorrhiza. III. Ultrastructural localisation of acid and alkaline phosphatase in onion roots infected by *Glomus mosseae* (Nicol. Gerd.) *New Phytologist* 82, 127–132.

Gianinazzi, S., Dexheimer, J., Gianinazzi-Pearson, V. and Marx, C. (1983) Role of the host arbuscule interface in the symbiotic nature of VA mycorrhizal associations: ultracytological studies of processes involved in phosphate and carbohydrate exchange. *Plant and Soil* 71, 211–215.

Gianinazzi-Pearson, V. and Gianinazzi, S. (1978) Enzymatic studies on the metabolism of vesicular-arbuscular mycorrhiza. II. Soluble alkaline phosphatase specific to mycorrhizal infection in onion roots. *Physiological Plant Pathology* 12, 45–53.

Gianinazzi-Pearson, V. and Gianinazzi, S. (1981) Role of endomycorrhizal fungi in phosphorus cycling in the ecosystem. In: Wicklow, D.T. and Carrol, G.C. (eds), *The Fungal Community: Its Organization and Role in the Ecosystem*. Marcel Dekker Inc., New York, pp. 637–652.

Gianinazzi-Pearson, V. and Gianinazzi, S. (1983) The physiology of vesicular-arbuscular mycorrhizal roots. *Plant and Soil* 71, 197–209.

Gianinazzi-Pearson, V. and Gianinazzi, S. (1986) Connaissances actuelles des bases physiologiques et biochimiques des effets des endomycorhizes sur le comportement des plantes. *Physiologie Végétale* 24, 253–262.

Gianinazzi-Pearson, V. and Smith, S.E. (1993) Physiology of mycorrhizal mycelia. In: Tommerop, I.C., Ingram, D.S. and Williams, P.H. (eds), *Mycorrhiza: A Synthesis, Advances in Plant Pathology* 8, Academic Press London (in press).

Gianinazzi-Pearson, V., Gianinazzi, S., Dexheimer, J., Bertheau, Y. and Asimi, S. (1978) Les phosphatases alcalines solubles dans l'association endomycorhizienne VA. *Physiologie Végétale* 16, 671–678.

Gianinazzi-Pearson, V., Bonfante-Fasolo, P. and Dexheimer, J. (1986) Ultrastructural studies of surface interactions during adhesion and infection by ericoid endomycorrhizal fungi. *NATO-ASI Series, Series H, Cell biology* 4, 273–282.

Gianinazzi-Pearson, V., Smith, S.E., Gianinazzi, S. and Smith, F.A. (1991) Enzymatic studies on the metabolism of vesicular-arbusuclar mycorrhizas V. Is H^+-ATPase a component of ATP-hydrolysing enzyme activities in plant-fungus interfaces? *New Phytologist* 117, 61–76.

Jeanmaire, C., Dexheimer, J., Marx, C., Gianinazzi, S. and Gianinazzi-Pearson, V. (1985) Effect of vesicular-arbuscular mycorrhizal infection on the distribution of neutral phosphatase activities in root cortical cells. *Journal of Plant Physiology* 119, 285–293.

Kough, J.L., Gianinazzi-Pearson, V. and Gianinazzi, S. (1987) Depressed metabolic activity of vesicular-arbuscular mycorrhizal fungi after fungicide applications. *New Phytologist* 106, 707–715.

Lemoine, M.C., Gianinazzi-Pearson, V., Gianinazzi, S. and Straker, C.J. (1992) Occurrence and expression of acid phosphatase of *Hymenoscyphus ericae* (Read) Korf & Kernan, in isolation or associated with plant roots. *Mycorrhiza* 1, 137–146.

Marx, C., Dexheimer, J., Gianinazzi-Pearson, V. and Gianinazzi, S. (1982) Enzymatic studies on the metabolism of vesicular-arbuscular mycorrhiza. IV. Ultra-cytoenzymological evidence (ATPase) for active transfer processes in the host-arbuscule interface. *New Phytologist* 90, 37–43.

Ollivier, B., Bertheau, Y., Diem, H.G. and Gianinazzi-Pearson, V. (1983) Influence de la variété de Vigna *unguiculata* dans l'expression de trois associations endomycorhiziennes à vésicules et arbuscules. *Canadian Journal of Botany* 61, 354–358.

Sen, R. and Hepper, C.M. (1986) Characterization of vesicular-arbuscular mycorrhizal fungi (*Glomus* sp.) by selective enzyme staining following polyacrylamide gel electrophoresis. *Soil Biology and Biochemistry* 18, 29–34.

Straker, C.J. and Mitchell, D.T. (1986) The activity and characterization of acid phosphatases in endomycorrhizal fungi of the Ericaceae. *New Phytologist* 104, 243–256.

Straker, C.J., Gianinazzi-Pearson, V., Gianinazzi, S., Cleyet-Marel, J.C. and Bousquet, N. (1989) Electrophoretic and immunological studies on acid phosphatase from a mycorrhizal fungus of *Erica hispidula* L. *New Phytologist* 111, 215–221.

Sylvia, D. (1987) Activity of external hyphae of vesicular-arbuscular mycorrhizal fungi. *Soil Biology and Biochemistry* 20, 39–43.

Tisserant, B., Gianinazzi-Pearson, V., Gianinazzi, S. and Gollotte, A. (1992) *In planta* histochemical staining of fungal alkaline phosphatase activity for analysis of efficient arbuscular mycorrhizal infections. *Mycological Research* 97 (in press).

45 Plant–Fungal Interface in VA Mycorrhizas: A Structural Point of View

P. Bonfante-Fasolo

Dipartimento di Biologia Vegetale dell'Università di Torino, Viale Mattioli 25 I-10125 Torino, Italy

The role of vesicular-arbuscular (VA) mycorrhiza in plant nutrition depends on the fungal ability to absorb specific nutrients from the soil and transfer them to the host plant, which, in turn, provides the fungus with photosynthates. This bidirectional exchange can be analysed not only at the ecological and organismic level (Fitter, 1991; Read, 1991), but also at the cellular and molecular one (Smith and Smith, 1990).

Here, the structures and the subcellular compartments which mediate the nutrient flux between the partners will be described.

Many morphological observations have shown that, during the establishment of the mycorrhizal symbiosis, contact between plant and fungus is mediated through their cell surfaces (Scannerini and Bonfante-Fasolo, 1983). Cell wall and cell wall-related molecules, together with the plasma membrane, have been hypothesized to play important roles in the short range interactions between the symbiotic partners. Nutrients as well as non-nutritional molecules – such as those involved in cell-to-cell signaling – must cross the interface formed by the contact of the partners' cell surfaces (Bonfante-Fasolo, 1988).

There are two different types of contact between partners, depending on whether the fungus occupies an intercellular or intracellular position. In the first case, as in ectomycorrhizas and in the intercellular phase of VA mycorrhizas, the cell walls of both partners are physically in contact. In the second case, mostly found in the intracellular phase of all endomycorrhizas (VA, ericoid, and orchid mycorrhiza), the thin fungal wall is separated from the host cytoplasm by the invaginated host plasmamembrane and by an interfacial material (Fig. 45.1).

This chapter will demonstrate that in VA mycorrhizas the zone of interface created during the intracellular phase of the fungus – and in particular during arbuscule development – is a compartment of high molecular complexity. As nutritional exchanges mostly occur through the plant–arbuscule interface, which is postulated to be a highly specialized structure (Cox and Tinker, 1976), understanding of the nature of such a compartment is crucial from a physiological point of view.

intercellular fungus

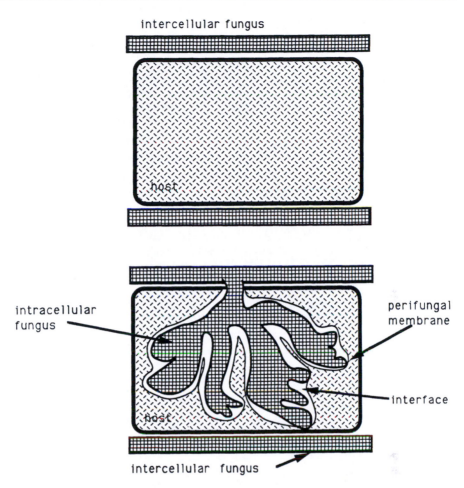

Fig. 45.1. The scheme represents the two different types of contacts or interfaces between plant and fungus in mycorrhizal associations, depending whether the fungus is intercellular (a) or intracellular (b). In the first case, the walls of both the partners are physically in contact. In the second case, the intracellular fungus is separated from the host cytoplasm by the invaginated host plasmamembrane (perifungal membrane) and by an interfacial material.

The Contribution of *in situ* Techniques

The recent development of affinity techniques, based on the specific non-covalent binding between molecules such as antigen-antibody, sugars–lectin, substrate–enzyme, has created a bridge between biochemical and morphological observations. The presence of specific molecules can be revealed in cell compartments, tissues and organs depending on the type of marker that has been used (colloidal gold for the electron and light microscope, fluorochromes for the fluorescence microscope) (Hall and Hawes, 1991 for a recent technical review). These *in situ* techniques have been demonstrated to be invaluable tools for the characterization of specific cell compartments and the understanding of biological processes. In plant biology, recent applications involve the synthesis of cell wall precursors and their assembly at the cell surface (Vian and Roland, 1991), pathogenic plant-fungal (Benhamou *et al.*, 1990) as well as analysis of symbiotic legume–*Rhizobium* interactions (Perotto *et al.*, 1991a). The use of these techniques to investigate the plant–fungus interface in different types of mycorrhizas has led to the identification of various carbohydrates, proteins and glycoproteins in the fungal wall, in the interfacial material and in the host perifungal membrane.

The Molecular Features of the Plant–Fungal Interface

VA mycorrhizal associations established on host plants as different as monocots and dicots (such as pea and *Ginkgo biloba*, or leek and maize) and a fungal symbiont (*Glomus versiforme* have been analysed with specific affinity probes. (Figs 45.2–45.6). The use of colloidal gold conjugated to cellobiohydrolase I (CBH I), an enzyme produced by *Trichoderma* spp. and specifically binding β (1–4) glucans, allowed us to localize the substrate not only over the host walls, but also around the branches of the fungal arbuscule. This probe was shown to be specific for the host wall, as the VA fungal wall was never labelled (Fig. 45.2). Such specificity indicates that the glucans found in the interface are of host origin.

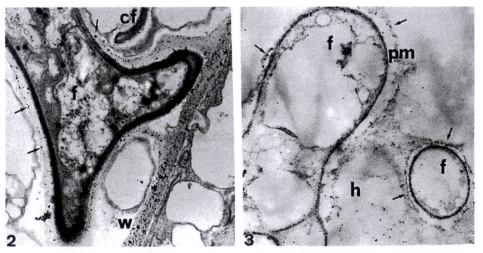

Fig. 45.2. Ultrastructural localization of β(1–4) glucans in a VAM root of *Ginkgo biloba* colonized by *Glomus* spp. The cellobiohydrolase-gold complex reveals the presence of glucans over the host wall (w) and the interface material (arrows). No labelling is present over the fungal wall (×24 000) (f: fungus, cf: collapsed fungus, h: host). **45.3.** Immunogold localization of HRGPs in a VAM root of *Pisum* colonized by *Glomus versiforme*. Colloidal gold granules are mostly present over the perifungal membrane (pm). Some granules are absorbed over the fungal wall (×30 500).

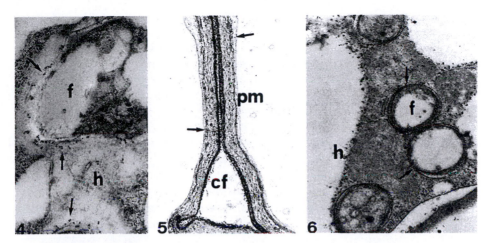

Figs 45.4., 45.5. and 45.6. Immunogold localization of pectins in a VAM root of *Pisum* colonized by *Glomus versiforme*.

Fig. 45.4. The monoclonal antibody JIM 5 reveals the presence of non-esterified pectins around the arbuscular branches (arrows) ($\times$ 30 000). **45.5.** Pectic material is abundant around the collapsed fungus (arrows) ($\times$ 24 000). **45.6.** No labelling is present around the arbuscular branches (f, arrows) when the monoclonal antibody JIM 7 is used. It mostly binds to methylesterified pectins ($\times$ 24 000).

In contrast, the use of the lectin wheat germ agglutinin (WGA), which is specific for oligomers of *N*-acetyglucosamine, has revealed that such residues are of fungal origin. They are present over the thin fungal wall, but they are not evident in the interface compartment (Bonfante-Fasolo *et al.*, 1990a).

The specificity of these components as plant or fungal products is not true of ericoid mycorrhiza, which are formed by the association between ericaceous plants and the ascomycete fungus *Hymenoscyphus ericae*. In this system, a colloidal gold-CBH I complex labelled not only the host cell walls, as expected, but also the fungal wall. Here fungal glucans are co-localized in the cell wall with chitin fibrils (Bonfante-Fasolo *et al.*, 1987). The cross-reactivity between plant and fungal glucans makes it impossible to define the origin of the material occurring in the interface that separates ericoid fungal partners from their hosts.

Non-esterified polygalacturonans were identified in the interfacial compartment in pea, leek and *Ginkgo* roots by using the monoclonal antibody, JIM 5 (Bonfante-Fasolo *et al.*, 1990b; Perotto *et al.*, 1990). Methylesterified polygalacturonans could also be identified by using another antibody, JIM 7, but this esterified form was rarely found in the interface (Figs 45.4, 45.5 and 45.6). Labelling with both these pectic molecules was present on the host cell walls and followed a precise distribution pattern, which appeared to be species and tissue specific (Bonfante-Fasolo and Perotto, 1992).

Cell wall proteins were also localized in the interfacial material. In particular, a polyclonal antibody raised against an extensin-like melon hydroxyproline rich glycoprotein, called HRGP$_{2b}$, (Mazau *et al.*, 1988) revealed related molecules over the walls of pea and leek plants and in the interface around the VAM fungus

(Fig. 45.3). The labelling was particularly evident over and near the host membranes (Bonfante-Fasolo *et al.*, 1991). The distribution of this protein component was analysed in maize mycorrhizal roots by using an antibody raised against a maize HRGP (Stiefel *et al.*, 1990). HRGP molecules from monocots are considered to be structurally different from the HRGPs occurring in dicots. However, the labelling pattern was very similar to that observed in dicots, gold granules being very abundant around the intracellular fungus.

The localization of HRGP molecules in the interface between plant and mycorrhizal fungus raises the question whether they represent structural cell wall molecules or are part of a plant defence response to the fungus.

It has been recently discovered that HRGP molecules are present over the peribacteroid membrane in the *Rhizobium*–legume symbiosis (Benhamou *et al.*, 1991) and some early nodulins are closely related to them in their primary structure (Van de Wiel *et al.*, 1990), but the extent of their involvement in the establishment of the symbiotic condition is not known.

In endomycorrhizas, the interfacial compartment is limited by a membrane of host origin which surrounds the fungus. The morphological continuity between the host plasmamembrane and the invaginated membrane suggests their relatedness, even if important differences exist concerning their functional activities (Smith and Smith, 1990; Gianinazzi-Pearson *et al.*, 1991). An invaginated membrane around a penetrating organism is a common pattern in heterologous interactions in the plant kingdom. An interesting approach for investigation of the perifungal membrane in endomycorrhizas is therefore to compare its molecular features with those shown by the invaginated membranes of other symbiotic associations.

A set of monoclonal antibodies (MAbs) was raised against molecular components of the peribacteroid membrane in pea nodules (VandenBosch *et al.*, 1989; Perotto *et al.*, 1991a). Some of the epitopes recognized by these MAbs were partially characterized and shown to be developmentally regulated during the formation of pea root nodules, probably being involved with different stages of interaction with *Rhizobium*. By using these MAbs as probes to study VA mycorrhizal pea roots it was possible to demonstrate that most components which were present on the peribacteroid membranes were also found to be expressed on the periarbuscular membrane. This suggests a strong relatedness between these two plant-derived membranes (Perotto in preparation). These data extend previous reports by Gianinazzi-Pearson *et al.* (1990) and confirm results by Wyss *et al.* (1990). The latter authors detected similar products in the membrane fractions of VA mycorrhiza and nodules by using antibodies to nodule specific gene products. These observations suggest that the molecular basis of the plant response during symbiosis may be based on the expression of similar proteins, irrespective of the penetrating organism.

Conclusions

The molecular dissection of the interfacial compartment in endomycorrhizas during arbuscule formation suggests the following conclusions.

1. The interface can be seen as an apoplastic compartment, as it shares common molecules with the host cell wall. However, the structure of the interface is morphologically different from the cell wall.
2. During tissue invasion by the VA fungus, a structure comparable with the infection thread produced by invading *Rhizobium* bacteria is formed. As in mycorrhizas, the infection thread wall possesses glucans and pectins (Rae *et al.*, 1992). However, the final stages of the interaction are different in endomycorrhizas and nodules.
3. The interface compartment may be an expression of compatible status, it being formed not only in mutualistic symbioses but whenever there is a biotrophic situation. In compatible associations between plants and biotrophic pathogenic fungi, an interface area is limited by the perihaustorial membrane containing molecules produced by both partners (Harder, 1989; Mackie *et al.*, 1991). In contrast, when an aggressive fungus, for example a pathogenic strain of *Rhizoctonia solani*, challenges a pea root, the host membrane is immediately broken, the interface area is not formed, there are no modifications in glucans or pectin deposition (Perotto *et al.*, 1991b), though there is an intense deposition of HRGP molecules.

Many questions concerning the structure of the interface, are still unanswered: How is this compartment built? How is the modification in the targeting of cell wall components regulated? What is the role of the plant cytoskeleton in the invagination of the plasmamembrane and the development of the interface? How is it related to the centralization of the nucleus? Is the cellulose synthase expressed and active on the periarbuscular membrane? Why is the structure of the interface different from the peripheral wall?

Answers to some of these questions may provide clues to a better understanding of the molecular nature of the interfacial compartment and to its function during plant–fungal exchanges.

References

Benhamou, N., Mazau, D. and Esquerré-Tugayé, M.T. (1990) Immunocytochemical localization of hydroxyproline-rich glycoproteins in tomato root cells infected by *Fusarium oxisporum* f. sp. *radicis-lycopersici*: study of a compatible interaction. *Phytopathology* 80, 163–173.

Benhamou, N., Lafontaine, P.J., Mazau, D. and Esquerré-Tugayé, M.T. (1991) Differential accumulation of hydroxyproline-rich glycoproteins in bean root nodule cells infected with a wild type strain or a C_4-dicarboxylic acid mutant of *Rhizobium leguminosarum* bv. *phaseoli*. *Planta* 184, 457–467.

Bonfante-Fasolo, P. (1988) The role of the cell wall as a signal in mycorrhizal associations. In: Scannerini S., Smith D., Bonfante-Fasolo P. and Gianinazzi-Pearson V. (eds), *Cell to Cell Signals in Plant, Animal and Microbial Symbiosis*, NATO ASI Series, H 17, Springer Verlag, Berlin, pp. 219–235.

Bonfante-Fasolo, P. and Perotto, S. (1992) Plants and endomycorrhizal fungi: the cellular and molecular basis of their interaction. In: D.P. Verma (ed.), *Molecular signals in Plant-Microbe communication*, CRC Press Boca Raton, USA, pp. 445–470.

Bonfante-Fasolo, P., Perotto, S., Testa, B. and Faccio, A. (1987) Ultrastructural localization of cell surface sugar residues in ericoid mycorrhizal fungi by gold-labeled lectins. *Protoplasma* 139, 25–35.

Bonfante-Fasolo, P., Faccio, A., Perotto, S. and Schubert, A. (1990a) Correlation between chitin distribution and cell wall morphology in the mycorrhizal fungus *Glomus versiforme*. *Mycological Research* 94, 157–165.

Bonfante-Fasolo, P., Vian, B., Perotto, S., Faccio, A. and Knox, J.P. (1990b) Cellulose and pectin localization in roots of mycorrhizal *Allium porrum*: Labelling continuity between host cell wall and interfacial material. *Planta* 180, 537–47.

Bonfante-Fasolo, P., Tamagnone, L., Peretto, R., Esquére-Tugaye, M.T., Mazau, D., Mosiniak, M. and Vian, B. (1991) Immunocytochemical location of hydroxyproline rich glycoproteins at the interface zone between a mycorrhizal fungus and host plants. *Protoplasma* 165, 127–138.

Cox, G. and Tinker, P.B. (1976) Translocation and transfer of nutrients in vesicular-arbuscular mycorrhizas. I. The arbuscule and phosphorus transfer: a quantitative ultrastructural study. *New Phytologist* 77, 371–378.

Fitter, A.H. (1991) Costs and benefits of mycorrhizas: implications for functioning under natural conditions. *Experientia* 47, 350–355.

Gianinazzi-Pearson, V., Gianinazzi, S. and Brewin, N.J. (1990) Immunocytochemical localisation of antigenic sites in the perisymbiotic membrane of vesicular-arbuscular endomycorrhiza using monoclonal antibodies reacting against the peribacteroid membrane of nodules. In: Nardon, P., Gianinazzi-Pearson, V., Grenier, A.M., Margulis, L. and Smith, D.C. (eds), *Endocytobiology IV*. INRA Press, Paris, pp. 127–131.

Gianinazzi-Pearson, V., Smith, S.E., Gianinazzi, S. and Smith, F.A. (1991) Enzymatic studies on the metabolism of vesicular-arbuscular mycorrhizas. V. Is H^+-ATPase a component of ATP-hydrolysing enzyme activities in plant-fungus interfaces? *New Phytologist* 117, 61–74.

Hall, J.L. and Hawes, C. (1991) *Electron Microscopy of Plant Cells*. Academic Press, London, New York.

Harder, D.E. (1989) Rust fungal haustoria–past, present future. *Canadian Journal of Plant Pathology* 11, 91–99.

Knox, J.P., Linstead, P.J., King, J., Cooper, C. and Roberts, K. (1990) Pectin esterification is spatially regulated both within cell walls and between developing tissues of root apices. *Planta* 181, 512–521.

Mackie, A.J., Roberts, A.M., Callow, J.A. and Green, J.R. (1991) Molecular differentiation in pea powdery – mildew haustoria. *Planta* 183, 399–408.

Mazau, D., Rumeau, D. and Esquerré-Tugayé, M.T. (1988) Two different families of hydroxyproline-rich glycoproteins in melon callus. *Plant Physiology* 86, 540–546.

Perotto, S., VandenBosch, K.A., Brewin, N.J., Faccio, A., Knox, J.P., and Bonfante-Fasolo, P. (1990) Modifications of the host cell wall during root colonization by *Rhizobium* and VAM fungi. In: Nardon, P., Gianinazzi-Pearson, V., Grenier, A.M., Margulis, M. and Smith, D.C. (eds), *Endocytobiology IV*, INRA Presse, Paris, pp. 115–117.

Perotto, S., VandenBosch, K.A., Butcher, G.W. and Brewin, N.J. (1991a) Molecular composition and development of the plant glycocalyx associated with the peribacteroid membrane of pea root nodules. *Development* 112, 763–774.

Perotto, S., Tamagnone, L., Knox, J.P., Faccio, A. and Bonfante-Fasolo, P. (1991b) Changes in the pea cell wall during interactions with a mutualistic and a pathogenic fungus. *Giornale Botanico Italiano* 125, 82–83.

Rae, A.L., Bonfante-Fasolo, P. and Brewin, N.J. (1992) Structure and growth of infection threads in the legume symbiosis with *Rhizobium leguminosarum*. *The Plant Journal* 2, 385–395.

Read, D.J. (1991) Mycorrhizas in ecosystems. *Experientia* 47, 376–391.

Scannerini, S. and Bonfante-Fasolo, P. (1983) Comparative ultrastructural analysis of mycorrhizal associations. *Canadian Journal of Botany* 61, 917–943.

Smith, S.E. and Smith, F.A. (1990) Structure and function of the interfaces in biotrophic symbioses as they relate to nutrient transport. *New Phytologist* 114, 1–38.

Stiefel, V., Ruiz-Avila, L., Raz, R., Vallés, M.P., Gomez, J., Pagés, M., Martìnez-Izquierdo, J.A., Ludevid, M.D., Langdale, J.A,. Nelson, T. and Puigdoménech, P. (1990) Expression of a maize cell wall hydroxyproline-rich glycoprotein gene in early leaf and root vascular differentiation. *Plant Cell* 2, 785–793.

VandenBosch, K.A., Bradley, D.J., Knox, J.P., Perotto, S., Butcher, G.W. and Brewin, J.B. (1989) Common components of the infection thread matrix and the intercellular space identified by immunocytochemical analysis of pea nodules and uninfected roots. *EMBO Journal* 8, 335–342.

Van de Wiel, C., Scheres, B., Franssen, H., VanLierop, M.J., VanLammeren, A., VanKammen, A. and Bisseling, T. (1990) The early nodulin transcript ENOD2 is located in the nodule parenchyma (inner cortex) of pea and soybean root nodules. *EMBO Journal* 9, 1–7.

Vian, B. and Roland, J.C. (1991) Affinodetection of the sites of formation and of the further distribution of polygalacturonans and native cellulose in growing plant cells. *Biology of the Cell* 71, 43–55.

Wyss, P., Mellor, R.B. and Wiemken, A. (1990) Vesicular-arbscular mycorrhizas of wild type soybean and non-nodulating mutants with *Glomus mosseae* contain symbiosis specific polypeptides (mycorrhizins), immunologically cross-reactive with nodulins. *Planta* 82, 22–26.

46 The Role of Ion Channels in Controlling Solute Exchange in Mycorrhizal Associations

M. Tester[1], F.A. Smith[1] and S.E. Smith[2]

[1]Department of Botany, University of Adelaide, Adelaide, South Australia 5001: [2]Department of Soil Science, Waite Agricultural Research Institute, University of Adelaide, Glen Osmond, SA 5064, Australia

Introduction

A key feature of the function of the interface between symbionts in mycorrhizal symbioses is an increased loss from each partner of particular solutes (Patrick, 1989; Smith and Smith, 1990); these solutes are then absorbed by the other symbiont. Cells can sometimes lose a wide range of solutes due to a general breakdown of the cell membrane, but in the case of mycorrhizal symbioses, the loss is at least partially specific. Generally speaking, the root cells lose photosynthate but not phosphate, and the fungal hyphae lose phosphate but not photosynthate. Unless the characteristics of the cells and their plasma membranes are radically different from 'ordinary' cells, it would be expected that the movement of both these solutes from cells would be passive. Due to the specificity of the loss of solutes from the partners in the symbiosis, a general increase in membrane 'leakiness' is unlikely; the loss most probably occurs through selective transport proteins, in particular through ion channels. In this chapter, we will speculate on the mechanisms for and control of the loss of phosphate from the fungus. This is of particular interest as cells do not normally lose phosphate, whereas the loss of photosynthate from plant cells is a normal feature of growth in intact plants. New methods are now available for the investigation of these possibilities.

Phosphate-coupled Transport

The loss of phosphate from the fungus is thermodynamically downhill, being driven by both an electrical gradient (with the cytoplasm negative relative to the wall) and a concentration gradient (say, 5 mM $H_2PO_4^-$ in the cytoplasm (Harold, 1962) and 1–10 μM in the wall (based on the affinity of plant phosphate transporters)). This suggests that efflux is most likely to occur by a passive transport rather than by an active extrusion. Nevertheless, the loss of phosphate from the fungus could be directly coupled to the uptake of or loss of another solute; this would be analogous to H^+-coupled solute transport which is widespread in plant and fungal cells, only phosphate is driving the transport, rather than H^+. Such phosphate-coupled transport has been described in bacteria (2 phosphate: 1 sugar phosphate exchange: Maloney *et al.*, 1984), chloroplasts (1 phosphate: 1 triose phosphate exchange: Flügge and Heldt, 1991) and mitochondria (a phosphate: dicarboxylate exchanger: Day and Wiskich, 1984). Depending on the degree of homology with a proposed phosphate-coupled transporter in the fungus, either inhibitors or antibodies could be used to investigate the existence and function of a phosphate-coupled transporter.

To our knowledge, no phosphate-coupled transport has been described in the plasma membrane of an animal, plant or fungus; nevertheless, such a mechanism should be seriously considered as a mechanism of phosphate efflux. It would have to act in parallel with other carbohydrate transporters, otherwise a restrictive stoichiometric linkage of phosphate and carbohydrate movement would result, but there is no reason why this could not occur. However, such transport would be an unusual (in fact, probably unique) mechanism to move phosphate out of a cell. A simpler mechanism, and one which has been described in the plasma membrane of fungi, plants and animals is via an ion channel.

Phosphate Efflux Through Ion Channels

Ion channels are proteins which catalyse rapid, passive electrogenic uniport of ions through pores spanning an otherwise poorly permeable lipid bilayer. Our current knowledge of fungal and plant channels is reviewed by Tester (1990). Although no ion channel known in any system is specific for the movement of phosphate, considerable efflux of phosphate can 'accidentally' occur through channels 'designed' to pass other ions, such as Cl^-. (This raises the question of the primary function of a particular channel protein which has been assumed to be designed for the ion of particular interest to the researcher.)

In a variety of systems, such as the outer membrane of all mitochondria (including those of fungi: Manella *et al.*, 1983), the sarcoplasmic reticulum of frog skeletal muscle (Hals *et al.*, 1989) and the plasma membrane of human T lymphocytes (Schlichter *et al.*, 1990), there exists a channel called the 'VDAC', which allows phosphate through at about half the rate of Cl^-. Although in the plasma membrane of some animal tissues, this particular channel has not been described in the ordinary plasma membrane of fungi or plants. Nevertheless, we know this channel is synthesized in fungal cells, but upon synthesis it would need to be redirected to the plasma membrane rather than the outer mitochondrial membrane, in response to the establishment of the mycorrhizal symbiosis.

Another low specificity anion permeable channel has been described in the plasma membrane of plants and fungi. In stomatal guard cells, an anion channel is reported to be almost as permeable to malate as to Cl^- (Hedrich *et al.*, 1990). In the plasma membrane of the water mould, *Blastocladiella emersonii*, an anion channel passes ions as diverse as NO_3^-, Cl^-, malate and the buffer, 1,4-piperazinediethanesulphonic acid (PIPES) and passes phosphate in significant amounts (Caldwell *et al.*, 1986). Thus, it is quite possible for phosphate to be released from the fungus through an ion channel which already exists in the plasma membrane and whose permeation characteristics do not need to be altered. Despite the apparent lack of specificity of the anion channel, the net flux from the fungus may be dominated by phosphate if its cytoplasmic free concentration is relatively high, or other ions may be recycled into the fungus by active transporters. In both cases, the transfer of anions would appear to be largely specific to phosphate (see Introduction).

Control of Ion Channel Activity

The most likely mechanism to stimulate the release of phosphate through an already existing channel would be to increase the probability of the channel being 'open' (i.e. actively transporting). Based on work with other cells, this can be done by a wide variety of mechanisms and provides almost endless possibilities for the plant to stimulate, actively, phosphate release from the fungus (see Table 46.1).

The anion channel of the water mould *Blastocladiella emersonii* is opened by an increase in cytosolic Ca^{2+} (Caldwell *et al.*, 1986), and an apparently similar channel found in the plasma membrane of guard cells appears to be opened by alterations in potential difference, cytosolic Ca^{2+} and nucleotides (Hedrich *et al.*, 1990). Another anion channel found in the giant celled alga, *Chara corallina* (and whose permeability to phosphate is still unknown), appears to be opened by a decrease in cytosolic pH (Smith and Reid, 1991).

Table 46.1. Known controls of ion channels in animals and plants, and examples of how the plant may stimulate fungal anion channels.

Controls of ion channels (in particular, opening)	How plant could open fungal channels
Ligands (e.g. acetylcholine) – not yet seen in plants or fungi	Release ligand
Potential difference (especially depolarization)	Increase wall K^+, decrease wall Ca^{2+}, decrease wall pH
Increase in cytosolic free Ca^{2+}	Stimulate opening of Ca^{2+} channels; inhibit Ca^{2+} pump (how?)
Inositol phosphates, ATP, AMP, GTP, etc. Phosphorylation	Release growth regulators, etc. which act by alteration of 2° messengers
Specific pharmacological agents	Synthesize or accumulate, and release upon infection
Stretch	Increase turgor of plant relative to fungus

Suggestions for Further Study

There are, therefore, many realistic possibilities for control of phosphate loss by the plant through ion channels in the fungal plasma membrane. These can be experimentally investigated using a variety of techniques, such as tracer efflux from cultured fungus (see Cairney and Smith, Chapter 47, this volume), patch clamp electrophysiology of fungal protoplasts (Tester, 1990) and scanning confocal fluorescence microscopy of the intact symbiosis with ion-sensitive ratio fluorochromes (Fricker *et al.*, 1992). Microscopic techniques using antibodies raised against transporters from other systems could also prove useful. For example, there exist antibodies for the VDAC raised against protein purified from yeast which recognize single mitochondrial proteins from *Neurospora* (Forte *et al.*, 1987); using these antibodies, it would be possible to check for the existence of the VDAC in the fungal plasma membrane at the mycorrhizal interface. The possibilities are many, and the prospects for useful results seem promising.

References

Caldwell, J.H., van Brunt, J. and Harold, F.M. (1986) Calcium-dependent anion channel in the water mold, *Blastocladiella emersonii*. *Journal of Membrane Biology* 89, 85–97.

Day, D.A. and Wiskich, J.T. (1984) Transport processes of isolated mitochondria. *Physiologie Végétale* 22, 241–61.

Flügge, U.-I. and Heldt, H.W. (1991) Metabolite translocators of the chloroplast envelope. *Annual Review of Plant Physiology and Plant Molecular Biology* 42, 129–144.

Forte, M., Adelsberger-Mangan, D. and Colombini, M. (1987) Purification and characterization of the voltage-dependent anion channel from the outer mitochondrial membrane of yeast. *Journal of Membrane Biology* 99, 65–72.

Fricker, M., Tester, M. and Gilroy, S. (1992) Fluorescence and luminescence techniques to probe ion activities in living plant cells. In: Mason, W.T. (ed.), *Fluorescent Probes for Biological Function of Living Cells – a Practical Guide*. Academic Press, London.

Hals, G.D., Stein, P.G. and Palade, P.T. (1989) Single channel characteristics of a high conductance anion channel in 'sarcoballs'. *Journal of General Physiology* 93, 385–410.

Harold, F.M. (1962) Depletion and replenishment of the inorganic polyphosphate pool in *Neurospora crassa*. *Journal of Bacteriology* 83, 1047–1057.

Hedrich, R., Busch, H. and Raschke, K. (1990) Ca^{2+} and nucleotide dependent regulation of voltage dependent anion channels in the plasma membrane of guard cells. *EMBO Journal* 9, 3889–3892.

Maloney, P.C., Ambudkar, S.V., Thomas, J. and Schiller, L. (1984) Phosphate/hexose-6-phosphate antiport in *Streptococcus lactis*. *Journal of Bacteriology* 158, 238–245.

Manella, C.A., Colombini, M. and Frank, J. (1983) Structural and functional evidence for multiple channel complexes in the outer membrane of *Neurospora crassa* mitochondria. *Proceedings of the National Academy of Science USA* 80, 2243–2247.

Patrick, J.W. (1989) Solute efflux from the host at plant-microorganism interfaces. *Australian Journal of Plant Physiology* 16, 53–67.

Schlichter, L.C., Grygorczyk, R., Pahapill, P.A. and Grygorczyk, C. (1990) A large, multiple-conductance chloride channel in normal human T-lymphocytes. *Pflugers Archiv* 416, 413–421.

Smith, F.A. and Reid, R.J., (1991) Biophysical and biochemical regulation of cytoplasmic pH in *Chara corallina* during acid loads. *Journal of Experimental Botany* 42, 173–182.

Smith, S.E. and Smith, F.A. (1990) Structure and function of the interfaces in biotrophic symbioses as they relate to nutrient transport. *New Phytologist* 114, 1–38.

Tester, M. (1990) Plant ion channels: whole cell and single channel studies. *New Phytologist* 114, 305–340.

47 Effect of Monovalent Cations on Efflux of Phosphate from the Ectomycorrhizal Fungus *Pisolithus tinctorius*

J.W.G. Cairney and S.E. Smith

Department of Soil Science, Waite Agricultural Research Institute, The University of Adelaide, Glen Osmond, SA 5064, Australia

Introduction

In mycorrhizal associations the rate of transfer of nutrients from fungus to plant will be controlled by the rate of release from the fungus into the apoplast of the fungus–plant interface and the subsequent rate of absorption by the host root. Calculations of phosphate flux across the symbiotic interface in vesicular-arbuscular (VA) mycorrhizas indicate rates of transfer of a similar order to measured rates of phosphate absorption by plant cells (Cox and Tinker, 1976; S.E. Smith *et al.*, unpublished results). This clearly implies that the rate of phosphate efflux from the fungus must be of a similar magnitude to the rate of absorption by the plant. Although no data are available at present, a similar situation is likely to apply in ectomycorrhizas.

Loss of phosphate from fungi is, under 'normal' circumstances, generally slight, being at most 10% of the rate of absorption (Beever and Burns, 1980). It has therefore been proposed that specific conditions favouring efflux of phosphate from the fungal symbiont, such as local ionic concentrations, might exist in the apoplast of the fungus–plant interface in mycorrhizas (Smith and Smith, 1990). We are currently investigating the effect of various exogenous factors on efflux of phosphate from mycelium of the ectomycorrhizal basidiomycete *Pisolithus tinctorius* and report here some preliminary findings on the effects of monovalent cations.

Materials and Methods

Pisolithus tinctorius (Pers.) Coker and Couch (isolate DI-15, Grenville *et al.*, 1986) was grown as unshaken liquid cultures on a medium containing (g l^{-1}): $(NH_4)_2HPO_4$, 0.50; KH_2PO_4, 0.30; $MgSO_4.7H_2O$, 0.14; $CaCl_2$, 0.05; NaCl, 0.025; $ZnSO_4$, 0.003; ferric EDTA, 0.038; thiamine, 0.1; glucose, 10.0 (adjusted to pH 5.5). Following 10–40 days incubation at 25°C mycelial mats from which agar inoculum plugs had been excised using a cork borer, were transferred to 0.1 M citrate buffer pH 5.1 containing 100 μM KH_2PO_4 and 500 μM $CaCl_2$ for 30 min. Mycelium was subsequently transferred to 100 ml of the same solution labelled with 370 Bq ^{32}P orthophosphate and incubated with stirring for 16 h.

Following loading with ^{32}P, mycelium was incubated in sequential 15 ml volumes of efflux solution over an 8-h period. Efflux periods were initially 1 min, increasing progressively to 1 h. At the conclusion of the 8 h elution period the mycelium was dried overnight at 80 °C, weighed and, as with each 15 ml volume of solution, ^{32}P content was determined by Cerenkov counting in a liquid scintillation counter. Loading of mycelium with ^{32}P and isotope elution were carried out at 18–24 °C. Mycelium was incubated in the following efflux solutions: distilled H_2O, citrate buffer pH 5.1 citrate buffer pH 5.1 containing 0.1 mM KH_2PO_4 and 0.5 mM $CaCl_2$ (uptake soln.), citrate buffer pH 5.1 (replaced with citrate buffer pH 2.4 after either 15 min (early) or 2 h (late)), 50 mM NaCl, 0.35 mM NaCl, 50 mM KCl, 0.35 mM KCl, 50 mM KCl + 0.05 mM $CaCl_2$, 50 mM KCl + 50 mM $CaCl_2$.

Control experiments, where the fungus was either killed or membrane transport inhibited, were conducted in order to determine the time required for removal of ^{32}P from cell walls and free space. In these experiments loss of ^{32}P from mycelium which had been treated with either 1 mM 2,4-dinitrophenol for 2 h before and during loading with ^{32}P, or fixed for 16 h in 2.5% glutaraldehyde or fumigated for 16 h with chloroform before loading was determined as described above.

Results and Discussion

Control experiments indicated that following the 16 h labelling period ten times less ^{32}P was associated with killed or inhibited mycelium than with the untreated fungus (*c.* 16 μmol g^{-1} and 213 μmol g^{-1} (DW) respectively). Ninety-nine percent of the ^{32}P associated with killed or inhibited mycelium (taken to represent ^{32}P in the hyphal apoplast) was removed during the first 40 min of the efflux. In the preliminary comparison of the effect of different efflux solutions on loss of ^{32}P from the *P. tinctorius* described here we therefore consider only that lost during the 40 min–8 h portion of the efflux period. It is important to note that some efflux of P from within the cytoplasm is likely to have occurred during the initial 40 min and the data may thus underestimate total efflux across the plasmamembrane.

Approximately 9% of the ^{32}P present in mycelium after the initial 40 min elution was effluxed into distilled H_2O during the period 40 min–8 h (Fig. 47.1a). Incubation in citrate buffer pH 5.1 significantly increased efflux to *c.* 19% of absorbed ^{32}P, however when citrate buffer pH 5.1 contained Ca^{2+} and PO_4^{3-}, efflux was similar to that into distilled H_2O (9%). Replacing citrate buffer pH 5.1 with the same buffer at pH 2.4 after 15 min elution similarly reduced efflux to the dist. H_2O value. Extending the elution period in citrate buffer pH 5.1 to 2 h before transfer to pH 2.4, significantly increased efflux to *c.* 15% of absorbed ^{32}P. A major difference between citrate buffer pH 5.1 and pH 2.4 is the concentration

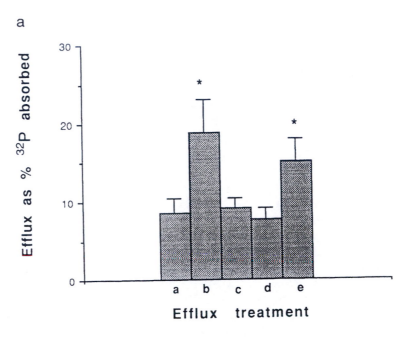

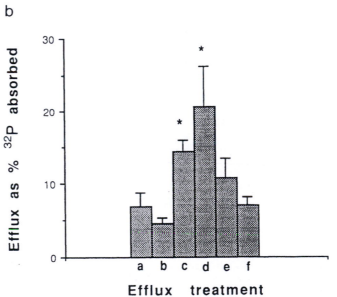

Fig. 47.1. Mean efflux (+s.e.) of ^{32}P from mycelium of *Pisolithus tinctorius* (expressed as % ^{32}P absorbed). Significantly differing treatments were identified from l.s.d. values computed following one-way ANOVA of log-transformed data. (**a**) a = distilled H_2O; b = citrate buffer (5.1); c = uptake solution; d = citrate buffer (5.1/2.4 early); e = citrate buffer (5.1/2.4 late). * denotes efflux significantly greater than distilled H_2O ($P < 0.01$). (**b**) a = 0.35 mM Na$^+$; b = 0.35 mM K$^+$; c = 50 mM Na$^+$; d = 50 mM K$^+$; e = 50 mM K$^+$ + 0.50 mM Ca^{2+}; f = 50 mM K$^+$ + 50 mM Ca^{2+}. * denotes efflux significantly greater than 0.35 mM Na$^+$ and 0.35 mM K$^+$ ($P < 0.01$).

of Na$^+$, *c.* 100 mM and 10 mM respectively. To investigate the putative stimulatory effect of monovalent cations further we assessed efflux into high and low concentrations of Na$^+$ or K$^+$.

Efflux of ^{32}P in the presence of 0.35 mM Na$^+$ or K$^+$ was similar to or lower than with distilled H$_2$O (Fig. 47.1b). Significantly greater efflux occurred into 50 mM Na$^+$ or K$^+$ (14% and 21% respectively). Inclusion of Ca^{2+} in the 50 mM K$^+$ efflux solution reduced efflux to the distilled H$_2$O level (Fig. 47.1b).

The results presented here clearly indicate that high external concentrations of monovalent cations can significantly increase efflux of P from mycelium of *P. tinctorius*. Under the experimental conditions we adopted, the mean rate of efflux was approximately doubled from the 'normal' distilled H$_2$O level of *c.* 9% to 21% in the presence of 50 mM K$^+$.

Electrophysiological studies have demonstrated that high concentrations of exogenous K$^+$ or Na$^+$ can result in membrane depolarization in higher plants and *Neurospora crassa* (Higinbotham, 1973), which may increase membrane permeability to a number of ions, including PO$_4$$^{3-}$. Inclusion of Ca^{2+} in the bathing medium can reduce the depolarizing effect of monovalent cations. The increased efflux from mycelium of *P. tinctorius* reported here may thus reflect increased membrane permeability via a similar mechanism. In this way localized concentrations of monovalent cations in the apoplast of the fungus–plant interface of the ectomycorrhiza might increase phosphate efflux from the fungal symbiont.

References

Beever, R.E. and Burns, D.J.W. (1980) Phosphorus uptake, storage and utilisation by fungi. *Advances in Botanical Research* 8, 128–219.

Cox, G. and Tinker, P.B. (1976) Translocation and transfer of nutrients in vesicular-arbuscular mycorrhizas I. The arbuscule and phosphorus transfer: a quantitative ultrastructural study. *New Phytologist* 77, 371–378.

Grenville, D.J., Peterson, R.L. and Ashford, A.E. (1986) Synthesis in growth pouches of mycorrhizae between *Eucalyptus pilularis* and several strains of *Pisolithus tinctorius*. *Australian Journal of Botany* 34, 95–102.

Higinbotham, N. (1973) Electropotentials of plant cells. *Annual Review of Plant Physiology* 24, 25–46.

Smith, S.E. and Smith, F.A. (1990) Structure and function of the interfaces in biotrophic symbioses as they relate to nutrient transport. *New Phytologist* 114, 1–38.

48 Comparative Analysis of IAA Production in Ectomycorrhizal, Ericoid and Saprophytic Fungi in Pure Culture

G. Gay, J. Bernillon and J.C. Debaud

Université Claude Bernard Lyon 1, Laboratoire d'Ecologie Microbienne du Sol associé au CNRS, bât. 405, 43 Boulevard du 11 Novembre, 69622 Villeurbanne, Cedex, France

Introduction

Numerous microorganisms, bacteria or fungi, are able to synthesize indole-3-acetic acid (IAA) when cultivated on a medium supplemented with a precursor containing the indole ring (e.g. tryptophan). This ability has been well documented in the case of soil bacteria (Katznelson and Sirois, 1961; Azcon and Barea, 1975), rhizospheric bacteria (Brown, 1972), symbiotic bacteria (Berry *et al.*, 1987; Atzorn *et al.*, 1988; Hunter, 1989) or pathogenic bacteria (Kaper and Velstra, 1958). Concerning fungi, Gruen (1959) analysed the results obtained with different fungi: 104 out of the 106 studied species released IAA when cultivated on a medium containing at least traces of tryptophan. However, numerous fungi were unable to release detectable amounts of IAA when cultivated in the absence of precursor containing the indole ring.

IAA production by fungi has been studied mainly in parasitic and symbiotic species which induce morphological and physiological modifications in the host plant which could be ascribed to fungal IAA. Thus, on the basis of the results of Slankis (see Slankis, 1973) who demonstrated that the typical dichotomous morphology of pine ectomycorrhizae could be induced by exogenous IAA, different investigators (Moser, 1959; Ulrich, 1960; Ek *et al.*, 1983; Gay and Debaud, 1987) showed that ectomycorrhizal fungi are able to synthesize and release IAA when cultivated in the presence of tryptophan. Indole-3-acetic acid production was also detected among ericoid endomycorrhizal fungi (Gay and Debaud, 1986; Berta *et al.*, 1988). All these studies were performed by measuring IAA in culture filtrates after culture durations ranging from some days to several months (Moser, 1959; Ulrich, 1960) and no data are available concerning the activity of IAA synthesizing enzymes, the dynamics of IAA production and its possible breakdown by these fungi. Moreover, although IAA production is not restricted to mycorrhizal fungi, a question remains open: is there any difference between IAA production in saprophytic and symbiotic fungi? Hence, the aim of the present study was to compare IAA production in three ectomycorrhizal, three ericoid endomycorrhizal and three saprophytic fungi for a better knowledge of the possible role of this hormone in the symbiosis.

Materials and Methods

Biological material

The fungal strains or isolates used in this work are presented in Table 48.1. Two isolates of ericoid endomycorrhizal fungus (n° 98 and XVIII) were isolated from mycorrhizas and were not identified whereas the other strains were isolated from sporocarp fragments.

Culture conditions

Mycelia were cultivated in 100 ml Erlenmeyer flasks containing 40 ml of the P_3 synthetic medium (Gay, 1990). The medium was prepared with and without a supplement of filter-sterilized tryptophan solution to give a final concentration of 1 mM. Mycelia were cultivated for 7 weeks at $22 \pm 1\,°C$, in the dark. The initial pH of the culture medium was 6.5; it was between 6.4 and 6.6 at the end of the culture period.

Table 48.1. Characteristics of the studied strains or isolates.

Species or isolate number	Source	Year/country of isolation	Host plant or habitat	Biological status
Hebeloma hiemale Bres.	G. Gay	1974, France	*Cistus laurifolius*	
Paxillus involutus (Batsch. ex Fr.) Fr.	J. Garbaye	1980, France	–	Ectomycorrhizal
Pisolithus tinctorius (Pers.) Cok. et Couch.	D.H. Marx	1975, USA	*Pinus* sp.	
XVIII	I. Vegh	1977, France	*Rhododendron* sp.	
98	M. Couture	1981, Canada	*Vaccinium* sp.	Ericoid endomycorrhizal
Hymenoscyphus ericae (Read) Korf and Kernan	D.J. Read	1974, UK	*Erica* sp.	
Collybia butyracea (Bull. ex Fr.) Kumm. (strain B 25)	F. Gourbière	1981, France	*Abies alba* litter	
Pleurotus eryngii (D.C. ex Fr.) Quél.	E. Jacquettan	1977, France	Stump of *Eryngium campestre*	Saprophytic
Agrocybe aegerita (Brig. ex Fr.) Sing. (strain 83–2)	G. Manachère	1983, France	Stump of *Populus nigra*	

Indole compounds identification and quantification

Indole compounds were extracted from culture medium or incubation solution with ethyl acetate at pH 3. They were purified by thin layer chromatography (TLC) and identified by high performance liquid chromatography (HPLC) according to Rouillon et al. (1986) and Gay et al. (1989). IAA was quantified according to Pilet and Collet (1962) by using the Salkowski reagent (Pilet, 1957), the specificity of which has been demonstrated by Rouillon et al. (1986). Indole-3-carboxylic acid (ICA) and indole-3-aldehyde (IAld), which originate from IAA breakdown, were quantified by HPLC with a RP 18 column (10 μm) 25 $\times$ 0.4 cm (Merck). The solvent system was acetonitrile, acetic acid, water (25: 0.1: 74.9 by volume) and the detection was UV absorbance at 280 nm.

Kinetics of IAA accumulation in culture filtrates

Mycelia were cultivated on the P_3 medium supplemented or not with 1 mM tryptophan. Indole-3-acetic acid was quantified every week in culture filtrates throughout the culture period. Sterile subsampling allowed kinetic studies of individual cultures. Indole-3-carboxylic acid and IAld were also measured in an attempt to estimate IAA oxidase activity of the mycelia.

Study of *in vivo* IAA synthesizing activity

Mycelia were cultivated as described for the kinetics of IAA accumulation. After a 7-week culture period, or after 2 weeks in the case of isolates 98 and XVIII, mycelia were collected and promptly rinsed with distilled water. The mycelium obtained from a single flask was incubated, at 22 $\pm$ 1°C, in the dark, in 4 ml of a solution containing 10 mM L-tryptophan and buffered with 3-(N-morpholino) propanesulphonic acid (MOPS) 10 mM (pH 9.0). IAA release in the incubation solution was measured for 24 h and *in vivo* IAA synthesizing activity was expressed as μmol IAA released mg^{-1} fungal protein assayed according to Lowry et al. (1951).

Study of *in vitro* enzyme activity

Mycelia were cultivated as described above. *In vitro* enzyme activity was studied by using the method of Gay (1986) modified as follows: mycelium was ground, at 4°C, in 10 mM tris (hydroxymethyl)aminomethane (Tris) buffer (pH 7.2). Supernatant obtained after centrifugation (27000g for 30 min) was diluted by half with Tris buffer and incubated in the presence of 10 mM (final concentration) tryptophan buffered with 10 mM Tris (final pH: 9.0). This solution was incubated at 22 $\pm$ 1°C, in the dark. Indole-3-acetic acid was regularly quantified in the incubation solution. *In vitro* IAA synthesizing activity was expressed as μmol IAA synthesized mg^{-1} protein.

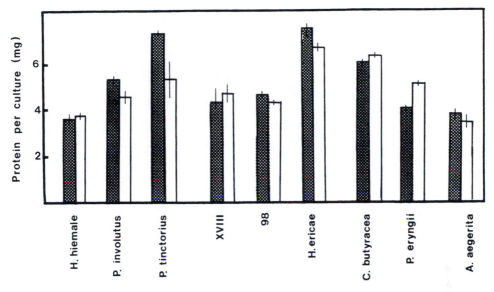

Fig. 48.1. Effect of 1 mM tryptophan on mycelial growth of 7-week-old mycelia. Open bars: mycelia cultivated on the P₃ medium without tryptophan; solid bars: mycelia cultivated on the P₃ medium supplemented with 1 mM tryptophan. Vertical lines represent standard deviation of the mean.

Results

Effect of one mM tryptophan on fungal growth

After a 7-week culture period, the mycelial growth of the studied fungi varied from 3.6 to 7.6 mg protein per flask (Fig. 48.1). The presence of 1 mM tryptophan in the nutrient solution only slightly affected the growth of the fungi. It should also be noted that *P. tinctorius* released high amounts of brown pigments when cultivated in the presence of tryptophan.

Kinetics of IAA accumulation in culture filtrate

All the studied fungi released IAA when cultivated on a medium supplemented with 1 mM tryptophan (Fig. 48.2). Two ectomycorrhizal fungi (*Hebeloma hiemale* and *Pisolithus tinctorius*) accumulated high amounts of IAA (8.5 µmol per culture) whereas the third one (*Paxillus involutus*) accumulated only 0.9 µmol IAA. The endomycorrhizal fungi accumulated IAA amounts comparable to those recorded in the case of ectomycorrhizal ones. IAA concentration in *Hymenoscyphus ericae* filtrates was 4 µmol per flask at the end of the culture period. The isolates 98 and XVIII accumulated higher amounts of IAA (respectively 10.5 and 6.9 µmol per flask) as soon as the second week of culture. The amount of IAA accumulated in culture filtrates of the three saprophytic fungi was c. ten times lower than the

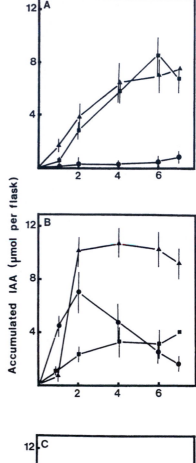

Fig. 48.2. Kinetics of IAA accumulation
in culture filtrates obtained in the presence of 1 mM
tryptophan.
(a) ectomycorrhizal fungi. ▲ = *Hebeloma hiemale*,
■ = *Pisolithus*
tinctorius, ● = *Paxillus involutus*;
(b) ericoid endomycorrhizal fungi. ▲ = isolate 98,
■ = *Hymenoscyphus ericae*, ● = isolate XVIII;
(c) saprophytic fungi. ▲ = *Agrocybe aegerita*,
■ = *Pleurotus eryngii*, ● = *Collybia butyracea*. Vertical
lines represent
standard deviation of the mean.

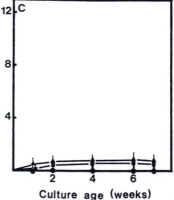

maximal amount recorded with mycorrhizal ones. It was however of the same
order as that recorded in the case of the ectomycorrhizal fungus *P. involutus*.

The decrease of IAA content in culture filtrates of isolate XVIII suggests that
IAA breakdown took place throughout the culture period. Therefore ICA and IAld

were quantified in 2-week-old culture filtrates of this isolate and of isolate 98 and in 7-week-old filtrates of the other fungi. ICA and IAld were detected in the filtrates of all the studied fungi (Table 48.2). The sum of these two compounds represented between c. 74% and 98% of total indoles in filtrates of saprophytic fungi. Indole-3-carboxylic acid and IAld were also abundant in the case of the ectomycorrhizal fungus *P. involutus* and of the endomycorrhizal fungi, particularly of isolate XVIII where they represented 84.2% of total indoles. By contrast, culture filtrates of the ectomycorrhizal fungi *P. tinctorius* and *H. hiemale* presented a very low ICA + IAld % which was respectively c. 14% and 23%.

Table 48.2 IAA synthesis and catabolism in mycorrhizal and saprophytic fungi. Accumulation in culture filtrates was quantified after a 7-week culture period (2-week in the case of isolates 98 and XVIII) on the P_3 medium supplemented with 1 mM tryptophan.

Fungus	Accumulated amounts (μmol mg^{-1} protein)			Relative proportion (as % of total indoles)	
	IAA	ICA	IAld	IAA	ICA + IAld
Hebeloma hiemale	2.04 ± 0.09	0.09 ± 0.004	0.53 ± 0.02	76.7	23.3
Paxillus involutus	0.18 ± 0.009	1.02 ± 0.05	0.03 ± 0.001	14.6	85.4
Pisolithus tinctorius	0.89 ± 0.09	0.14 ± 0.01	0.01 ± 0.001	85.6	14.4
XVIII	0.43 ± 0.03	2.18 ± 0.02	0.12 ± 0.003	15.8	84.2
98	0.97 ± 0.02	2.05 ± 0.02	0.04 ± 0.004	48.5	51.5
Hymenoscyphus ericae	0.54 ± 0.01	0.39 ± 0.01	0.006 ± 0.0004	57.7	42.3
Collybia butyracea	0.23 ± 0.008	0.18 ± 0.006	0.48 ± 0.02	25.9	74.1
Pleurotus eryngii	0.13 ± 0.005	0.08 ± 0.003	5.39 ± 0.02	2.3	97.7
Agrocybe aegerita	0.13 ± 0.009	1.27 ± 0.009	0.03 ± 0.002	9.1	90.9

Abbreviations. IAA: indole-3-acetic acid; ICA: indole-3-carboxylic acid; IAld: indole-3-aldehyde.

In vivo IAA synthesizing activity

Only two fungi yielded a high activity (Fig. 48.3). These are the ectomycorrhizal fungus *H. hiemale* and the saprophyte *Agrocybe aegerita*. When obtained in the absence of tryptophan, these two fungi presented a comparable IAA synthesizing activity (respectively 0.81 and 0.64 μmol IAA synthesized) but mycelia of both fungi grown on a tryptophan supplemented medium presented a lower IAA synthesizing activity. All the other studied species presented a very low activity whereas some of them (e.g. endomycorrhizal fungi or *P. tinctorius*) accumulated high amounts of IAA in their culture filtrates.

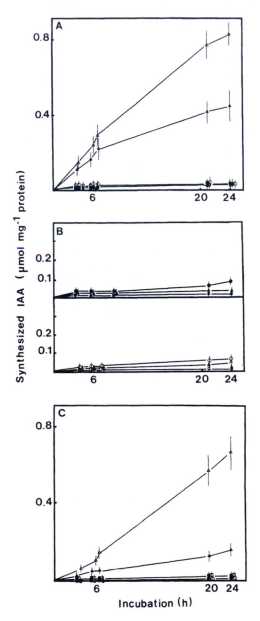

Fig. 48.3. *In vivo* IAA synthesizing activity of 7-week-old (2-week-old in the case of isolates 98 and XVIII) mycelia obtained on the P_3 medium with or without 1 mM tryptophan, supplement. **(a)**: ectomycorrhizal fungi. Triangles = *Hebeloma hiemale*, squares = *Pisolithus tinctorius*, circles = *Paxillus involutus;* **(b)** ericoid endomycorrhizal fungi. Triangles = isolate 98, squares = *Hymenoscyphus ericae*, circles = isolate XVIII; **(c)** saprophytic fungi. Triangles = *Agrocybe aegerita*, squares = *Pleurotus eryngii*, circles = *Collybia butyracea*. Open symbols indicate mycelia obtained on the P_3 medium without tryptophan and filled ones indicate mycelia obtained on the P_3 medium supplemented with 1 mM tryptophan. Vertical lines represent standard deviation of the mean.

In vitro IAA synthesizing activity

The two fungi presenting the highest *in vivo* enzyme activity, namely *H. hiemale* and *A. aegerita*, also had the highest *in vitro* IAA synthesizing activity (Fig. 48.4). Mycelia of these two fungi obtained on a medium containing no tryptophan synthesized respectively 0.34 and 0.35 μmol IAA mg^{-1} protein. Their activity was only slightly affected by the presence of tryptophan in the culture medium. Two other fungi, *P. involutus* and *Collybia butyracea* which had a low *in vivo* enzyme activity presented a rather high *in vitro* activity (respectively 0.09 and 0.19 μmol IAA synthesized mg^{-1} protein). As previously recorded in the case of *in vivo* enzyme activity, the three endomycorrhizal fungi presented a low *in vitro* IAA synthesizing activity. It was also the case for *P. tinctorius* and *Pleurotus eryngii*.

Discussion

All the fungi studied, irrespective of their biological status, were able to metabolize exogenously supplied tryptophan to IAA, but none of them was able to synthesize IAA in the absence of this precursor (data not shown). This confirms the conclusion of Gruen (1959) that most fungi are able to synthesize IAA. However, on the basis of the reported results, the ability to accumulate a high amount of IAA in culture filtrate seems to be characteristic of symbiotic (ecto-or endomycorrhizal) fungi.

Five out of the nine studied species accumulated large quantities of IAA (over 4 μmol per flask) in culture filtrate, but only two of them (*H. hiemale* and *A. aegerita*) presented a high IAA synthesizing activity. Thus, the amount of IAA accumulated in culture media does not always reflect IAA synthesizing activity of mycelia. Indole-3-acetic acid accumulation in culture filtrate of some species was in agreement with their IAA synthesizing activity. By contrast, in some others, no correlation could be detected between their IAA synthesis and IAA accumulation. These fungi are characterized either by a high IAA production associated with a low accumulation (e.g. *A. aegerita*) or by a poor synthesis associated with a high accumulation in culture media (e.g. *P. tinctorius* or isolate 98). It must be emphasized that IAA accumulation in culture filtrate is the result of IAA production and of its breakdown by fungal IAA oxidases. Thus the poor IAA accumulation in filtrate, which appeared as being a characteristic of saprophytic fungi, was mainly due to their high IAA oxidase activity.

The very limited IAA degradation in *P. tinctorius* filtrate can probably be ascribed to the presence of a high amount of phenolic pigments that can act as IAA protectors by inhibiting IAA oxidases (Tomaszewski and Thimann, 1966). The ability to release polyphenolic pigments is characteristic of numerous ectomycorrhizal fungi and is correlated with the ability to accumulate IAA in culture filtrates (Tomaszewski and Wojciechowska, 1975). For example a strain of *Amanita* sp. that did not release pigment degraded IAA more rapidly than it was able to synthesize it and was therefore unable to accumulate this metabolite in culture filtrate. By contrast, mycelia of saprophytic Basidiomycetes are frequently white coloured in pure culture and do not release pigments. The role of these metabolites in

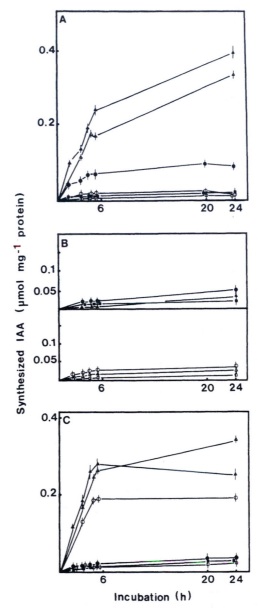

Fig. 48.4. *In vitro* IAA synthesizing activity of 7-week-old (2-week-old in the case of isolates 98 and XVIII) mycelia obtained on the P_3 medium with or without 1 mM tryptophan, supplement. (**a**): ectomycorrhizal fungi. Triangles = *Hebeloma hiemale*, squares = *Pisolithus tinctorius*, circles = *Paxillus involutus;* (**b**) ericoid endomycorrhizal fungi. Triangles = isolate 98, squares = *Hymenoscyphus ericae*, circles = isolate XVIII; (**c**) saprophytic fungi. Triangles = *Agrocybe aegerita*, squares = *Pleurotus eryngii*, circles = *Collybia butyracea*. Open symbols indicate mycelia obtained on the P_3 medium without tryptophan and filled ones indicate mycelia obtained on the P_3 medium supplemented with 1 mM tryptophan. Vertical lines represent standard deviation of the mean.

mycorrhizas has never been clarified, however, it must be emphasized that ectomycorrhizal fungi considered as being very infective (e.g. *P. tinctorius* or *Suillus* sp.) are frequently pigment producers whereas fungi as *Amanita* sp. which do not release pigments are generally less infective. Hence, the role of these pigments as possible chemical mediators of the infectivity of ectomycorrhizal fungi should be considered.

Acknowledgements

The authors are grateful to Professors D.J. Read and G. Manachere and to Drs M. Couture, J. Garbaye, F. Gourbiere, E. Jacquettan, D.H. Marx and I. Vegh for providing fungal strains or isolates, also to C. Raffier for her valuable technical assistance.

References

Atzorn, R., Crozier, A., Wheleer, C.T. and Sandberg, G. (1988) Production of gibberellins and indole-3-acetic acid by *Rhizobium phaseoli* in relation to nodulation of *Phaseolus vulgaris* roots. *Planta* 175, 532–538.

Azcon, N.R. and Barea, J.M. (1975) Synthesis of auxins, gibberellins and cytokinins by *Azotobacter vinelandii* and *Azotobacter beijerinckii* related to effects produced on tomato plants. *Plant Soil* 43, 609–619.

Berry, A.M., Kahn, K.S. and Booth, M.C. (1987) Identification of indole compounds secreted by *Frankia* sp HFPAR13 *in vitro*. *Plant Physiology* 83, 72.

Berta, G., Gianinazzi-Pearson, V., Gay. G. and Torri., G. (1988) Morphogenetic effect of endomycorrhiza formation on the root system of *Calluna vulgaris* (L.) Hull. *Symbiosis* 5, 33–44.

Brown, M.R. (1972) Plant growth substances produced by microorganisms of soil and rhizosphere. *Journal of Bacteriology* 35, 443–451.

Ek, M., Ljungquist, P.O. and Stenström, E. (1983) Indole-3-acetic acid production by mycorrhizal fungi determined by gas chromatography-mass spectrometry. *New Phytologist* 94, 401–407.

Gay, G. (1986) Effect of glucose on indole-3-acetic acid production by the ectomycorrhizal fungus *Hebeloma hiemale* in pure culture. *Physiologie Végétale* 24, 185–192.

Gay, G. (1990) Effect of the ectomycorrhizal fungus *Hebeloma hiemale* on adventitious root formation in de-rooted *Pinus halepensis* shoot hypocotyls. *Canadian Journal of Botany* 68, 1265–1270.

Gay, G. and Debaud, J.C. (1986) Preliminary study on IAA synthesis by ericoid endomycorrhizal fungi. In: Gianinazzi-Pearson, V. and Gianinazzi, S. (eds), *Physiological and Genetical Aspects of Mycorrhizae*, Proceedings of the 1st European symposium on mycorrhizae, Dijon, France, July 1–5 1985. Institut National de la Recherche Agronomique, Paris. pp. 677–682.

Gay, G. and Debaud, J.C. (1987) Genetic study on indole-3-acetic acid production by ectomycorrhizal *Hebeloma* species: inter- and intra-specific variability in homo- and dikaryotic mycelia. *Applied Microbiology and Biotechnology* 26, 141–146.

Gay, G., Rouillon, R., Bernillon, J. and Favre-Bonvin, J. (1989) IAA biosynthesis by the ectomycorrhizal fungus *Hebeloma hiemale* as affected by different precursors. *Canadian Journal of Botany* 67, 2235–2239.

Gruen, H.E. (1959) Auxin and fungi. *Annual Review of Plant Physiology* 10, 405–440.

Hunter, W.J. (1989) IAA production by bacteroids from soybean root nodules. *Physiologia Plantarum* 76, 31–36.

Kaper, J.M. and Velstra, H. (1958) On the metabolism of tryptophan by *Agrobacterium tumefasciens*. *Biochimica et Biophysica Acta* 30, 333–376.

Katznelson, H. and Sirois, J.C. (1961) Auxin production by species of *Arthrobacter*. *Nature* 91, 1323.

Lowry, O.H., Rosebrough, N.J., Farr, A.L. and Randall. R.J. (1951) Protein measurement with the Folin phenol reagent. *Journal of Biological Chemistry* 193, 265–275.

Moser, M. (1959) Beiträge zur Kenntnis der Wuchsstoffbeziehungen im Bereich ektotropher Mykorrhizen. *Arkiv für Mikrobiologie* 34, 251–269.

Pilet, P.E. (1957) Dosage photocolorimétrique de l'acide β-indotylacétique: application à l'étude des auxine-oxydases. *Revue Générale de Botanique* 64, 106–122.

Pilet, P.E. and Collet, G. (1 962) Méthodes d'analyse du catabolisme auxinique. Zwahlen Lausanne, 20pp.

Rouillon, R., Gay, G., Bernillon, J., Favre-Bonvin, J. and Bruchet, G. (1986) Analysis by HPLC-mass spectrometry of the indole compounds released by the ectomycorrhizal fungus *Hebeloma hiemale* Bres. in pure culture. *Canadian Journal of Botany* 64, 1893–1897.

Slankis, V. (1973) Hormonal relationship in mycorrhizal development. In: Marks, G.C. and Kozlowski, T.T. (eds), *Ectomycorrhizae*. Academic Press, London, New York, pp. 231–298.

Tomaszewski, M. and Thimann, K.V. (1966) Interaction of phenolic acids, metallic ions and chelating agents on auxin-induced growth. *Plant Physiology* 41, 1443–1454.

Tomaszewski, M. and Wojciechowska, B. (1975) The role of growth regulators released by fungi in pine mycorrhizae. In: Carr, D.J. (ed.), *Plant Growth Substances*. Hirokawa Publishing Company, Tokyo, pp. 217–227.

Ulrich, J. (1960) Auxin production by ectomycorrhizal fungi. *Physiologia Plantarum* 13, 429–443.

Part Seven
Posters

The Survival of Transplanted Seedlings of Ectomycorrhizal Rain Forest Legumes in Relation to Forest Composition

I.J. Alexander[1], J.A. Rother[1] and D.Mc. Newberry[2]
[1]*Department of Plant and Soil Science, University of Aberdeen, Aberdeen AB9 2UE, UK:*
[2]*Unit of Applied Tropical Ecology, Department of Biological and Molecular Sciences, University of Stirling, Stirling, FK9 4LA, UK.*

In the lowland rainforest of Korup National Park, Cameroon, ectomycorrhizal legumes are not uniformly distributed, and three large emergent species, *Microberlinia bisulcata*, *Tetraberlinia bifoliolata* and *T. moreliana* appear to form groves in which they constitute up to 50% of the basal area. In order to test the hypothesis that survival and growth of seedlings of these species is greatest under an existing canopy of ectomycorrhizal trees a transplant experiment was carried out. Seeds of *M. bisulcata*, *T. bifoliolata* and *T. moreliana* were germinated in river sand, remote from the forest. These non-ectomycorrhizal seedlings were then planted into plots in the forest with low (<15%) and high (>45%) basal area abundance of ectomycorrhizal trees. Survival and height growth were monitored for 500 days and then the seedlings were harvested for determination of mycorrhizal infection, biomass and nutrient content. Suitable ectomycorrhizal inoculum appears to be present in the forest outwith the *Tetraberlinia/Microberlinia* groves and only one species. *T. bifoliolata*, showed greater survival in the plots with high (>45%) basal area abundance of ectomycorrhizal trees. The results are discussed in relation to the factors likely to determine seedling establishment in the forest.

Effects of Nitrogen on the Mycelial Extension of Four Different Ectomycorrhizal Fungi Grown in Symbiosis with *Pinus sylvestris*

K. Arnebrant and B. Söderström
Department of Microbial Ecology, University of Lund, Helgonavägen 5, S-223 62 Lund, Sweden.

It is well known that forest fertilization influences ectomycorrhizal fungi. A reduction in the total number of ectomycorrhizal root tips has often been reported as an immediate response to nitrogen amendments. Long-term effects on species composition have also been reported. One reason for the initial reduction of ectomycorrhizal root tips found after nitrogen fertilization could be reduced growth, and/or lower infection potential of the external mycelium. The influence of nitrogen on the extension of ectomycorrhizal mycelium has been studied using perspex microcosms. Peat amended with different concentrations of nitrogen (0–4 mg N g^{-1} DW peat) either as $(NH_4)_2 SO_4$, $NaNO_3$ or as a complete nutrient solution (Ingestad) was used as substrate. *Pinus sylvestris* or *P. contorta* seedlings infected with four ectomycorrhizal fungi were planted in the microcosms. The extension rate and total extension area of the mycelium were recorded. Mycelial colonization was estimated using a semi-quantitative method, whereby fungal growth was classified into three categories according to the spread of the mycelium: extensive, medium or no or very little growth. Mycelial extension rate was registered through measuring the mycelial front at different time intervals.

Four different ectomycorrhizal fungi were used, *Paxillus involutus*., *Suillus bovinus* and two unidentified species (vg 1 87.10 and vgk 2 89.10) originally isolated from root tips. These fungi responded differently to the nitrogen treatments. One of the unidentified fungi (vg 1 87.10) was very sensitive, and showed very little growth on any of the nitrogen treatments. This type had previously been shown to disappear as a result of forest fertilization. In contrast, the growth of *P. involutus* was shown to be virtually unaffected by the nitrogen, since the fungus colonized the peat extensively, even at the highest nitrogen concentrations of the different treatments. However, the extension rate of *P. involutus* was reduced to approximately 80% of the control. The other two fungi, *S. bovinus* and the unidentified isolate (vgk 2 89.10), both appeared to be negatively affected by the nitrogen amendments since the number of microcosms with extensive mycelial growth was lower in the treatments with nitrogen than those without. The growth rate was also reduced, to approximately 60% of the control in *S. bovinus* and 50% of the control in the unidentified fungus.

Transformation of a Mutant of *Pisum sativum* cv. 'Sparkle' by *Agrobacterium rhizogenes* – A Possible Plant Partner for VA Fungi and *Rhizobium*

A.M. Ba[1], S. Chabot[1], T. La Rue[2], D. Tepfer[3] and Y. Piché[1]
[1]*Centre de Recherche en Biologie Forestière, Faculté de Foresterie et de Géomatique, Université Laval, Québec, G1K 7P4, Canada:* [2]*Boyce Thompson Institute for Plant Research, Tower Road, Ithaca, NY 14853–1801, USA:* [3]*Laboratoire de Biologie de la Rhizosphère, INRA, F-78026 Cedex Versailles, France.*

Transformation assays of one mutant of *Pisum sativum* cv. 'Sparkle' [sym 13 sym 13 (nod + fix −)] by six strains of *Agrobacterium rhizogenes* (L. Moore A4, ATCC 11325, ATCC 15834, 8196, 2659 and 1855) have been investigated. The epicotyls of some seedlings of *P. sativum* cultivated in Petri dishes were wounded with a syringe needle and injected with a liquid culture of *A. rhizogenes*. Other epicotyls of *P. sativum* seedlings were cut into segments and submerged for 30 s in an *A. rhizogenes* liquid culture and then subcultured in an inverted position on White's medium in magenta vessels. All *A. rhizogenes* strains induced the formation of callus on pea epicotyls, in both of the inoculation procedures tested. Roots only appeared later on callus induced by *A. rhizogenes* strains ATCC 15834, 2659 and 1855. Under the same experimental conditions, uninoculated control plants did not initiate any root growth at the wounded site. Only explanted roots induced by *A. rhizogenes* strain 1855 were successfully cultivated on a hormone-free White sucrose medium. Strain 2659 was less virulent than strain 1855. Roots induced by strain 1855 displayed the typical 'hairy root' phenotype. The growth rate of *A. rhizogenes*-induced roots (0.8–1.3 cm per day) was higher than that of control roots, excised from uninoculated seedlings (less than 0.3 cm per day). They were also profusely branched, while control roots had little lateral branching. Thus roots induced by *A. rhizogenes* behaved like transformed roots. However, the transformation should be confirmed by the detection of specific opines and/or by southern blot analysis.

Early Events in Ectomycorrhiza Formation Studied by Electron Microscopy

A.M. Ba[1, 2, 3]*, F.F. Lapeyrie[2], J. Garbaye[2] and J. Dexheimer[3]
[1]*Laboratoire de Microbiologie du Sol, ORSTOM B.P. 1386, Dakar, Sénégal:* [2]*Laboratoire de Microbiologie Forestière, INRA, Champenoux, 54280 Seichamps, France:* [3]*Laboratoire de Biologie des Ligneux, Université de Nancy I, B.P. 239, 54506 Vandoeuvre-Les-Nancy, Cedex, France.* *Present address: CRBF, Faculté de Foresterie et de Géomatique, Université Laval, Québec, G1K 7P4, Canada.*

Differences in the colonization of *Eucalyptus camaldulensis* and *Acacia holosericea* root systems by *Pisolithus tinctorius* (445), *Scleroderma dictyosporum* (ORS. 7731), *Amanita* sp. (ORS. 7735) and *Collybia platyphylla* have been observed *in vitro*. The *P. dictyosporum* isolate formed a mantle and Hartig net with *A. holosericea* and *E. camaldulensis*. *Amanita* sp. formed a mantle, but no Hartig net with *E. camaldulensis* but did not colonize *A. holosericea*. *C. platyphylla* was pathogenic towards *E. camaldulensis*. The variations in mycorrhizal anatomy with these fungi suggested differences in the precolonization stages between the four isolates and *E. camaldulensis* and between three of them (*P. tinctorius*,

S. dictyosporum and *Amanita* sp.) and *A. holosericea* seedlings. Using histochemical tests, ultrastructural comparison of the pre-colonization stage of both host plants by these fungal isolates showed substantial differences. On day 2, the PATAg and SWIFT tests revealed glycoprotein fibrils in the space between the hyphae and cortical cells on the ectomycorrhizal associations *E. camaldulensis/P. tinctorius*, *A. holosericea/P. tinctorius* and *A. holosericea/ S. dictyosporum*. Although the interaction between *E. camaldulensis* and *S. dictyosporum* did not result in ectomycorrhiza development, comparable fibrils reacted positively to PATAg and SWIFT tests. However, this fibrillar material was absent in the interactions between *E. camaldulensis* and *Amanita* sp. which later formed a mantle but no Hartig net. Fibrillar material was absent between *Acacia holosericea* and *Amanita* sp. and between *E. camaldulensis* and *C. platyphylla*. Our results suggest that glycoprotein fibril production differs between mycorrhizal and non-mycorrhizal fungi and that their production by mycorrhizal fungi can occur in the presence of host and non-host plants. On day 2, plasmalemmal acid phosphatase activities were also detected on hyphae of *P. tinctorius* and *S. dictyosporum* associated with the *E. camaldulensis* and *A. holosericea* root surfaces, respectively. These activities were only detected on intercellular hyphae of *P. tinctorius* in *A. holosericea* root systems, and were not detected on fungal cells of *S. dictyosporum*, *Amanita* sp. or *C. platyphylla* associated with *E. camaldulensis* or between *Amanita* sp. and *A. holosericea*. In contrast to fibrillar material, plasmalemmal acid phosphatase activities on hyphae were only observed in the presence of the host plant. These activities could be considered to be an early physiological consequence of host recognition and may serve as an indication of compatibility.

Dual *in vitro* Rhizobial and Ectomycorrhizal Colonization of *Acacia holosericea*

A.M. Ba and Y. Piché
Centre de Recherche en Biologie Forestière, Faculté de Foresterie et de Géomatique, Université Laval, Québec, G1K 7P4, Canada.

The complex interactions that occur in systems having more than one type of symbiosis were studied using one isolate of *Bradyrhizobium* sp. (ORS. 928) and one of the ectomycorrhizal fungus *Pisolithus tinctorius* (445) inoculated on to the roots of *Acacia holosericea in vitro*. After inoculation of *A. holosericea* by the *Bradyrhizobium* sp. strain, small white protuberances which typically represent an early stage of the nodulation process were observed after 5 days. Uninoculated plants did not form these. Symbiotic bacteria entered the plant roots by infection threads in the root hairs and sometimes by intercellular penetration between root hairs at lateral root junctions. After 15 days, intercellular penetration of bacteria was seen deeper in the cortex and in the nodular meristem. Intracellular release of bacteria occurred in the nodular meristem. After inoculation by *P. tinctorius*, young lateral roots of *A. holosericea* often exhibited a sparse mantle after 2 days and showed early intercellular penetration after 4 days. In general, on day 7, typical ectomycorrhizas were observed with a conspicuous mantle and a Hartig net. After dual inoculation, successful formation of nodule-like structures and ectomycorrhiza were frequently observed on the lateral roots of *A. holosericea*. It was possible under the same experimental conditions to find adventitious roots colonized by both microorganisms on the stem of *A. holosericea*. After 15 days, nodule-like structures on lateral roots were surrounded by a hyphal envelope without fungal penetration into the host roots. Sporadically, a non-infected zone occurred between nodule-like structures and ectomycorrhizal colonization. In the nodule-like structures covered by hyphae, important nodulation processes were observed: thread infection in the root hairs, intercellular penetration of bacteria in the cortex and in the nodular meristem, invagination of intercellular infection threads and intracellular release of bacteria in the nodular meristem. However, no attempts were made to measure the efficiency of these nodule-like structures. Paradoxically, in some sections, the initiation and development of nodule-like structures were observed without bacteria. This suggests that *Bradyrhizobium* sp. induced, at distance, the formation of a nodular meristem probably by diffusible signals. Bacteria and hyphae followed the same infection pattern during the colonization process of the root system. The disappearance of root hairs and fungal sheath formation could be an obstacle to colonization of the root system by bacteria. Nevertheless, in the ectomycorrhizas, bacteria were present in the mantle and, more rarely, in the Hartig net. Dual inoculation did not affect the percentage of mycorrhizal infection or the number of nodule-like structures on the root system, suggesting that there was no competition for infection sites.

The Influence of Scots Pine Needle and Humus Extracts on the Growth of Some Ectomycorrhizal Fungi

J. Baar, B. van der Laan, I.L. Sweers and Th. W. Kuyper
Biological Station of the Agricultural University, Kampsweg 27, 9418 PD Wijster, The Netherlands.

> *Laccaria bicolor*, *Laccaria proxima*, *Paxillus involutus*, *Rhizopogon luteolus*, and *Xerocomus badius* were grown on agar with Scots pine needle extracts from three provenances at two concentrations. The growth rate of *L. proxima* and *R. luteolus* was significantly negatively affected by the needle extracts. Only the high concentration had a significantly inhibitory effect on the growth rate of *P. involutus* and *X. badius*. The growth of *L. bicolor* was significantly enhanced by the needle extracts. In a subsequent experiment *L. proxima* and *L. bicolor* were grown on agar with needle and humus extracts from old pine forest, again at two concentrations. The needle extracts had a significant negative effect on the growth rate of *L. proxima*, but a significantly positive effect on the growth rate of *L. bicolor*. The growth of *L. proxima* was not affected by the humus extracts whereas the growth of *L. bicolor* was slightly inhibited by the humus extracts. The species-specific sensitivity to litter extracts was discussed in relation to ectomycorrhizal succession.

Interactions between Indigenous VAM Fungi and Soil Ecotype in *Terminalia superba* in the wet Tropics (Ivory Coast)

B. Blal and V. Gianinazzi-Pearson
Laboratoire de Phytoparasitologie, INRA-CNRS, Station de Génétique et Amélioration des Plantes, INRA, BV 1540, 21034 Dijon cédex, France.

> Production of *Terminalia superba* one of the primary tree species used in reafforestation programmes in the Ivory Coast, can vary between plantation sites and soil type. We have investigated whether VA mycorrhizas could contribute to this variability by studying the infectivity and efficiency of indigenous VA fungal populations from different plantation sites, and their behaviour following inoculation into the different soils.
>
> Soil was collected from three plantation sites of the Centre Technique Forestier Tropical: Mopri (clay loam, pH 5.2, 17.4 p.p.m. available P), Anguédédou (sandy clay, pH 4.1, 52.4 p.p.m. available P) and Yapo (sandy loam, pH 5.3 13.1 p.p.m. available P).
>
> VA infection, plant growth and P contents were compared in *T. superba* seedlings after 10 weeks' growth under nursery conditions in non-disinfected soil, disinfected soil, or disinfected soil inoculated with indigenous fungi in the form of fine mycorrhizal roots taken from 4–5-year-old *T. superba* trees at the three sites.

Plants in unsterile soil from Mopri and Anguédédou sites were highly mycorrhizal (91% and 85% infection frequencies) and had higher fresh mass (2.3- and 3.2-fold) and P contents (2.6- and 3.2-fold) than those in Yapo soil, where VA infection was low (27%).

Growth and P contents of seedlings were greatly reduced by disinfection of all three soils: 2.3- and 4.9-fold in Mopri, 3.6- and 4.5-fold in Anguédédou and 1.4-fold in Yapo. The more important effect of soil disinfection in the Anguédédou and Mopri soils suggests an essential contribution of the symbiotic microflora to the better plant development in these two soils.

Reintroduction of VA fungi from Mopri and Anguédédou sites into their original disinfected soil restored VA infection, plant growth and P contents to levels comparable to those in the corresponding unsterile soils.

In contrast, reinoculation of Yapo soil was not beneficial to seedling development and VA infection was as poor as in unsterile soil. Since no parasites were found in roots, low VA fungal infectivity may be responsible for the poor ecological adaptation of *T. superba* to Yapo soil.

VA fungi from Mopri and Anguédédou sites infected well in disinfected Yapo soil (91 and 75%) and considerably improved seedling growth (2.3- to 3.9-fold) and P uptake (2.3-fold). Yapo root inoculum infected poorly in the other two soils (20–27%), although it slightly increased plant growth in Mopri soil (1.6-fold) and P uptake (1.9-fold) in Anguédédou soil. VA fungi from the Mopri site were also efficient in infecting (86%) and improving seedling development (2- to 3-fold) in soil from Anguédédou, while the inverse was not true.

These results show that (1) *T. superba* is a highly VA-dependent species, (2) VA fungal populations vary in infectivity and effectivity, (3) differences exist in the adaptability of VA fungi to different soil ecotypes and (4) introduction of efficient VA fungi is a guarantee for plant production under nursery conditions.

Influence of Artificial Substrata on Mycorrhization of Micropropagated Fruit Trees in a Horticultural System

B. Branzanti[1], V. Gianinazzi-Pearson[2], S. Gianinazzi[2], S. Predieri[3] and R. Baraldi[3]

[1]*Dipartimento di Protezione e Valorizzazione Agroalimentare, Università degli Studi di Bologna, Italy:* [2]*Laboratoire de Phytoparasitologie, INRA-CNRS, Station de Génétique et d'Amélioration des Plantes, INRA, Dijon, France:* [3]*Centro Tecnica Frutticola – CNR, Università degli Studi di Bologna, Italy.*

Potting mixes used in nurseries can be incompatible with the development of vesicular-arbuscular endomycorrhiza. As part of a bilateral research programme on the use of vesicular-arbuscular endomycorrhiza in the production of micropropagated fruit tree species, we have studied the effect of different horticultural substrates on infection and mycorrhizal growth responses in micropropagated pear (OHF 51) and almond × peach hybrid (GF 677) clones.

Micropropagated plants were inoculated at outplanting from *in vitro* with inoculum consisting of clover roots infected with a *Glomus* species, isolated from a peach orchard. Potting substrates (600 g) containing various combinations of soil, sand, peat, vermiculite, perlite or expanded clay were used and plants were grown in a glasshouse. Plant development (height and number of internodes) and mycorrhizal infection were recorded after one year's growth.

Fungal inoculation enhanced plant growth as compared with uninoculated plants in all cases but *Glomus* infection and mycorrhizal growth responses varied depending on the micropropagated species and potting mix used. Mycorrhizal infection of the almond ×

peach hybrid varied depending on the substratum from 30% in a soil, peat, sand, expanded clay (1:1:1:1) mix to 70% in a peat, sand (1:1) mix, while that of pear plants was less variable (30–50%).

The almond × peach hybrid grew best in an inoculated soil, peat, vermiculite (1:1:1) mix but mycorrhizal responses were maximum (six-fold) in the sand–peat substratum. For pear plants, growth and mycorrhizal effects were optimum (four- to six-fold) in the inoculated mix of soil, sand, peat and expanded clay or of soil, peat and perlite (2:1:1). Number of internodes per plant was generally more homogeneous in the mycorrhizal plants. There was no apparent relationship between infection levels and mycorrhizal growth effects.

These results illustrate species variability in the response of plants to mycorrhizal infection in different horticultural substrates and underline the need for a controlled mycorrhizal weaning system for optimizing production of micropropagated plants.

Occurrence of Vesicular-Arbuscular Mycorrhiza on Douglas Fir and Western Hemlock Seedlings

E. Cázares and J.E. Smith
Department of Botany and Plant Pathology and United States Department of Agriculture, Forest Service, Pacific Northwest Research Station, Forestry Sciences Laboratory, 3200 Jefferson Way, Corvallis, OR 97331, USA.

Vesicular-arbuscular (VA) mycorrhiza develop on 60–80% of the world's plant species, including bryophytes, pteridophytes, gymnosperms and angiosperms. Douglas fir and Western hemlock typically form endo- and ectomycorrhizal (EM) associations but have not been reported to form VA mycorrhizas. Seedlings of *P. menziesii* (PSME) and *T. heterophylla* (TSHE) were grown in a glasshouse for 6 and 12 months in field soil that had been collected from young (12-year-old) and rotation-aged (60-year-old) Douglas fir forests in the Oregon Coast Range.

Root samples were randomly selected from root systems that had been washed free of soil in running tap water, cleared, stained, and the extent of VA colonization was determined. One hundred and eighteen seedlings (61 PSME, 57 TSHE) were examined. Infection was distinguished by the presence of vesicles, arbuscules and non-septate hyphae in the cortical cells of both secondary and feeder roots; however, vesicles were more common than arbuscules. VA fungi were often the only fungi seen in the cortical cells, but EM and dark-walled septate endophytes were commonly present in the same root. Low levels ranging from 1 to 25% of VA colonization were found in 46% of the PSME seedlings and 25% of the TSHE. Levels ranging from 51 to 75% of VA colonization were found in one PSME seedling. Higher levels of VA colonization did not occur in either Douglas fir or Western hemlock. This study reports for the first time the presence of VA in EM colonized seedlings of *P. menziesii* and *T. heterophylla*. The occurrence of both EM and VA in tree genera has been previously reported in *Acacia*, *Casuarina*, *Cupressus*, *Juniperus*, *Populus*, *Salix*, *Tilia* and *Ulmus*. Succession from VA to EM in the same root system has been described for *Helianthemum*, *Eucalyptus* and *Alnus*. We have no evidence that VA/EM succession occurred in our Douglas fir and western hemlock seedlings, but we suppose that VA fungi penetrated the cortical cells before the EM fungi produce mantles on the root surface. Douglas fir and western hemlock are generally considered obligate EM hosts. EM root tips are relatively easy to assess without differential staining and by direct observation with a stereomicroscope. The presence of VA and dark septate endophytes would be overlooked with this procedure. The serendipitous discovery of VA in Douglas fir and western hemlock generates many questions about their presence in different habitats, associations with plants at certain ages, occurrence with other species of Pinaceae, the evolution of mycorrhizal symbiosis, and the ecological importance of different mycorrhizal types in forest ecosystems.

Effects of Cadmium on Ectomycorrhizal Pine (*Pinus sylvestris*) Seedlings

J. Colpaert and J. Van Assche
Plantkunde Instituut, K.U. Leuven, K. Mercierlaan 92, 3001 Heverlee, Belgium.

Many investigators have found an ameliorating effect of mycorrhizal infection on the growth of host plants cultivated at high metal concentrations. We studied the effects of Cd on *Pinus sylvestris* seedlings, inoculated with nine mycorrhizal strains (Table 1). Plants were grown in root chambers with perlite. Over 2 months a Cd solution (44.5 μmol) was supplied to 5-month-old seedlings. Plants were harvested 1 month after cessation of the treatment. Cd concentrations in needles and stems were determined by atomic absorption spectrophotometry (AAS). The percentage of perlite colonized with living extramatrical (EM) mycelium could be calculated from the root observation chambers by planimetry (Table 1). All mycorrhizal fungi tested were able to reduce Cd uptake by their host plant. A treatment effect on the growth of the plants was not detected. However, a direct effect on the EM mycelium was obvious. The colonization percentages of the perlite show that many fungal strains were inhibited in the polluted substrate. Species with a high density of EM mycelium are least affected by the Cd addition. A considerable amount of the added Cd may be complexed and thus retained by the mycorrhizal biomass, but Cd retention by the fungi can only be efficient over a long time-course if the fungi are able to maintain growth of their EM mycelium.

Table 1. Cadmium concentration (μg g^{-1} DW ± s.e.) in non-mycorrhizal and mycorrhizal *Pinus sylvestris* seedlings, supplied with a 44.5 μmol Cd solution or not (co). Effect of this Cd solution on the colonization percentage of the substrate by the extramatrical mycelium (n = 12).

Fungal associate	Treatment	% living mycelium mean (range)	[Cd] needles	[Cd] stem
Non-mycorrhizal plants	co	0	0.29 ± 0.04	0.61 ± 0.05
	Cd	0	1.23 ± 0.31	9.02 ± 0.71
Thelephora terrestris	co	95	0.22 ± 0.04	0.51 ± 0.04
	Cd	5 (0–10)	0.43 ± 0.04	3.82 ± 0.36
Laccaria laccata	co	86	0.32 ± 0.09	0.45 ± 0.12
	Cd	26 (16–43)	0.89 ± 0.11	3.54 ± 0.49
Scleroderma citrinum	co	84	0.48 ± 0.04	0.72 ± 0.07
	Cd	21 (5–33)	0.52 ± 0.04	2.06 ± 0.09
Paxillus involutus	co	95	0.17 ± 0.03	0.40 ± 0.05
	Cd	5 (1–8)	0.46 ± 0.05	2.10 ± 0.13
Suillus luteus 1	co	100	0.20 ± 0.01	0.40 ± 0.04
	Cd	72 (38–87)	0.40 ± 0.03	2.36 ± 0.22
Suillus luteus 2	co	100	0.30 ± 0.04	0.79 ± 0.07
	Cd	98 (91–100)	0.38 ± 0.02	2.33 ± 0.24
Suillus bovinus 1	co	90	0.29 ± 0.03	0.21 ± 0.04
	Cd	24 (10–31)	0.37 ± 0.05	3.73 ± 0.34
Suillus bovinus 2	co	97	0.17 ± 0.02	0.34 ± 0.04
	Cd	97 (95–100)	0.33 ± 0.03	1.17 ± 0.08
Suillus bovinus 3	co	86	0.19 ± 0.03	0.46 ± 0.04
	Cd	48 (31–92)	0.32 ± 0.03	1.76 ± 0.18

Effects of Simulated Acid Rain, Soil Contamination and Mycorrhizal Infection on *Picea abies* Seedlings

P. Cudlín[1], A. Bystrican[2], P. Zvára[3], P. Siffel[3], J. Santrucek[3], P. Kindlmann[4] and Z. Braunová[3]
[1]*Institute of Landscape Ecology, Acad. Sci., Na sádkách 7, 370 05 České Budejovice:* [2]*PYRUS, Dlouhá 3096, 400 01 Ustí n.L.:* [3]*Institute of Molecular Biology of Plants, Acad, Sci., Branisovská 31, 370 05 České Budejovice:* [4]*Biomathematical Laboratory, Acad. Sci., Branisovské 31, 370 05 České Budejovice, Czechoslovakia.*

One-year-old *Picea abies* seedlings, grown hydroponically in a forest nursery, were planted in February 1988 into large containers of natural soil from Sumava Mts (less polluted) and Krusné hory Mts (heavily polluted) as well as into a γ-irradiation sterilized peat–perlite substrate. In November seedlings were transplanted in the following arrangement: 60 plants grown in Sumava Mts soil and 60 from peat–perlite substrate were transferred to fresh soil from the same locality; seedlings from Krusné hory Mts were treated in a similar way. They were grown in a heated greenhouse and from September 1989 to June 1990 were acidified by simulated acid rain (mixture of H_2SO_4 and HNO_3, pH 3) in total quantity of 405 mm. Controls were moistened by the same quantity of distilled water of pH 6.3. In June 1990 seedlings were harvested and evaluated for basic shoot and root characteristics including some photosynthetic parameters and mycorrhiza development.

Simulated acidification adversely influenced both shoots (dry weight of needles, root collar diameter) and roots (dry weight of roots with diameter 1–2 mm). The most significant differences in root tip and mycorrhiza numbers appeared in seedlings transplanted to the Sumava Mts soil from sterile peat–perlite. No significant effects of acidification on chlorophyll content and contents of individual chlorophyll proteins were found, however, partial interruption of energy transfer from light harvesting antenna to photosystem 2 was observed in all acidified seedlings. These effects were analogous to the decrease of growth and physiological activity of seedlings in polluted soil from Krusné hory Mts compared with the not so degraded soil from Sumava Mts.

The changes caused by long-term man-made deterioration of soils were comparable with those induced by the short-term experimental acidification. Simulated acid precipitations resulted in increased physiological activity (higher photosynthesis rate, increasing of extension of root system) of seedlings grown under relatively non-stressed conditions in Sumava Mts soil.

Mycorrhizal Amelioration of Metal Toxicity to Plants

H. Denny and I. Ridge
Department of Biology, Open University, Walton Hall, Milton Keynes, UK.

Mycorrhizal fungi ameliorate metal toxicity to plants. They reduce the concentration of toxic metals in the aerial parts of the plants. Relatively high concentrations of zinc are found in the cell walls and/or slime of the extramatrical hyphae, of ectomycorrhizal birch. This suggests that the amelioration mechanism entails sequestration of the metal in the hyphal wall/slime layers of the fungus, thereby lowering the concentration of metal in the soil solution and concomitantly the metal concentration to which the plant is exposed. There is some evidence to suggest that the slime is important as the metal binding agent.

The aim of this project is to investigate the part played by the hyphal slime in the amelioration mechanism; three strains of some ericaceous mycobionts, which exhibit a range of fibrillar slime production, are being compared with respect to their ability to bind zinc, tolerate zinc and ameliorate its toxicity to higher plants.

The first objective was to discover whether the ericaceous fungi behaved like their ectomycorrhizal counterparts. An isolate of the ericaceous mycobiont *Hymenoscyphus ericae* Korf and Kernan was grown in liquid culture at a range of zinc concentrations. The fungus grew well at 0 and $0.5 \, mmol \, dm^{-3}$ zinc, was slightly inhibited at $1.5 \, mmol \, dm^{-3}$ and severely inhibited at $3.0 \, mmol \, dm^{-3}$ zinc. Samples of mycelium were prepared for quantitative X-ray microanalysis. Preliminary results indicate that this ericaceous mycobiont does indeed behave like the ectomycorrhizal *Paxillus involutus* Fr. in that zinc accumulates within the hyphal wall/slime, and does not cross the plasmalemma in significant quantities unless the external zinc concentration is toxic to the fungus.

The second objective was to relate the ability of the individual strains of ericaceous mycobiont to tolerate zinc to how much fibrillar slime they produce. The three strains, denoted: highly slimy Fib + +, slightly slimy Fib + and not at all slimy Fib −, were grown in liquid culture at a range of zinc concentrations between 0 and $6 \, mmol \, dm^{-3}$ zinc. All three grew well in the zinc free control, Fib − grew very little at any of the other concentrations, Fib + grew only slightly better than Fib −, but Fib + + grew well at all concentrations except the highest. Therefore there is some evidence to suggest that zinc tolerance is proportional to the capacity to produce fibrillar slime.

Given that most fungal strains are more tolerant of metals than the associated higher plant, these results are compatible with the hypothesis that the fungal mycelium may colonize and detoxify new areas of soil before its exploitation by the roots of the higher plant and that the extrahyphal slime may have an important role in this process.

Effect of 10 Years of Low-input Sustainable Agriculture upon VA Fungi

D.D. Douds[1], Jr., R.R. Janke[2], and S.E. Peters[2]
[1]*USDA-ARS ERRC, 600 E. Mermaid Lane, Wyndmoor, PA 19118, USA:* [2]*Rodale Research Center, 611 Siegfriedale Rd., Kutztown, PA 19530, USA.*

Spore populations and colonization of maize and soyabean roots by VA fungi were studied in a field experiment in which a conventional agriculture rotation and two low-input sustainable (LISA) systems were maintained for 10 years. Farming systems included: a conventional maize–soyabean rotation with chemical fertilizers and weed control, a LISA rotation with animal manure as fertilizer and an emphasis on the production of hay as well as grains, and a LISA system with green manure and small grain cover crops which produced grain for income. Soil samples were obtained at planting and after harvest for two growing seasons. Maize and soyabean root systems were sampled in early summer.

Low-input plots tended to have higher populations of spores and greater diversity of VA fungi than conventionally farmed plots. *Gigaspora gigantea* was uncommon under conventional management (1 spore per ten $50 \, cm^3$ samples) yet very common in LISA systems (up to 30 spores per $50 \, cm^3$ sample). *Glomus* spp. also tended to be more numerous in the LISA systems. No spores which could be identified as *Acaulospora* or *Entrophospora* were found.

Colonization of maize and soyabean in the field by VA fungi was not consistently reflective of spore populations. Soil collected for sampling for spores was used for colonization bioassays in the greenhouse with maize or bahiagrass (*Paspalum notatum*) as test plants. Results showed greater colonization in soil from LISA plots than from conventional plots in both years studied.

These results indicate that the conventional practice of the soil lying bare from autumn harvest through spring sowing may be detrimental to obligate symbionts such as VA fungi. The LISA fields were covered with live plants 70% of an average year compared with 40% for the conventionally farmed plots. Further, the application of chemical fertilizers may have resulted in lower levels of VA fungi in conventionally farmed soils. More research needs to be done, however, to determine the contribution to crop growth and yield by mycorrhizas in LISA relative to that obtained by creative nitrogen and soil cover management.

Preferential Cycling of Phosphorus: The Role of Mycorrhizas

W.R. Eason and E.I. Newman*
AFRC Institute of Grassland and Environmental Research, Aberystwyth SY23 3EB, UK.
(*Present address: Department of Botany, University of Bristol, Woodland Road, Bristol BS8 1UG, UK)

The work is led by the hypothesis that arises if interplant phosphorus transfer occurs via mycorrhizal hyphal links. If there is nutrient cycling between plants by mycorrhizal hyphae one would anticipate that there would be a preferential cycling between species that share the same type of mycorrhiza. This has been tested in pot experiments.

Each pot intially contained three plants each forming its normal type of mycorrhiza. The combinations were:

A. *Fraxinus excelsior* (VA), *Lolium perenne* (VA), *Acer pseudoplatanus* (VA)
B. *F. excelsior* (VA), *L. perenne* (VA), *Larix eurolepis* (EC)
C. *A. pseudoplatanus* (VA), *Lolium perenne* (VA), *Larix eurolepis* (EC)
D. *Calluna vulgaris* (ER), *C. vulgaris* (ER), *Molina caerulea* (VA)

(where VA = vesicular-arbuscular mycorrhizal, EC = ectomycorrhizal and ER = ericoid mycorrhizal)

The three plants were in the pot in a row, in the order listed, with the central one as the donor. So in A both receivers had the same type of mycorrhiza as the donor, but in other combinations one receiver formed a different mycorrhiza from the donor (and so was not able to form links with it). Radiophosphorus (^{32}P) was either supplied directly to the donor after which its roots were detached. Alternatively the donor roots were detached but ^{32}P was supplied by injecting it into the soil. The ability of the receivers then, to acquire P either from the soil pool or from dying roots was compared.

In A, where all species were VA infected, the distribution of ^{32}P between *Fraxinus* and *Acer* was the same regardless of how the ^{32}P was supplied. In all other combinations however, the species that had the same type of mycorrhiza as the donor acquired a higher proportion of the ^{32}P when it was provided via the donor roots than when it came only from the soil. These results are in agreement with the original hypothesis that P transfer can occur via direct links, resulting in a preferential cycling of nutrients between species with the same type of mycorrhiza. The evidence is far from conclusive however: further work involving the use of non-mycorrhizal controls and the use of single-strain sources of inocula is planned. If established the implications of this work for nutrient cycling and species balance in plant communities are profound.

Nitrogen Translocation Through a Root-free Soil Mediated by VA Fungal Hyphae

B. Frey and H. Schüepp
Swiss Federal Research Station, Department of Phytopathology and Soil Microbiology, CH-8820 Wädenswil, Switzerland.

In a first experiment the objective was to determine whether vesicular-arbuscular fungi can enhance nitrogen uptake from a root-free soil. Maize (*Zea mays* L.) was grown for 12 weeks in a system with two compartments, one for root and one for hyphal growth. Compartmentation was accomplished by a 40-μm nylon net. Plants were inoculated with *Glomus fasciculatum* or uninoculated. After 45 days of growth a solution of $(^{15}NH_4)_2SO_4$ at 10 atom% excess ^{15}N was applied with a syringe at 6 cm distance from the root compartment.

Mycorrhizal inoculation greatly increased N concentration and the percentage of N as ^{15}N above natural abundance in plants (Table 1). Mycorrhizal plants contained an average of 40% of the ^{15}N applied, while the control plants derived an average of only 2.6%. Plant biomass was not affected by the mycorrhizal fungus. Hyphal length (1.5–3.0 m g^{-1} soil) was determined in the hyphal compartment.

In summary we propose that the external hyphae of *G. fasciculatum* can take up N and transport it into the rooting zone and/or the fungi can affect nitrification to increase movement of NO_3^+ to roots by mass flow and diffusion.

In a second experiment $^{15}N_2$ was used to investigate the transfer of fixed nitrogen from clover (*Trifolium alexandrium*) to maize. Plants were separated by a soil-bridge (2 cm) allowing hyphal growth to pass and were grown for 11 weeks in a cuvette-system. Clover was inoculated with *Glomus fasciculatum* and rhizobia or with rhizobia alone. After 50 days of growth the cuvettes were sealed in plastic bags, while the aerial environment remained undisturbed. $^{15}N_2$ gas was applied through a septum and the soil was incubated for 2 days.

Inoculated clover fixed nitrogen containing more ^{15}N than the non-legume. Dual-inoculated clover did not fix significantly more $^{15}N_2$ than single inoculated plants. Atom % ^{15}N excess and ^{15}N content in maize did not differ significantly between treatments, although the ^{15}N content in the roots tended to be higher in the presence of the fungus (Table 2).

In this study mycorrhizal hyphae did not improve transfer from labelled fixed nitrogen in clover to maize.

Table 1. Concentrations of N and ^{15}N in plants. (−M = without mycorrhizal inoculation: +M = with mycorrhizal inoculation).

Treatment	% N		Atom % ^{15}N excess	
	Shoot	Root	Shoot	Root
−M	0.7	0.77	0.049	0.078
+M	0.92	1.27	0.534	0.513

Table 2. Atom % ^{15}N enrichment and total ^{15}N content in shoots and roots of donor and receiver plants.

Treatment	Atom % excess		^{15}N content (μg)	
	Shoot	Root	Shoot	Root
Donor plants: *Trifolium alexandrinum*				
−M/ + R	0.0152	0.0112	23.71	5.32
+M/ + R	0.0139	0.0101	24.21	6.18
Receiver plants: *Zea mays*				
−M/ + R	0.0033	0.0031	1.10	0.48
+M/ + R	0.0033	0.0035	1.09	0.58

Early Events of VA Infection in Host and Non-host Plants

M. Giovannetti[1] L. Avio[1], C. Sbrana[2], and A.S. Citernesi[1]
[2]*Istituto di Microbiologia Agraria, Centro di Studio per la Microbiologia del Suolo, Pisa, Italy:*
[2]*Scuola Superiore di Studi e Perfezionamento S. Anna, Pisa, Italy.*

The establishment of mycorrhizal symbiosis is a multi-step process involving different degrees of recognition between plants and fungi. The utilization of host and non-host plants allows the study and definition of the early events involved in the process. It has been shown that two species of vesicular-arbuscular (VA) fungi were not able to infect *Brassica* roots, although they could adhere to them and produce swellings resembling appressoria. In this work we studied the non-host genus *Lupinus*. We investigated: (1) the effect of lupin roots on germination and hyphal elongation of *Glomus mosseae*; (2) the effect of excised roots of lupin and lucerne (*Medicago sativa*) on appressorium formation by *G. mosseae*.

Lupin roots did not inhibit spore germination and hyphal growth of *G. mosseae*. The hyphae grew on and around excised lupin roots, forming swellings resembling appressoria. Such enlarged structures sometimes produced thin hyphae, growing along and perpendicular to root cell walls. These rapidly aborted, retracting their cytoplasm and forming consecutive septa which isolated hyphal tips. In the presence of excised roots of the compatible host plant lucerne, hyphae extending from appressoria were capable of penetrating adjacent epidermal cells, forming coils, but they soon retracted their cytoplasm and produced septa. Consequently the fungus was not able to spread further into dying roots. When both host and non-host root systems were killed by liquid nitrogen treatment, the fungus did form swellings, but no appressoria and was unable to penetrate even the host roots.

Our results suggest that in lupin roots hyphal adhesion and appressorium formation are inhibited by a factor associated with intact living plants. The failure of fungal infection at a different stage confirms that mycorrhizal infection is a multi-step process, during which signals may initiate consecutive recognition events between host and symbiont, eventually leading to the formation of a functional symbiosis.

Soil Solution Chemistry of Ectomycorrhizal Mat Soils

R.P. Griffiths[1], B.A. Caldwell[1], and J.E. Baham[2]
[1]*Department of Microbiology, Room 220 Nash Hall,* [2]*Department of Crop and Soil Science, Oregon State University, Corvallis, OR 97331, USA.*

Ectomycorrhizal fungal mats are distinct features of Pacific northwest coniferous forests and other forests throughout the world. Organic acids produced by these fungi may play an important role in nutrient availability and mineral weathering within the soil ecosystem. The dissolved chemical elements in soil solutions isolated from two ectomycorrhizal fungal

Table 1. Concentrations (μmol) of dissolved constituents in soil extracts.

	Hysterangium		Soil	Gautieria		Soil
H$^+$	74.1	***	11.0	31.4	***	2.1
K	470	***	120	1 800	***	210
Ca	710	***	120	2 000	***	110
Mg	410	***	51	1 500	***	480
Zn	2.67	*	1.13	12.6	**	1.23
Mn	10.3	***	2.00	1 220	***	9.1
Cu	0.56	***	0.20	1.83	**	0.18
Fe	39	*	16	320	**	6.3
Al	353	**	56	8 044	***	55
Oxalate	59	*	5.2	11 638	***	35
DOC	67 000	**	7 300	188 000	***	5 200
Total N	550	**	270	1 011	***	183
CN$^+$	122		32	186		28

*, **, *** = significance of difference between mat and non-mat soil solutions; $P < 0.05$, 0.01 and 0.001 respectively. $n = 5$ (Wilcoxin non-parametric test). † molar ratio for DOC/Total N.

(*Hysterangium setchellii* and *Gautieria monticola*) mat soils were compared with those from adjacent soils with no visible mat development. The concentrations of dissolved constituents were greater, in all cases, for the mat soils (Table 1).

Concentrations of cations, oxalate, dissolved organic carbon (DOC) and total N were always greatest in mat soils with concentrations in *G. monticola* mat solutions usually being higher than in *H. setchellii* mat solutions. The chemical constituents showing the largest differences between mat and non-mat soils for both mat types included: Al, Fe, Mg, Mn, oxalate and DOC. The reduced pH and elevated oxalate and DOC concentrations in the mat soils suggest that organic acids produced by the fungal mat community accelerated weathering of the soil mineral phase causing an increase in pore-water cation concentrations. The elevated C:N ratios in mat soil solutions may be caused by the input of organic acids by the ectomycorrhizal fungus and/or by the selective removal of dissolved organic nitrogen by the fungus resulting in an enrichment of DOC. This process may be responsible for releasing plant nutrients to be transported by the fungi to the host tree.

Fungicide Interactions with VA Fungi in *Ananas comosus* Grown in a Tropical Environment
J.P. Guillemin and S. Gianinazzi
Laboratoire de Phytoparasitologie, INRA–CNRS, Station de Génétique et d'Amélioration des Plantes, INRA, BV 1540, 21034 Dijon Cédex, France.

The aim of the study was to determine the effects of several fungicides (fosetyl-Al, etridiazole, captan and maneb) on VA infection development and plant growth in pineapple (*Ananas comosus*).

Micropropagated plantlets of Queen and Smooth Cayenne (clone CY0) were inoculated with *Glomus* sp. (LPA 21) in an acid soil (pH 5.0) and growth in a growth chamber with artificial tropical conditions ($300 \, \mu E \, s^{-1} \, m^{-2}$, 12 h day, 29/25°C). Previous studies showed that under these conditions, endomycorrhizal pineapple plants have better growth than non-mycorrhizal plants. Fungicides were applied at field rates recommended for root disease control: fosetyl-Al ($10 \, g \, m^{-2}$), etridiazole ($3 \, g \, m^{-2}$), captan ($1.2 \, g \, m^{-2}$) and maneb ($1.5 \, g \, m^{-2}$).

Fungicide application did not negatively influence endomycorrhizal growth effects on shoot and root fresh mass, except for etridiazole, which caused a decrease in shoot growth. Captan positively affected non-mycorrhizal and endomycorrhizal plants and in certain cases it reinforced the endomycorrhizal effect (+34% shoot and +68% root fresh mass of Queen, +25% root fresh mass of Smooth Cayenne). This was also true for fosetyl-Al and maneb treated endomycorrhizal plants of the Queen variety (+46% root fresh mass) and Smooth Cayenne variety (+48% root fresh mass), respectively.

The root:shoot ratio was not modified by fungicide application, except for Queen variety plants treated with fosetyl-Al (+30%) and etridiazole (+41%); plants treated with the two fungicides showed increases in root production.

The tested fungicides were not harmful to the endomycorrhizal infection by *Glomus* sp., except in relation to arbuscule frequency of plants treated with captan and maneb, which were reduced respectively by 24% and 32% for the Queen variety and of 37% and 34% for Smooth Cayenne. However, this decrease in infection did not affect the growth of plants.

These results show that certain fungicide treatments are compatible with the positive effect of controlled endomycorrhization of micro-propagated plantlets of pineapple during the post-*vitro* weaning period, and they open interesting perspectives for combining the positive effect of endomycorrhizas in micropropagated plant production with the chemical control of fungal pathogens.

Native Populations of the Glomales Influenced by Terracing and Fertilization under Cultivated Potato in the Tropical Highlands of Africa

J. Heinzemann, E. Sieverding and C. Diederichs
Institute of Agronomy in the Tropics (IAT), University of Göttingen, Grisebachstr. 6, 3400 Göttingen, Germany.

The present study was conducted in the northern part of Rwanda, Commune Giciye, where soil erosion is a severe problem. We looked at the occurrence of Glomales associated with potatoes (*Solanum tuberosum*). Three factors were examined regarding the number of glomalean spores:

1. erosion control with top soil replacement on terraces of 1 m height versus non-erosion control;
2. fertilizer application (500 kg ha^{-1} rock phosphate, 10 t ha^{-1} sheep dung, combination of both);
3. four land sites with similar climatic and pedologic conditions.

The objective was to determine whether these management factors influenced the density of fungal propagules (spores cm^{-3} soil) and the distribution of species of the Glomales.

In total, 13 species of Glomales were found of which nine belonged to *Glomus*. The mean density of spores (311 spores per 100 cm^3 soil) and the numbers of fungal species (8–11 per sample) were similar to those known from other tropical soils. The number of Glomalean spores and species were not correlated.

The mean spore density in the soil did not differ significantly among the four locations, but at one site the cumulative species number was lower. Analysis of the spore number of each determined fungal species, suggests that the quantitative composition of the fungal community is site specific. However, the composition may have been influenced by the preceding crop and the agricultural management methods used in the years before sampling. Further investigations should examine the dynamics of the fungal populations over several crop cycles.

The erosion control system did not have a significant influence on the total spore density nor on the cumulative fungal species number. It was observed that the formation of terraces enhanced the presence of some species, i.e. of *G. callosum* and *G. aggregatum* and suppressed others (*G. fasciculatum*, *G. occultum*). Yet, no information is available on the effectiveness of these different fungal species for plant growth. Nonetheless, this erosion control system appears to preserve the biological fertility of tropical hill sites, possibly as a result of top soil replacement.

The combined application of rock phosphate with animal manure increased the spore number per unit soil volume by 45%, this effect was mainly due to improved reproduction of *G. fasciculatum*, *G. aggregatum* and *G. geosporum*. Animal dung and slow releasing P from rocks may enforce the proliferation of the Glomales in tropical soils.

The study demonstrates the need to investigate the successional behaviour of fungi at the species level.

Mycorrhizas in African Miombo Savanna Woodlands

P. Högberg
Section of Forest Science, Department of Ecological Botany, University of Umeå, S-901 87 Umeå, Sweden.

Deciduous miombo woodlands occupy vast areas in East and South-Central Africa. The soils are poor in organic matter (<1% org-C), N (<0.1% tot-N) and available P. There are up to 30 woody species per hectare. The dominant trees are ectomycorrhizal (notably

Caesalpinioideae, but also Dipterocarpaceae and Euphorbiaceae.). VA mycorrhizal and nodulated, potentially N_2-fixing, VA mycorrhizal species are also present. Foliar analyses indicate that N-fixation by the latter is limited by the low supply of P, whereas non-nodulated VA mycorrhizal species sometimes accumulate P, but have low concentrations of N. Ectomycorrhizal species appear to have the most balanced N and P nutrition. They have 0.5% (of dry mass) higher foliar concentration of N than VA mycorrhizal species. Preliminary studies show that their ^{15}N abundance is different from that of VA mycorrhizal species, which may indicate that they utilize different sources of N. There is no indication from data on ^{15}N and N% to suggest that non-nodulated VA mycorrhizal species receive N from nodulated VA mycorrhizal species, although they are likely to share a common mycelial network.

Ectomycorrhizal Fungi in Kenya
M.H. Ivory and F.M. Munga
Oxford Forestry Institute, South Parks Road, Oxford, OX1 3RB, UK.

Pines in East Africa were known to be ectomycorrhizal in habit before 1977, but few of their symbionts had then been identified. It was later realized that some indigenous trees in small areas of coastal forest, such as *Afzelia quanzensis*, were also ectomycorrhizal.

Since 1978 brief surveys of ectomycorrhizal fungi have been carried out by the authors in indigenous coastal forests and various pine plantations. Several attempts were also made to introduce exotic pine ectomycorrhizal fungi into Kenya for field testing. Several were successfully introduced on to nursery plants, but only one, *Pisolithus tinctorius*, is known to have been successfully established in field planting. Probable ectomycorrhizal fungi now known to occur in Kenya, or which were introduced (1978–1987) are listed in Table 1.

Table 1. Probable ectomycorrhizal fungi in Kenya.

Lactarius pandani [1]	*Rhizopogon nigrescens* [3]
Russula annulata f. *lutea* [1]	*Rhizopogon rubescens* [2]
Russula cfr. *cinerella* [1]	*Geastrum dissimile* [1]
Russula cfr. *cyanoxantha* [1]	*Hebeloma crustuliniforme* [2]
Russula sp. (sect. *Ingratae*) [1]	*Inocybe lanuginella* [2]
Cantharellus longisporus [1]	*Laccaria laccata* [2]
Thelephora terrestris [2,3]	*Lepista lentiginosa* [2]
Pisolithus tinctorius [1,3]	*Suillus granulatus* [2,3]
Scleroderma bovista [3]	*Suillus luteus* [2]
Scleroderma texense [3]	*Telamonia* sp. [2]
Scleroderma verrucosum [1]	*Tuber* cfr. *rapaedorum* [2]
Rhizopogon luteolus [2,3]	*Wilcoxina mikolae* [2]

[1] Indigenous forest; [2] pine associate; [3] introduction 1978–87

Axenic Sand Culture for the Study of Mycorrhizal Root Systems and their Rhizospheres
G. Jentschke, H. Schlegel and D.L. Godbold
Forstbotanisches Institut, Universität Göttingen, D–3400 Göttingen, Germany.

In investigations of the toxicity of metals and nutrient uptake by mycorrhizal roots, it is necessary to control the element levels in the culture system. In most axenic systems for the culture of mycorrhiza this is not possible. In this work an axenic sand–nutrient solution culture system has been developed in which element levels in the system can be controlled. Three types of system have been developed, all working on the same basic principle. Small aliquots of nutrient solution (1–10 ml) are added periodically (0.5–5 h) to the rooting substrate. The nutrient solution is forced into the system using sterile filtered air. Excess nutrient solution is removed using ceramic lysimeter candles under negative pressure. The supply of elements to the plants can be controlled by determining the concentration of elements in the eluent and by subsequently adjusting the input rate of the nutrient solution. Thus, depletion of the nutrient solution by the plants can be minimized. Three systems have so far been developed. The basic system uses a 2 litre culture vessel in which up to ten spruce seedlings may be grown. The seedlings are fully enclosed in the culture vessel. In the two other systems the aerial plant parts are not enclosed. One system uses 60 cm^3 glass tubes containing one plant per tube, and the other uses 30 × 30 × 2 cm rhizotron plates. The rhizotron system allows easy access to the roots and rhizospheres, and permits continuous measurement of root growth. In all systems seedlings of spruce (*Picea abies*) and or pine (*Pinus sylvestris*) have been inoculated using agar plugs or liquid cultures. Mycorrhizas with *Paxillus involutus*, *Lactarius rufus*, *Lactarius theiogalus* and *Pisolithus tinctorius* have successfully been established with effective colonization.

This system permits investigations at environmentally relevant levels of nutrients and toxic metals, and may have many applications in assessing the role of mycorrhizal fungi in plant nutrition and stress physiology.

Do Ectomycorrhizas Affect Uptake and Toxicity of Metals in Roots of Norway Spruce?

G. Jentschke, H. Schlegel and D.L. Godbold
Forstbotanisches Institut, Universität Göttingen, D-3400 Göttingen, Germany.

Experimental evidence indicates that metal toxicity is reduced in mycorrhizal plants. Mycorrhizas may prevent metal ions from being taken up by the roots. Using the dye cellufluor as an apoplasmic tracer, in excised mycorrhizas no penetration of the dye through the hyphal sheath into the root cortex was observed. This suggests that the fungus may isolate ectomycorrhizal roots from the soil and thus from toxic ions in the soil solution. However, direct evidence of reduced uptake of metals into the cortex of ectomycorrhizal roots is lacking. In the experiments presented here, mycorrhizal and non-mycorrhizal spruce seedlings (*Picea abies*) were grown in an axenic quartz sand culture system with frequently renewed nutrient solution, containing either 1 μmol PbCl$_2$ or 800 μmol Al(NO$_3$)$_3$ for 13 to 19 weeks. Using X-ray microanalysis, the distribution and content of Pb and Al in mycorrhizal and non-mycorrhizal root tips have been compared. Independent of mycorrhizal colonization and fungal species (*Lactarius rufus* and *L. theiogalus* and two strains of *Paxillus*

Table 1. Effect of fungal colonization on distribution of Pb and Al (X-ray microanalysis) in root tips of *Picea abies* seedlings.

Mycobiont	Cell wall content of Pb or Al, mmol dm^{-3}			
	Hyphal sheath	Cortex	Endodermis	Stele
Pb				
Non-mycorrhizal	–	6.1	5.1	1.4
P. involutus 533	0.8	5.1	4.2	1.1
P. involutus 537	0.7	2.8*	3.9	0.7
L. rufus	n.d.	3.6	4.8	0.7
Al				
Non-mycorrhizal	–	93	110	25
L. theiogalus	72	177	97	22
L. rufus	63	89	113	16

*, significantly different (*P* = 0.1) from non-mycorrhizal root tips; n.d., not detected.

involutus) the highest contents of Pb and Al were found in the apoplast of the root cortex (Table 1). Except for seedlings colonized with *Paxillus involutus* strain 537 and exposed to Pb, no significant difference in Pb or Al contents in cell walls of the root cortex of mycorrhizal and non-mycorrhizal root tips was found. In the mycorrhizal seedlings exposed to Al, content of Mg in needles and roots were strongly reduced (Table 2). This was attributed to displacement of Mg in the root cortex cell walls. The results indicate that the fungal sheath does not prevent Al and Pb uptake into the root.

Table 2. Effect of a 13 (17) week treatment with 800 μmol l^{-1} Al(NO$_3$)$_3$ on Mg contents in needles (N), roots (R) and cell walls of the root cortex (Cwi) of spruce seedlings colonized with *L. rufus* (*L. theiogalus*).

Mycobiont		Tissue content of Mg, mmol kg DW^{-1} or mmol dm^{-3} (Cwi)		
		N	R	Cwi
L. rufus	−Al	26	27	9.4
	+Al	9*	13*	3.7*
L. theiogalus	−Al	36	27	6.8
	+Al	12*	13*	1.6*

*, significantly different (*P* = 0.05) from corresponding control.

Nitrogen Transport and Depletion of Soil Nitrogen by External Hyphae of VA Mycorrhizas

A. Johansen, I. Jakobsen and E.S. Jensen
Plant Biology Section, Environmental Science and Technology Dept, Risø National Laboratory, DK-4000 Roskilde, Denmark.

Transport of soil N to plants via external hyphae of VA mycorrhiza has been demonstrated by ^{15}N techniques. However, reliable estimates of the potential role of mycorrhizas in N transport may be established only if precautions are taken to minimize direct uptake of ^{15}N by the roots. The objective of this work was to demonstrate hyphal transport of ^{15}N and the subsequent depletion of inorganic N from soil treated with a nitrification inhibitor.

Cucumber was grown with or without *Glomus fasciculatum* in containers subdivided into a root compartment (RC) and two hyphal compartments (HC$_A$ and HC$_B$) by 20-μm nylon mesh. HC$_A$ was subdivided into a buffer compartment and a labelling compartment. Nitrogen-15 as (^{15}NH$_4$)$_2$SO$_4$ (25 mg N kg^{-1} soil) and N-serve (5 mg N kg^{-1} soil) were mixed into the labelling compartment after 34 days, when hyphae from colonized roots were observed in the labelling compartment. Plants were harvested after 65 days and their ^{15}N content measured. Soil cores taken from HC$_A$ and HC$_B$ at 1, 3 and 5 cm distance from the RC were used for measuremens of length and ^{15}N content of hyphae and of KCl^{-} extractable NH$_4^+$ and NO$_3^-$.

Mycorrhizal plants had 63% of their root length colonized. Plant dry weight and N concentrations in plant tissues were unaffected by *G. fasciculatum*, which in contrast had a major effect on ^{15}N in the plants (Table 1). The ^{15}N in mycorrhizal plants corresponds to a 30% recovery of applied N. Length of hyphae in HC$_A$ and HC$_B$ of mycorrhizal plants were in the range 7.6–12.9 m g^{-1} soil. Hyphae from HC$_A$ contained significant amounts of ^{15}N (1.1–1.5 atom % ^{15}N, excess), whereas hyphae from HC$_B$ contained no ^{15}N. The soil was efficiently depleted of inorganic N by the external mycelium at all distances from the RC. Nitrogen depletion was considerably less in controls.

N-serve and a hyphal buffer compartment proved useful for minimizing the movement of the applied N source to roots by mass flow and diffusion. Consequently, the observed

[15]N enrichment of mycorrhizal plant provides an unequivocal demonstration of hyphal N transport. The absence of ^{15}N in hyphae from HC$_B$ indicates a low capacity for N transport from host to microsymbiont. The efficient uptake of inorganic nitrogen from the hyphal compartments, by the external hyphae, suggests the ecological importance of VA-mycorrhizal fungi in the conservation of soil nitrogen.

Table 1. Concentration of N and ^{15}N in plants.

	% N		Atom % [15]N (excess)	
	shoot	root	shoot	root
Control	0.91	1.54	0.01	0.01
G. fasciculatum	0.99	1.66	0.22	0.56

Effects of Nitrogen Application on Ericoid Mycorrhiza of *Calluna vulgaris* on a Danish Heathland

M. Johansson
University of Copenhagen, Institut for Sporeplanter, Øster Farimagsgade 2D, DK-1353 Copenhagen K, Denmark.
During the past few decades there has been a widespread conversion of heathland into grassland due to an increased atmospheric N-input. In order to clarify the effect of the addition of nitrogen on ericoid mycorrhizal colonization, field experiments have been carried out on a Danish heathland.

Root samples of *Calluna vulgaris* were obtained from two localities, Melby and Kongenshus, with N-deposition levels of 10 and 30 kg N ha^{-1} yr^{-1}, respectively. Ingrowth cores made of nylon bags containing sieved mor and bleached sand were established on each plot in cylindrical holes near the centre of individual *Calluna* plants. The ingrowth cores were taken into the laboratory in November after 6 months' growth. During this period ammonium nitrate was applied monthly to the Melby plot (total 35 kg N ha^{-1}). Roots from the mor layer and bleached sand were washed and cut into 5 mm lengths. The infected root lengths were determined in subsamples by the line-intersect method. The infection levels were determined in oven dried subsamples by the colorimetric assay of chitin.

The root biomass was significantly ($P < 0.05$) higher in the ingrowth cores from Kongenshus than Melby. The edaphic conditions at the two localities are very similar. Melby is, in contrast to the homogeneous *Calluna* cover on Kongenshus, a mosaic of *Calluna* and grasses. Vegetation structure influences the root competition and could be the basis of differences in root production. Different levels of atmospheric N-deposition or infection levels could influence the root biomass as well. The largest part of the root biomass is confined to the mor layer.

The mean infection level estimated as percentage of mycorrhizal roots was significantly higher in roots from Kongenshus than at Melby ($P < 0.05$). The mean infection was higher in the roots from the mor layer as compared with the roots from bleached sand and infection measured by the chitin assay followed the same trend.

The root biomass and infection both increased significantly at Melby following nitrogen application.

It is concluded that nitrogen has a stimulating effect on root growth, and that increases of mycorrhizal infection reflect vigour of the host plant. However, it is possible that even higher levels of nitrogen application would reduce the infection level. This needs further investigation.

Enhanced Growth of External VA Mycorrhizal Hyphae in Soil Amended with Straw

E.J. Joner and I. Jakobsen
Plant Biology Section, Environmental Science and Technology Department, Riso National Laboratory, DK-4000 Roskilde, Denmark.

Organic matter in soil appears to be an important factor for the development of vesicular-arbuscular (VA) mycorrhiza but only few authors have looked at the direct effect of organic matter on VA mycorrhizal hyphae. The objective of this study was to investigate how straw incorporation into soil affects root colonization and hyphal growth of VA mycorrhizal fungi.

Soil amended with chopped wheat straw (0, 10, 30 or 90 g kg^{-1} soil) and NH$_4$NO$_3$ (9.8 mg N kg^{-1} straw) was incubated indoors or outdoors for 14 months. Surplus precipitation was allowed to drain from soil incubated outdoors. Cucumber plants were grown with *Glomus caledonium* (RI 42) in irradiated (10 kGy) soil in containers consisting of a central root compartment separated from 12 hyphal compartments (HC) by 37-μm nylon mesh. Each HC contained 170 cm^3 irradiated soil from one of six incubation treatments. Length of hyphae in all HC (three replicates) was measured after 31 days by a membrane filter/grid intersect method.

Straw amendment significantly increased C:N ratio and organic C in soil. The highest straw level increased the CN ratio from 9.2 to 13.4. Root colonization by *G. caledonium* was 85%. Colonization did not differ among plants grown directly in soil of different straw contents. Length of external hyphae was increased by straw amendment from *c.* 17 m cm^{-3} soil at zero amendment to *c.* 28 m cm^{-3} at 90 g kg^{-1}. Inorganic N increased to a maximum of 233 mg kg^{-1} at the highest straw addition in indoor incubated soil, but was unaffected by straw amendment (50–53 mg kg^{-1}) outdoors. It is concluded that the increase in length of external hyphae did not result from increases in inorganic N. Organic matter appears to influence the external mycelium directly as root colonization was not affected.

Effect of Mycorrhizal Inoculation in Forest Nurseries

K. Kropáček
University of Agriculture, Faculty of Forestry, 281 63 Kostelec nad Cernymi lesy, Prague, Czechoslovakia.

Inoculation can be used to increase stress tolerance of seedlings after outplanting at sites disturbed by human activity. The aim of this study was to evaluate the effects of various fungal symbionts. The host species *Picea abies*, *Pinus sylvestris* were inoculated with *Hebeloma crustuliniforme*, *Hygrophorus pustulatus*, *Inocybe lacera*, *Laccaria laccata*, *Laccaria proxima*, or *Suillus luteus*. The substrate was sterilized peat–bark mixture and the inoculum carriers were alginate, or peat–vermiculite. Bareroot seedlings were grown in a cold greenhouse, and inoculum was applied immediately before or 1 and 2 months after sowing. After the first season, seedlings were containerized and cultivated for 2 years in open beds. During a 3-year period the mycorrhiza effect and occurrence of indigenous *Thelephora terrestris* mycorrhiza were evaluated. Mycorrhiza formation on the inoculated *Picea abies* seedlings cultivated in the laboratory in sterilized and non-sterilized substrate sampled from the reforestation sites was also evaluated.

L. laccata and *L. proxima* were found to be the best symbionts both for *P. abies* and *P. sylvestris*. At the end of the first season there was an almost complete infection on the tested seedlings. The vermiculite–peat inoculum was characterized by a rapid development of mycorrhizal infection at the beginning of the growth season. The alginate inoculum induced slower but more intensive infection at the end of the season and the mean dry weight of the seedlings was significantly greater than in the vermiculite–peat treatments or non-mycorrhizal controls. Differences between inoculation simultaneously with sowing or 1–2 months after sowing were not found. During the 3-year period the effect of artificial mycorrhization was manifested primarily as a more balanced growth of seedlings. Increased contamination of substrate by the indigenous mycorrhizal fungus *T. terrestris* was observed during cultivation in outdoor beds.

Effects of Organic Matter Removal on Fruitbody Production of Ectomycorrhizal Fungi in Stands of *Pinus sylvestris*

Th.W. Kuyper, B.W.L. de Vries and J. Baar
Biological Station of the Agricultural University, Kampsweg, 27, 9418 PD Wijster, The Netherlands.

During the last decade numbers of species of ectomycorrhizal fungi have declined in Europe. This decline is especially evident in older stands in polluted regions and is more severe than that seen during forest succession. Field observations suggest that accumulation of organic material in the forest floor might explain this decline in ectomycorrhizal diversity. At the beginning of this century Kallenbach noted that removal of litter had positive effects especially on hydnaceous fungi. It has also been shown to stimulate fruiting of *Russula* spp. but to decrease that of *Lactarius*.

An experiment was set up in which litter and humus layers were completely removed in three old stands, ranging from 40–80 years, in N.E. Netherlands. In each stand two plots of 100 m^2 were used as controls, and in two other plots of each stand the litter and humus layer were completely removed (OM removal) manually in 1985. The fruitbody production was investigated from 1985 to 1989. OM removal had a significantly positive effect on both species richness and fruitbody abundance. The effects of the experimental treatment became progressively more clear with time. Most species were clearly stimulated by OM removal.

The mechanisms involved in the increase in diversity and abundance of ectomycorrhizal fungi are unknown. The positive effect of OM removal on ectomycorrhizal diversity (and most likely on ectomycorrhizal functioning) might be relevant for restoration of nitrogen-saturated environments. Even though removal of OM leads to a decrease in cation availability and a lower water holding capacity, positive effects on mycorrhizal functioning after removal of the nitrogen excess might well prevail. In order to understand better the role of the OM horizons in influencing the mycorrhizal community a new large-scale experiment has been established in young, intermediate and old stands of *Pinus sylvestris*. OM has been removed and also added to other plots in young and intermediate stands.

Results of Ectomycorrhizal Inoculation of Pine Species with *Pisolithus tinctorius* and *Thelephora terrestris* in Korea

K.J. Lee[1] and C.D. Koo[2]
[1]*Department of Forestry, Seoul National University, Suwon, Kyonggido, 441–744 Korea:* [2]*Forestry Research Institute, Cheongryangri-dong, Tongdaemun-Ku, Seoul, 130–101 Korea.*

Pinus koraiensis (Pk), *P. rigida* (Pr), and *P. rigida* × *taeda* (Pr.t) seedlings in a bare-root nursery were artificially inoculated with *Pisolithus tinctorius* (Pt) and *Thelephora terrestris* (Tt) to test long-term effects of ectomycorrhizal inoculation on host growth and survival of introduced mycorrhizal fungi in the field.

Mycelial inocula of Pt and Tt were mass-cultured in vermiculite–peatmoss mixture and introduced into fumigated nursery soil before seed sowing. Bare-root, inoculated seedlings were outplanted to the field at 1–4 years of age.

At the time of outplanting, Pk seedlings (4 years old), Pr seedlings (2 years old), and Pr.t seedlings (1 year old), all infected by Pt, were significantly taller by 28%, 26%, and

77%, respectively, than control seedlings infected by natural populations of mycorrhizal fungi in non-fumigated plots. Ten years after inoculation, and 6–9 years after outplanting to the field, Pk seedlings inoculated with Pt were significantly taller by 9%, Pr.t seedlings significantly taller by 18%, and Pr slightly taller by 2% (not significant) than control seedlings, suggesting that the stimulatory effect of Pt on host growth gradually declined or became minimal after outplanting. Tt failed to stimulate host growth either in the nursery or in the field, and survival rate of outplanted seedlings was not different among fungal treatments.

Loss of the infected roots during lifting the seedlings for outplanting could be the primary cause of the reduced effect of Pt in the field. More than 90% of the fine roots were Pt infected in the fumigated nursery during the first growing season, but Pt mycorrhizas disappeared completely in the field. A secondary cause of the reduced effect could be that it failed to compete successfully with the natural population of ectomycorrhizal fungi. It is necessary to select those mycorrhizal fungi which adapt well to both nursery and field.

Ecology of Ectomycorrhiza and Ectomycorrhizal Fungi in Norway Spruce Forest Ecosystems of Sumava Mts, Czechoslovakia

A. Lepsová[1], R. Kocourek[2] and R. Král[3]
[1]*Inst. Landscape Ecol. CSAV, Na Sádkách 7, Ceské Budejovice:* [2]*Husova 465, Námest nad Oslavou:* [3]*Inst. Inorg. Chem. CSAV, Rez u Prahy, Czechoslovakia.*

Research on ectomycorrhizal (EC) fungi and fine roots of Norway spruce at 1300 m in the Sumava Mts is reported. Permanent plots in an old growth mountain spruce forest were studied.

The biomass of EC carpophores varied from 0.72 to 7 kg ha dry weight (1987, 1988). Most biomass was formed by *Russula ochroleuca* (about 50%). Fine root material consisted mostly of senescent roots. The biomass of mycorrhizas varied from 300 to 1000 kg ha^{-1}. Most of the fine root biomass was found in layers rich in very acidic organic matter (pH 3.2–4.4). Spruce needles showed slight deficiency of Mg, Cu and Zn. *Russula ochroleuca* accumulated Zn, Cu and Mg and excluded Fe, Mn and Al and Ca slightly. Mycorrhiza accumulated Fe, Ca, Mg, Cu and Zn and Al slightly. The role of fungi and mycorrhiza in spruce nutrition could be elucidated by measurement of needle:carpophore and needle:mycorrhiza element ratio.

Present data suggest that mycorrhiza do not affect the uptake of Mn and Ca by *Picea abies*, but EC fungi may enhance the uptake of Cu and Zn while providing a barrier against Al and Fe. The role of EC fungi in Mg uptake is not clear. Zinc accumulation by *R. ochroleuca* is about five times greater than by other fungi.

Production of Siderophores by Ectomycorrhizal Fungi

C. Leyval[1], F. Watteau[1], J. Berthelin[1] and C.P.P. Reid[2]
[1]*Centre de Pédologie Biologique, UPR 6831 du CNRS associé à l'université B.P.5, 54501, Vandoeuvre-les-Nancy, Cedex France:* [2]*Department of Forestry, 118 Newins-Ziegler Hall, University of Florida, Gainesville, FL 32601, USA.*

In iron-deficient conditions, most bacteria and fungi are known to release iron-chelating siderophores. We investigated the production of siderophores by three ectomycorrhizal fungi (*Laccaria bicolor, Pisolithus tinctorius, Suillus granulatus*), in liquid or solid nutrient medium using a colorimetric assay and a bioassay (Arthrobacter JG9). The colorimetric assay was also tested with young mycorrhizal slash pine seedlings in non-axenic conditions. Two other experiments are also reported to illustrate the influence of ectomycorrhizal fungi on iron mobilization from minerals and on iron nutrition of mycorrhizal plants, that could be attributed to siderophore production.

On Petri plates the bioassay is a more sensitive method than the chrome azurol method. The latter cannot be used to detect siderophores from filtrates of fungal cultures, because of its low sensitivity. In liquid fungal cultures without added iron, siderophores were detected with the bioassay after 2 days with *L. bicolor* and *S. granulatus*. The production was delayed in presence of increasing concentrations of $FeCl_3$. Above 10 μmol which was the optimum concentration for fungal growth, no more siderophores were detected with *L. bicolor*. A maximum concentration of 0.5 to 1 μg ml^{-1} DFOB (deferrioxamine B) equivalent was found with *S. granulatus* and 0.2 μg ml^{-1} with *L. bicolor*. Grown on agar plates in presence of chrome azurol, *S. granulatus* released more iron chelates than the two other fungi. On this buffered (pH 6.8) medium *P. tinctorius* did not seem to release siderophores. But the production varied greatly with the nutrient medium.

Both methods show that ectomycorrhizal fungi (*P. tinctorius*, *S. granulatus* and *L. bicolor*) produce siderophores in pure culture. The precise structure of these compounds has not been analysed.

In a flask experiment, *S. granulatus* released iron from goethite, an iron oxyhydroxide. The solubilization of iron seemed to be related to the release of siderophores rather than to organic acids.

The colorimetric method indicated that mycorrhizal (*P. tinctorius*) pine roots released siderophores.

Such results show that fungal siderophores may be involved in iron mobilization in soils and in iron nutrition, and that their effect in a mycorrhizal symbiosis needs further investigation.

Fungal Mass in Sporophores, Mycorrhizas and Living Mycelia in Scots Pine Stands Along a Pollution Gradient in the Oulu Region of Finland

A.M. Markkola, R. Ohtonen, U. Ahonen-Jonnarth, O. Tarvainen and A. Ohtonen
Department of Botany, University of Oulu, Linnanmaa, SF-90570 Oulu, Finland.

The benefits of mycorrhizal symbiosis to forest trees need to be assessed in areas with heavy deposition of air pollutants and reduced photosynthesizing biomass. We studied the effect of air pollution on fungi and at the same time tried to estimate the cost of symbiosis to the trees by measuring the total fungal mass in sporophores of both mycorrhizal and decomposing fungi, pine mycorrhizas and living mycelium in humus from Scots pine stands on a pollution gradient of N, S and heavy metals around the city of Oulu, Northern Finland.

The combined fungal mass measured as sporophore production, Scots pine mycorrhizas (assayed as chitin) and mycelium were about equal at both ends of the pollution gradient. About 76% of the fungal mass was in the mycorrhizas at the less polluted sites, 18% in mycelium and 6% in the sporophores. At the more polluted sites the amount of mycelium was doubled to 35% of the total fungal mass and the proportion of mycorrhizas and sporophores was decreased to 63% and 2% respectively. The sporophore production of mycorrhizal fungi was reduced to one-third and that of decomposer species increased fourfold at these sites compared to the less polluted sites.

It was not possible to distinguish the mycelia of mycorrhizal and decomposer fungi in the soil. The increase in the production of sporophores of decomposing species at the more polluted sities, however, is probably attributable to an increase of mycelia of that group in the forest humus.

The decreased fungal mass in Scots pine mycorrhizas and sporophores of mycorrhizal fungi in polluted sites could be partly due to the toxic effects of pollutants deposited in the humus layer, but partly to both the reduced amount and changed accumulation pattern of carbohydrates in the root-fungus systems, which might in turn favour decomposer fungi.

Effect of Lead on the Growth of Ectomycorrhizal Fungi

P. Marschner and D.L. Godbold

Forstbotanisches Institut, Universität Göttingen, D-3400 Göttingen, Germany.

The lead tolerance of two different species and five isolates of ectomycorrhizal fungi has been tested. Using normal culture methods Pb^{2+} concentrations to which the fungi are exposed may be decreased by precipitation as Pb phosphates and by complexation with agar. A technique was developed to asses the effect of metals on the growth of fungi in nutrient solution.

In this system the effect of Pb on the growth of ectomycorrhizal fungi can be estimated without formation of Pb phosphates or complexation of Pb by agar. This allows assays to be carried out at known Pb concentrations.

The fungi were grown on Petri dishes divided into three equal parts. The composition of the nutrient solution used was based on soil solutions from the Solling forest (Germany) amended with glucose at $10 \text{ g } 1^{-1}$. One compartment of the Petri dish was filled with the complete nutrient solution with malt extract solidified with agar. The second was filled with the nutient solution without phosphorus (P) and in the third with the nutrient solution and with 1–500 μmol Pb as $PbCl_2$. An agar plug from the edge of an actively growing colony was placed near the middle of the dish on the agar side.

The Pb tolerance varies greatly between and within species (Table 1). Increasing the P concentration on the agar side strongly increased the Pb tolerance of *Paxillus involutus* Nau. Supplying P and Pb in the same nutrient solution strongly reduces the toxicity of Pb. This work shows that when assessing the effects of Pb in the environment on ectomycorrhizal fungi the differential tolerance between and within species and the effect of the P supply on the Pb tolerance must be taken into account.

Table 1. Surface area of the mycelium of different ectomycorrhizal fungi on the Pb side of split plates at different Pb concentrations. Expressed as % of the control.

	Pb conc (μmol)			
	1	10	100	500
Paxillus involutus				
COU	83 ± 10	75 ± 6	56 ± 6	56 ± 14
NAU	117 ± 7	84 ± 7	16 ± 2	10 ± 1
533	109 ± 9	93 ± 3	66 ± 11	30 ± 1
031	102 ± 26	97 ± 14	29 ± 6	5 ± 3
Pisolithus tinctorius				
(Marx)	94 ± 35	85 ± 6	38 ± 4	22 ± 3

The Role of VA Mycorrhiza and Soil Phosphate in the Early Life History of the Bluebell, *Hyacinthoides non-scripta*

J.W. Merryweather and A.H. Fitter

Department of Biology, University of York, York, YO1 5DD UK.

The bluebell *Hyacinthoides non-scripta* (Liliaceae) can dominate the ground flora in some English woodlands. Preliminary findings of a study on bluebell mycorrhizas, illustrate an increasing VA dependence of individual bulbs as they mature and migrate downward to a very infertile soil horizon where they survive for many years as adults.

At the study site in North Yorkshire, adult bulbs and their roots occur in the highly-leached sandy *Bs* soil horizon (*c.* 15 cm downward). The short, thick unbranched roots are strongly mycorrhizal, as might be expected of a plant with so inadequate a root system. Available P (water extract) is plentiful in the upper horizons, especially in the litter (mean [P] 35 μg ml^{-1}) where seeds fall and the young plantlets establish themselves. However, P concentration decreases to values as low as 0.10 μg ml^{-1} at a depth of 15–25 cm.

In July bluebell seeds are deposited onto a bed of dry litter composed of dead leaves of both their parents and the canopy species (bracken, oak, sycamore etc.). When covered by wet, rotting leaves in November they germinate, producing a small primary root which exploits this rich substrate where nutrients are plentiful. Few of these primary roots encounter a fungal inoculum, and by spring very few seedlings have become mycorrhizal. In subsequent years successful plantlets (mortality is high, up to 97% dying before maturity) descend, by means of contractile roots and downward bulb growth for about 5 years, until they reach the Bs horizon where they proliferate to form a colony. As the bulbs progress downward through the soil, ageing from seed to adult plant and passing from a high to a very low P environment, the roots become progressively more colonized by VA endophytes.

The principal benefit to the plant partner of a VA mycorrhizal fungus is the enhanced uptake of phosphorus, and this is particularly important to the bluebell, which inhabits a soil where available P is scarce. Seed reserves and nutrients in decomposing leaf litter provide initial supplies as bulbs descend into less fertile horizons. As bulbs descend they begin to encounter inocula of VA fungi, eventually establishing a permanent relationship: since the adults' new roots follow the paths of the previous years' dead roots, they are immediately reinfected, maintaining the mycorrhiza established previously. The bluebell provides not only a model for the obligate mycotrophic system, but it also illustrates the way in which a mycorrhiza may be maintained for almost the entire life of an individual, being intimately involved throughout its life history.

Inoculation with *Glomus intraradix* Improves Growth of *Acacia nilotica* under Non-sterile Nursery Conditions in Ethiopia

A. Michelsen
Institute of Plant Ecology, University of Copenhagen, Øster Farimagsgade 2 D, DK-1353 Copenhagen K, Denmark.

The experiment was designed to measure the response of the multipurpose tree *Acacia nilotica* to inoculation with a VA fungus and to P fertilization under nursery conditions in Ethiopia. Seedlings of *A. nilotica* were grown for 5 months in pots in a tree nursery at Jima, W. Ethiopia. The climate is subhumid and the soil a highly P-sorbing dystric nitosol. The experimental design was completely randomized 2*2 factorial: ± mycorrhizal inoculation and ± phosphorus fertilization, with nine replicates. One part inoculum consisting of an Italian isolate of the VA fungus *Glomus intraradix* was added to 30 parts of a standard, non-sterile nursery soil mixture. In the P treatments 15 mg P kg^{-1} soil was added three times (as $Ca(H_2PO_4)_2$), in weeks 2, 7 and 14.

Inoculation with *G. intraradix*, P fertilization and the combination of the two treatments improved shoot dry weight by 41, 45 and 40%, respectively. Height and leaf number were significantly increased by infection from week 7 and by P from week 22 and 14, respectively. Inoculation increased colonization of the roots from 19 to 33% in non-fertilized plants and from 18 to 27% in fertilized plants. Total root length increased accordingly, by 99 and 52%, whereas shoot P uptake was not significantly affected.

The results obtained in the nursery emphasize the potential of inoculation of tropical tree nursery soils which often have suboptimal populations of indigenous VA fungi. They contrast with earlier results obtained in a greenhouse. The influence of *G. intraradix* could be due to improved water relations as well as to enhanced P uptake. The increased root length to shoot dry weight ratio of inoculated seedlings may facilitate recovery from drought stress and improve soil moisture extraction at low soil water potentials. No additive effect on shoot dry weight was seen when combining P fertilization and VA inoculation. Other factors may thus limit seedling growth when available P or water is sufficient.

VA – *Rhizobium* interaction in productivity and nutrient content of yard long bean (*Vigna unguiculata sesquipedalis*)

M.A.U. Mridha[1], M. Mitra[1] and Z. Parveen[2]
Department of Botany, University of Chittagong, Bangladesh: [2]Department of Soil Science, Dhaka University, Bangladesh.

Yard long bean (*Vigna unguiculata sesquipedalis*), harvested as green pods, is an important vegetable crop of Bangladesh, and is commonly gown in the hilly and plain areas of Chittagong. Its response to dual inoculation with vesicular-arbuscular mycorrhiza (VAM) and *Rhizobium* has been studied in Bangladesh.

The experiment was carried out on a loamy soil in pots under field conditions with the host inoculated with a local isolate of *Rhizobium*. The soil was amended with root segments of *Mimosa pudica* infected with mycorrhizal fungi. There were four treatments: uninoculated control, *Rhizobium* inoculated, VA inoculated, and *Rhizobium* + VA inoculated.

The growth, yield and nutrient content were all markedly increased by dual inoculation with VA and *Rhizobium*. Plants inoculated with either mycorrhiza or *Rhizobium* produced significantly higher growth, yield and nutrient content compared with non-inoculated controls.

Rhizobium inoculation enhanced shoot and root growth, nodule number and nodule dry wight, to a greater extent than mycorrhizal inoculation. Increased plant height, fruit length and fruit number were observed with mycorrhizal inoculation compared with *Rhizobium* inoculation.

These results are in agreement with other studies which have shown that VA endophytes can have important effects on nodulation and N-fixation in legumes. Practical application of VA and *Rhizobium* is important because in cropping systems of Bangladesh inputs of P and N fertilizers are limited by economic constraints and shortfalls in supply.

The Development of Ectomycorrhizal Infection, and its Relationship to Seedling Growth

A.C. Newton
Institute of Terrestrial Ecology, Bush Estate, Penicuik, Midlothian, EH26 0QB, UK.

In order to assess the importance of nutrient uptake and mycorrhizal infection on the establishment of plants of *Quercus robur* and *Betula pendula*, seedlings were grown on two nutrient-poor soils in two woodlands sites, and in pots of soil collected from each field site.

Sequential harvests were taken throughout the first growing season. The onset of ectomycorrhizal infection of field-grown plants occurred 2–4 weeks before that of pot-grown plants; intial infection of field-grown seedlings was observed 3 weeks after outplanting. Percentage infection of field-grown seedlings was higher than in pots throughout the growing season, by up to 40%. Different mycorrhizal types displayed contrasting patterns of development: *Paxillus involutus* tended to increase rapidly, to maximum percentage infections of over 30%; whereas *Cenococcum geophilum* was consistently recorded at <10% infection.

When a single root system of pot-grown oak was analysed in detail, seven mycorrhizal types were differentiated. The relative abundance of these types varied depending on whether infection was assessed by counting branched mycorrhizas, or individual mycorrhizal tips. This reflected variation in the pattern of branching of the different mycorrhizal types. Of the 41 mycorrhizal types differentiated in the entire survey of 1800 seedlings, five were consistently recorded as unbranched, and six were found to produce second-order branches.

Different mycorrhizal types displayed contrasting patterns of abundance, both with respect to site and host species. For example, 21 types were recorded infecting oak, whereas only 15 were found on birch; only three types were found infecting both oak and birch on both sites.

In this survey, seedlings were dominated by between one and three mycorrhizal types. Two types were particularly dominant: those formed by *Scleroderma citrinum* and *P. involutus*, which accounted for up to 61% and 92% of the mycorrhizas of oak and birch respectively. These fungi were exceptional among those encountered in the present analysis, in producing both highly branched mycorrhizas and mycelial strands.

The relationship between ectomycorrhizal infection and seedling growth was assessed by correlation. Total seedling dry mass was in general found to be positively correlated with the total number of ectomycorrhizal tips, but poorly correlated with the extent of ectomycorrhizal infection, except where seedling growth had previously been shown to be limited by phosphate availability. Seedling dry mass was also in general poorly correlated with number of mycorrhizal types.

Auxin Production and Mycorrhizal 'Virulence'

J.-E. Nylund and H. Wallander
Department of Forest Mycology and Pathology, SUAS, P.O. Box 7026, S-75007 Uppsala, Sweden.
Some fungal species known to be vigorous ectomycorrhiza formers, such as *Pisolithus tinctorius* and *Laccaria bicolor*, excrete large amounts of indole acetic acid (IAA) in pure culture. Considering suggestions of a key role of IAA in regulating the symbiosis, a potential 'super mycobiont' may be sought among fungi with high IAA production.

Mycorrhiza (M) formation by two very low IAA producing wildtype isolates of *Hebeloma cylindrosporum* and four hyper-IAA mutants were compared using *Pinus pinaster* as a natural host and *P. sylvestris* as a non-natural but compatible host plant. Previous results of IAA and indole butyric acid (IBA) application to ectomycorrhiza in hydroponic culture suggested the mutants caused a greater sugar drain from the host, a larger fungal biomass in the mycorrhiza, but a reduced host growth compared with the wildtypes, and also a difference in root morphogenesis. However, the mutants did not differ from the wildtypes in two hydroponic culture experiments with *P. pinaster*. In one experiment with *P. sylvestris*, the two mutants stimulated growth rate by 7–10%, while in other respects the lack of difference was as notable as with *P. pinaster*.

The IAA levels of M (*L. bicolor*) and NM roots were examined with GS-SIS-MS, as increased IAA has been reported in the former. Our results were again contrary to this view. In both short roots and complete root systems, IAA levels were 50% lower with mycorrhiza. Others have reported that high nitrogen (N) levels reduce IAA in mycorrhiza. However, here with high N availability, there was an *increase* in IAA levels, particularly in M seedlings, in which fungal biomass was suppressed by the high N levels.

Thus, our findings provide little evidence in favour of the view that high IAA production would be an important characteristic for a vigorous mycorrhiza former. Yet, there are indications of IAA being of importance, although the mechanisms may be other than those expected. One such feature is the observed growth stimulation; another is the occasionally observed enhanced mycorrhiza development around agar plugs containing IBA.

Another question is whether 'virulence' is desirable. Some results suggest that vigorous inoculated mycobionts may prevent natural species from colonizing roots after plantation, thus being disadvantageous except under extreme circumstances.

Immunological Aspects of the Characterization of *Tuber magnatum* and *Tuber albidum*

G. Papa[1], C. Polimeni[1], P. Mischiati and G. Cantini Cortellezzi[2]
[1]*Centro di Studio sulla Micologia del Terreno-CNR c/o Dipartimento di Biologia Vegetale–Università v/le Mattioli 25–10125 Torino, Italy:* [2]*Instituto Zooprofilattico, via Bologna 148–10154-Torino, Italy.*

Tuber magnatum, the typical Italian truffle characteristic of some areas of Piedmont (NW Italy), is an extremely valuable agricultural product. For this reason, the practice of preparing pre-infected seedlings, usually of *Quercus* and *Populus*, is gaining in importance; where possible, these are planted in areas which are already known to be suitable for the development of the truffle. The mycorrhization process of seedlings of *Quercus* and *Populus* is more difficult with *T. magnatum* than it is with *T. melanosporum*, which is much more widespread in Europe.

The production of seedlings that are successfully infected, but free from polluting agents such as *Sphaerosporella brunnea* or some other predominant *Tuber* species, requires sterility of the glasshouses to be rigorously controlled. One of the most common contaminant mycorrhizas of *T. magnatum* is *T. albidum*, which forms a more competitive symbiosis, but which has little commercial value.

These two mycorrhizas (*T. magnatum* and *T. albidum*) are morphologically very similar, making an unambiguous method to distinguish them essential. A routine, high-sensitivity immunological method was set up for this purpose: three different antisera were obtained in rabbits, one by injecting an electrophoretically-pure characteristic component of the carpophore, the other two by injecting an extract of root apexes mycorrhized with *T. magnatum* (b) and, in a second phase of the experiment, an extract with concentraded protein content (b').

Immunodiffusion experiments in 1% agarose showed that the mycorrhiza from *T. magnatum* can be distinguished from that of *T. albidum* by antiserum (b'). Antiserum (a) had a partially positive reaction against both mycorrhizas.

Competitive absorption experiments of the antisera with extracts from non-mycorrhizal root apexes of *Q. robur* and those mycorrhizal with *T. albidum* confirmed the reliability of this immunological detection method. Preliminary results with mycelia obtained *in vitro* from the two *Tuber* species do not show specific reactions.

Comparison of the Ectomycorrhizas Formed by *Russula ochroleuca* on Several Tree Species

A. Pillukat
Institut für Systematische Botanik, Menzinger Straße 67, D-8000 München 19, Germany.

A general supposition is that the fungal features of the ectomycorrhizas of a defined fungal species are independent of the autobiont involved, but that features originating from the latter may differ. Evidence in favour of this view was found in a comprehensive examination and comparison of the ectomycorrhizas of *Russula ochroleuca* on the following tree species: *Abies alba*, *Betula pendula*, *Carpinus betulus*, *Fagus sylvatica*, *Larix decidua*, *Picea abies*, *Pinus sylvestris* and *Quercus robur*.

The ectomycorrhizas were isolated from soil samples taken directly under fruitbodies of *Russula ochroleuca*. The identification of the fungus and tree species was by comparison with published descriptions. Immediately after isolation from the soil, mycorrhizal features which could change following fixation of the ectomycorrhizas in formalin–acetic acid (FAA) (e.g. colour, auto-fluorescence and staining reactions) were recorded. The pattern of ramification and the structure of the mantle, in both plan view and when seen in section, was examined and recorded for each tree species. The dimensions of the ectomycorrhizal systems, the mantle and cortical cells were compared.

In all mycorrhizas examined, the mantle structure was very similar, irrespective of the tree species involved. However, there were host dependent differences in the type of ramification, shape, dimensions, occurrence of tannin cells, shape and orientation of cortical cells and development of the Hartig net. Sections of the ectomycorrhizas revealed that all angiosperms lacked tannin cells whereas there was always a tannin layer present in ectomycorrhizas on gymnosperms. Moreover, the first layer of cortical cells of angiosperms was generally tangentially elongated, whereas those of gymnosperms were more or less round to rectangular shaped. Usually the Hartig net in angiosperms surrounded only the first layer of cortical cells, in contrast to gymnosperm mycorrhizas, where it penetrated further.

Independent of the tree species, at the distal end of the ectomycorrhizas, within the second and third upper mantle layers, there were cells with smaller dimensions and thinner cell walls than in the subapical region. The origin of these cells appears to be by division of larger thicker walled cells, which produces a characteristic pattern. These pseudoparenchymatous cells are cautiously interpreted as being part of an initiation zone of the mantle.

Identification of Ectomycorrhizal Fungi by Use of Immunological Techniques

I. Plattner[1], T. Grabher[2], I. Hall[3], K. Haselwandter[1] and G. Stöffler[2]
[1]*Department of Microbiology, Faculty of Science, University of Innsbruck,* [2]*Department of Microbiology, Medical School, University of Innsbruck, A-6020 Innsbruck, Austria,* [3]*Ministry of Agriculture and Fisheries, Invermay Agricultural Centre, Mosgiel, New Zealand.*

The objective of this study was to develop specific and rapid identification methods for *Tuber* spp. which would aid investigations of their ecology and taxonomy. The experiments involved use of antisera and monoclonal antibodies in various immunological assays. As immunoassays are rapid and easy to perform they have potential for use in field studies, such as those investigating the spread and distribution of mycorrhizas along the rootsystem or monitoring competition with other soil fungi.

Soluble extracts from fruitbodies of *Tuber melanosporum* were used to raise antisera in New Zealand white rabbits. The antisera were first characterized for their reactivity with extracts from *Tuber melanosporum* fruitbodies and corresponding mycorrhizal root-tips. The degree of cross-reaction with *Tuber magnatum* and four other ectomycorrhizal isolates (*Suillus variegatus, Suillus luteus, Amanita muscaria, Rhizopogon roseolus*) was then examined using ELISA, immunofluorescence, dot blot and Western blot.

The antisera were found to react with both fruitbodies and root-tips formed by *Tuber melanosporum*. They exhibited high titres in ELISA where dilutions up to 1:64 000 gave positive results. The amount of antigen present in a single root-tip was detectable in dot blot. Using indirect immunofluorescence, single hyphae and the ectomycorrhizal mantle could be visualized. Both results indicate the high sensitivity of the antisera.

Cross-reaction of the antisera with *Tuber magnatum* was almost complete in ELISA (91%) but in Western blot differentiation of the two *Tuber* spp. was possible. With the other ectomycorrhizal fungi cross-reaction was very low: in ELISA approximately 10% with *Suillus grevillei* and only 5% with *Suillus luteus, Amanita muscaria* and *Rhizopogon rubescens*. These results indicate that the specificity of the antisera is confined to the genus *Tuber*.

Recently, monoclonal antibodies specific for *Tuber melanosporum* were produced. From two fusion experiments 28 culture supernatants were reactive with *Tuber melano-*

sporum in ELISA. Four clones were selected for growth in large culture dishes. Three of these were also reactive in Western-blot and stained protein bands of molecular weights with 66, 40 and 36 kDa, respectively.

Monoclonal antibodies will be used to develop specific tests which should allow us to discriminate between different *Tuber* spp. and between members of the genus *Tuber* and other ectomycorrhizal fungi.

Mycorrhizal Status of *Quercus* and *Fagus* in Latium (Central Italy)

G. Puppi and R. Isopi

Dipartimento di Biologia Vegetale, Università 'La Sapienza', Roma, Italy.

A mycorrhizal association is unequivocally defined only by the identification, at the species level, of the fungal partner. Since however the presence of fruiting bodies may not reflect the mycorrhiza actually formed and one of the most commonly observed detrimental effects of atmospheric pollution on forest trees is reduction of mycorrhizal frequency, with alterations in morphology and reduction in types, the need for description of mycorrhizal types is now recognized. A survey of the mycorrhizal types of some important tree species in Latium (Central Italy) is reported.

Quercus ilex, *Q. cerris* and *Fagus sylvatica* were sampled in Mediterranean coastal forests on a range of soil types. Samples of *Q. ilex* were taken in April 1990, all others were collected in early July. Mycorrhizal types were determined, on a morphological basis (colour, ramification pattern, surface characters, mantle structure) after extraction from subsamples by wet sieving and decanting.

Twenty-two distinct mycorrhizal types were recognized, in 45 soil cores. One to four different types could be found in each core; 13 types were found associated with only one site-tree combination, while only one type was common to all sites and plant species. The highest diversity in types was recorded on a poor soil at Castelporziano as well as the lowest percentage infection; out of 12 types, nine were found on *Q. ilex* and seven on *Q. cerris*, four were shared; on a good quality soil at Tolfa, out of nine distinct types, seven were observed on *F. sylvatica*, and only four on *Q. cerris*, two were shared; on an organic soil at Vico, out of 11 types, seven were found on *F. sylvatica*, and five on *Q. cerris*, only one was in common. *Q. cerris* at Tolfa and at Vico exhibits the same, few, mycorrhizal types, which however have very different frequency in the two sites. Mycorrhizal types found on *F. sylvatica* at Tolfa and at Vico are rather distinct; percentage mycorrhizal infection is higher at Tolfa, and dominated by one type.

While some relationships were found between frequency of mycorrhizal infection and the stress indicator peroxidase it is difficult to relate mycorrhizal types to environmental conditions. A more extended survey, as well as identification at the species level of the symbiont and controlled experiments will contribute to the understanding of the relationships between mycorrhizal types and optimal mycorrhizal conditions in Mediterranean areas.

Effects of Liming and N-fertilization on Ectomycorrhizas in a Mature Beech Stand in the Solling Area (Germany)

C. Rapp

Institut F Waldbau, Büsgenweg 1, D-3400 Göttingen, Germany.

The application of fertilizers in forest ecosystems can stimulate or compensate for the ecological consequences of high element input, e.g. soil acidification. Although fertilizers are widely used in forestry practice knowledge of their effects on ectomycorrhizas is sparse.

A field experiment was carried out in a mature beech stand on acid forest soil. Soil acidification was counteracted by application of dolomitic limestone or accelerated by annual application of ammonium sulphate. Consequent effects on mycorrhizal biomass, distribution and species composition of ectomycorrhiza were investigated.

Surface fertilization of plots was carried out as shown in Table 1.

Table 1. Surface fertilization of plots.

Plot	Date	Fertilizer	Amount (kg ha^{-1})
BO	control	–	
BK	1982	limestone	30 000
BN	1983–1988	$(NH_4)_2SO_4$	660

Four years after treatment (1986/1987) root samples were obtained by soil coring.

Mean annual mycorrhiza biomass from surface layers and mineral soil (20 cm depth) was 100–120 kg ha^{-1} on the control plot. Liming reduced mycorrhizal biomass slightly, whereas N-application induced a distinct decrease. The surface layers of the control plot comprised 40–70% of mycorrhizal biomass, whereas corresponding values after N-treatment were below 10%.

Mycorrhiza of *Lactarius subdulcis* were the most frequent of five ectomycorrhiza species identified in the surface layer O_F on the control plot. Liming reduced the abundance of *L. subdulcis*, whereas percentage of *Laccaria amethystina* increased threefold. N-treatment reduced abundance of *L. amethystina*, but did not alter percentage of *L. subdulcis*. Both treatments reduced total numbers of different ectomycorrhizal types.

Examination of Single Spore Cultures of VA Fungi by Isoenzyme Patterns after Polyacrylamide Gel Electrophoresis (PAGE)

I. Raschen and H. von Alten
Institut für Pflanzenkrankheiten und Pflanzenschutz, Universität Hannover, Herrenhäuser Str. 2, D-3000 Hannover 21, Germany.

From a pot culture of *Glomus intraradix* on *Tagetes erecta* originating from three morphologically identical spores 11 single spore cultures were raised on *Allium porrum* in autoclaved sand. When mycorrhiza had developed root systems were removed from the substrate causing minimal damage and cut back to approximately one-third of the original size before the plants were repotted in sterilized sand to conserve the original cultures. For further propagation surface sterilized seeds of *A. schoenoprasum* were sown either into 5% of the remaining root/sand or into a pot with one mycorrhizal plant from subsequent generations in sterilized sand. Isoenzyme distribution pattern analysis was carried out on 5 g of fresh root material homogenized in protein extraction buffer with polyvinylpolypyrrolidone (PVPP) on ice and centrifuged for 20 min. Aliquots of the supernatant containing 10 mg protein were loaded per well and separated electrophoretically in a discontinuous polyacrylamide gel constituting a stacking gel of 3.5% T, 5% C_{bis}, pH 6.8 and a separating gel of 15% sucrose, 6.5% T, 5% C_{bis}, pH 8.8. After electrophoresis at 4°C, 150 V, 30 mA for 90 min gels were stained for non-specific esterase activity (EC3.1.1.1).

Using both morphological characters of the resting spores and characteristic isoenzyme distribution patterns of total protein extracts from mycorrhizal roots after PAGE, the continuity of genotypic traits of one culture was examined over four generations of propagation.

Additionally, the remaining single spore cultures were cross examined for species specific properties.

All spores observed were of uniform appearance with no significant morphological differences between those of different single spore cultures or generations. However, whereas the examination of one single spore culture over four generations yielded identical isoenzyme distribution patterns for non-specific esterase, a cross examination comparing ten single spore cultures resulted in three categories of bands, those common to VA in general, those common to *G. intraradix* only and those specific for single spore cultures. These results indicate a continuity of the characters expressed in single spore cultures under glasshouse conditions, extending over several generations. Isoenzyme distribution patterns were thus shown to be isolate specific and are suggested as a tool to aid the assessment of purity of VA strains by the use of objective parameters. Additionally the method will be developed further to obtain quantitative measurements allowing research on the competitive abilities of two or more VA mycorrhizal isolates in experiments on joint colonization of root systems.

The Role of Mycorrhiza in the Transfer of Nitrogen from White Clover to Perennial Ryegrass in Pasture Ecosystems

J.B. Rogers, P. Christie and A.S. Laidlaw
Department of Agricultural Botany, Queen's University, Newforge Lane, Belfast BT9 5PX, UK.

One possible route for transfer of fixed nitrogen from the legume to associated grass in pasture ecosystems is *via* the hyphae of root infecting fungi. Interspecific root connections can be formed by the external hyphae of vesicular-arbuscular (VA) mycorrhiza, and stable or radioactive isotopes have been observed to pass from 'donor' to 'receiver' plants. However the net flow of these isotopes from one plant to another has not been demonstrated. The possibility of net flow of ^{15}N between white clover (*Trifolium repens*) and perennial ryegrass (*Lolium perenne*) was investigated using a split root technique and selective mesh barriers of varying pore size, which allow control over the degree of root and hyphal mixing. Infection was achieved using mycorrhizal roots from the field.

The growth subtrate was sterilized 3:1 sand: soil mixture. N was applied directly to the portion of the root system of the donor plant separated from the pot containing the two plant species as 99.7 atom % $(NH_4)_2SO_4$. Both plant species were used as donors or receivers in order to determine net flow of ^{15}N. The experiment comprised 60 experimental units each composed of two pots, one which isolated the roots of the donor and the other with the remainder of the donor roots and the receiver roots, separated by barriers.

VA mycorrhizal inoculation significantly increased dry matter production and total N yield of clover shoots as receivers by 28% and 57% respectively. The contribution of the labelled source of N to the total N of the receiver shoots was not affected by the presence of mycorrhiza irrespective of barrier type, although an effect of barrier pore size was seen where grass was the receiver. This result indicates that any transfer of ^{15}N from clover to grass occurred by mass flow and/or diffusion in this experimental system. Unfortunately, breakdown around the edges of the complete barriers resulted in some root or hyphal mixing. Therefore the 5 μm pore barrier could be taken instead to represent minimum interaction between root systems.

The increase in the growth and total N yield of the clover shoots was probably due to increased uptake of phosphorus by mycorrhizal clover. However, infection did not appear to have any significant effect on the growth or N yield of the grass shoots or to facilitate transfer of N to the grass plants in the grass–clover association. This indicates that the observed net flow of ^{15}N from clover to grass was not a mycorrhizal effect, and implies that direct hyphal transfer of N may not be a significant process in grass–clover swards.

Seasonal Variation in Occurrence of VA Mycorrhizal Infection Types in a Danish Grassland Community

S. Rosendahl[1] and C.N. Rosendahl[2]
[1]*University of Copenhagen, Institut for Sporeplanter, Øster Farimagsgade 2D, DK-1353 Copenhagen K. Denmark:* [2]*Plant Protection Division, Novo-Nordisk Novo Allé 1, DK-2880 Bagsværd, Denmark.*

The occurrence of vesicular-arbuscular (VA) mycorrhizal infection types in four plant species: *Hieracium pilosella, Trifolium repens, Agrostis capillaris* and *Plantago lanceolata* were examined from April to October at five sample dates. *T. repens* and *H. pilosella* had high levels of infection at all harvests, whereas the infections in *P. lanceolata* and *A. capillaris* were lower at the first harvest, but increased during the season.

Several distinct infection types were found on each host plant, but the frequency of the infection types was strongly dependent on the harvest time. *T. repens* had coarse, medium and fine endophytes in April and June, but only the coarse type at the last harvest in October. Almost no coarse endophytes were found in *P. lanceolata*. The endophyte flora of this plant species was dominated by two medium endophytes. In *H. pilosella* a coarse endophyte was found most frequently at the last harvest in October, whereas the fine endophyte was most abundant in April. Three types of medium endophytes could be distinguished in this species. *A. capillaris* had two medium endophytes of which one was found exclusively in April, whereas the other medium endophyte which had smaller vesicles was found at all harvests. A coarse endophyte was the dominant infection type at the last harvest.

The richness of infection types found in the host plants decreased during the period from April to October, in all plants except *H. pilosella*.

The fluctuations in the endophytic flora which occurred during the period of investigation, could be the result of competition between the endophytes. In temperate perennial grasses a more rapid turnover rate of roots has been found in spring compared with late summer. This may favour endophytes that are able to spread and infect rapidly. Moreover, the low soil temperature in April could favour endophytes which are able to grow at low temperatures. These endophytes gradually disappear as other endophytes are favoured later in the season, during which the temperature is higher and the root turn-over rate lower.

In the present work no attempts were made to compare infection types found on different plant species. It is highly doubtful whether anatomical characters of VA infection are independent of the host plant species. The observed seasonal fluctuations found in the endophyte flora do not eliminate the possibility that some endophytes can infect more than one plant species.

Studies of the Extension of Individual Mycelia of VA Mycorrhizal Fungi in Natural Vegetation

S. Rosendahl, A. Michelsen and C.N. Rosendahl
Institut for Sporeplanter, University of Copenhagen, Øster Farimagsgade 2D, DK-1353 Copenhagen K Denmark.

The aim of the present study was to determine the extension of individual mycelia of vesicular-arbuscular (VA) mycorrhizal fungi by comparing isozyme patterns of VA endophytes isolated from the same root system. Roots of one specimen of chive (*Allium schoenoprasum* L.) from a Danish grassland and roots from a maize pot culture of *Glomus* sp. (E3 strain from Rothamsted) were cut into 1 cm pieces which were used as inoculum for chive and leek (*A. porrum*) respectively. Plants were inoculated with 1 cm root pieces each, and grown in a multipot system with autoclaved soil. The endophytes recovered from these plants were characterized by their isozyme pattern of malatedehydrogenase (MDH) and peptidase (PEP) after polyacrylamide gel electrophoresis (PAGE). The VA endophytes from chive all showed different isozyme patterns of MDH, whereas the PEP pattern was similar. Some differences in the MDH pattern of the E3 endophytes were also found, although the variation was less distinct.

The results suggest that the individual mycelia have a very limited extension. However, it is more likely that they arise as a result of fragmentation of a heterokaryotic mycelium. Spores of VA fungi are known to contain a large number of nuclei, and the nuclei are also regularly distributed in the mycelium which therefore contains a population of nuclei. When the mycelium is fractured the fragments, containing different populations of nuclei, appear different with respect to banding pattern. The isolates found in the present study could then still be connected on the host plant.

Effect of Specific Ectomycorrhizal Fungi on Growth of Beech Seedlings in Damaged Stands

D. Schmitz
Research Institute for Mushroom Cultivation, Germany.

Wind damage and forest decline necessitates reafforestation in Northrhine-Westfalia. Therefore ectomycorrhizal fungi which are beneficial to trees in damaged areas are being sought. A field trial was started using beech inoculated with specific ectomycorrhizal fungi on clearfelled sites. Seedlings of *Fagus sylvatica* were planted in Rootrainers, containing a mixture of vermiculite and peat. They were inoculated with *Paxillus involutus*, strains W 50 and F 56; *Pisolithus tinctorius* strain 301, or were uninoculated (controls), and planted in Bergisches Land, near Wuppertal, the field design being a two factorial (three mycorrhizal fungi and control, with and without lime) with eight plots, each with 24 plants, in six blocks. The response of plants was determined by measuring shoot length and diameter of stem base. Two-way analysis of variance and least significant difference tests were carried out. After the first growth season beeches with mycorrhiza had significantly larger stem diameters relative to those without mycorrhiza. There was no effect on stem length. Significant differences in diameter and length arising from mycorrhizal infection were also seen after the second and third growth season. An influence of liming on growth in length and diameter was also observed after the third growing season. *Paxillus involutus* was more effective than *Pisolithus tinctorius*. *Scleroderma aurantium* also gave good results on air polluted stands when inoculated on spruce and beech.

Occurrence of Ecto- and Ericoid Mycorrhizas on *Gaultheria shallon* and *Rhododendron macrophyllum* Seedlings Grown in Soils from the Oregon Coast Range

J.E. Smith[1], D.A. Perry[2] and R. Molina[1]
[1]*USDA, Forest Service, Pacific Northwest Research Station, 3200 Jefferson Way, Corvallis, OR 97331.* [2]*Department of Forest Science, Oregon State University, Corvallis, Oregon, 97331, USA.*

Ericaceous plants are integral components of forest ecosystems in the Pacific northwestern US, yet their mycorrhizal symbionts and ecosystem functions are poorly understood. Ericaceae typically form ericoid mycorrhiza, but may also form ectomycorrhiza. Because the dominant forest trees in this region are ectomycorrhizal, it is critical to evaluate the ability of ericaceous plants and overstorey conifers to share compatible mycorrhizal fungi. This study was undertaken to explore root and mycorrhizal fungus interactions when ericaceous and coniferous seedlings were grown in mixtures in microcosms.

Seedlings of two ericaceous species, *Gaultheria shallon* (GASH) and *Rhododendron macrophyllum* (RHMA), and two coniferous species, *Pseudotsuga menziesii* (PSME) and *Tsuga heterophylla* (TSHE), were grown together for approximately 1 year in pots containing field soil that had been collected from three sites in young, managed forests

(12-year-old PSME plantations) in the Oregon Coast Range. From each site, 10 soil samples were collected along two randomly selected transects at 20 m intervals. Soil samples were kept separate so that within site comparisons could be made. Enough soil to fill two 2-gallon pots when mixed with perlite (3:1 volume ratio field soil: perlite) was collected at each sample point. An autoclaved mixture of soil from the ten sample points on each site was used for controls. A total of 66 pots (264 seedlings) were grown in the greenhouse. At harvest, root systems were examined for the presence of ectomycorrhizas and ericoid mycorrhizas.

Two ectomycorrhizal types were encountered on 24% of the ericaceous plants in trace amounts. One type was dark brown with emanating dark brown cystidia that were *Rhizopogon*-like in morphology. Cystidia were 1.5–2.5 μm wide, tapered to a point and often bent at the midway point. The other type was formed by a creamish-white basidiomycete with clamp connections and hyaline hyphae similar in width, 2.5–5(6) μm, to *Laccaria laccata* or *Thelephora terrestris*. A total of nine ectomycorrhizal types, including these three, were characterized for PSME and TSHE.

Ericoid mycorrhiza, characterized by septate hyphae forming a loose weft around the root and extensive hyphal coils within the epidermal cells, were common in GASH and RHMA. Hyphae were hyaline to dark and ranged in width from 1–2.5 μm. Darkly pigmented arthroconidia attached to hyphae were observed inside root cells. Arthroconidia have been reported in the anamorphic state of two genera that contain species known to form ericoid mycorrhiza, *Myxotrichum* and *Gymnascella*.

The function of ericaceous plants in forest ecosystems is not clear. Ericoid fungi degrade complex organic substrates and provide the host with a nitrogen source that otherwise would be unavailable. Laboratory studies indicate that plants of the same and different species are linked by their mycorrhizal hyphae and may be sharing nutrients. Although the extent of this linkage function in natural systems needs further examination, the existence of hyphal connections between plants of different species illustrates the complexity of the below-ground structure of forest ecosystems. Formation of ectomycorrhiza by ericaceous plants in this study, though not extensive, indicates that hyphal links between ericoids and conifers are at least possible. More information is needed about the connections between plant species and their mycorrhizal symbionts in order to understand better the ecological role of understorey species.

Reaction of Flax *(Linum usitatissimum)* to Different Stress Factors after Mycorrhizal Infection

U. Steiner, U. Drüge, H. von Reichenbach and F. Schönbeck
Institut für Pflanzenkrankheiten und Pflanzenschutz, Universität Hannover, Herrenhäuser Str. 21, D-3000 Hannover 21, Germany.

Flax plants, growing on fields differing in soil type and nutrient supply, were found to be highly infected by VA-mycorrhiza fungi, but the significance of the mycorrhiza (M) for growth and development of the plants is unknown. Compared with other crops, the development of the flax root system is poor which suggests the possibility that benefits may arise from the symbiosis.

Flax grown in a nutrient rich compost showed a strong growth response to infection. The effects were especially large under short-day length. Compared with non-infected (NM) flax, M plants showed an enhanced biomass and shoot length, an increase of leaf area and of chlorophyll content. Inflorescence number increased and flowering occurred in short-day length conditions. Improved uptake of N, P and K occurred. Especially after drought stress shoot dry mass was regularly higher for M plants than for NM controls.

Growth responses of inoculated plants suggest that photosynthesis and leaf water status are changed. Under well-watered conditions and even under drought stress inoculated flax showed increased transpiration associated with higher CO_2-assimilation rates.

Enhancement of cytokinin content in M plants has been described for other plant species. The content of zeatinriboside (ZR) in flax plants was therefore investigated using monoclonal antibodies (ELISA).

In the early phase of infection the ZR content of roots was distinctly decreased by symbiosis. In contrast the ZR content of the shoot increased. From 4 weeks after inoculation onwards the roots of mycorrhizal plants revealed a higher ZR content than that of the controls. The lower ZR content of mycorrhizal roots in the early period of infection may be due to a disturbed production by the plant in the beginning of the symbiosis. The simultaneously increased ZR content of the shoot indicates enhanced transport into this organ. The higher ZR content of mycorrhizal roots during further development may be the result of either cytokinin production by the fungus or a stimulated production by the plant.

In view of these responses to VA mycorrhiza infection, flax can be characterized as mycorrhiza dependent. The alterations in the cytokinin content may be involved in the increase in growth of shoot and roots, because they were always preceded by a higher ZR content of the respective organ. The involvement of phytohormones is also indicated by the accelerated flowering.

Dissolution and Immobilization of Phosphorus and Cadmium from Rock Phosphates by Ectomycorrhizal Fungi

T. Surtiningsih, C. Leyval and J. Berthelin
Centre de Pédologie Biologique, LP 6831 du CNRS associé à l'Université, 17 rue N.D. des Pauvres BP5 54501 Vandoeuvre-les-Nancy, Cedex, France.

A number of soil microorganisms, bacteria and fungi, have been reported as solubilizers of insoluble inorganic phosphates. The solubilization could be related to the production of acid, complexing or reducing alkaline compounds or to enzymatic processes. Mobilization of P from insoluble phosphates by ectomycorrhizal fungi has been demonstrated. It depends on the fungal strains and on the nature (chemical composition, specific surface . . .) of the phosphates.

Some rock phosphates used to produce fertilizers contain cadmium, and since some fungi are known to accumulate trace elements such as Cd the bioavailability (immobilization, solubilization) of Cd and P in these rock phosphates was studied using ectomycorrhizal fungi. The results were compared with those obtained with a phosphate-solubilizing bacterium (*Agrobacterium radiobacter*).

Pisolithus tinctorius and *Suillus granulatus* were grown in flasks in liquid medium supplemented with 1 g of rock phosphate (from North Carolina, Senegal or Togo) in a dialysis bag. P and Cd were determined in the solution and in the biomass after acid digestion (65% nitric acid) using inductively coupled plasma mass spectrometry (ICP).

After 20 days *P. tinctorius* mobilized more P and less Cd from N. Carolina and Senegal phosphates than *S. granulatus*. But it mobilized less P and more Cd from Togo phosphate than *S. granulatus*. The percentage of P solubilized by the fungi and the bacteria was higher than the immobilized P (less than 20% was taken up), but the immobilization of Cd by the fungi was greater (>65%) than the solubilization. In contrast the Cd was mainly solubilized (immobilization <3%) by the bacterium.

The bioavalilability (solubilization + immobilization) of P from the rock phosphates with the fungi and the bacteria was in the following order: N.Carolina > Togo > Senegal.

These results are in accordance with the physical and chemical properties of the phosphates (formic acid P extraction, specific surface). The bioavailability of Cd however did not give the same order suggesting specific biological processes. The orders were:

- for *P. tinctorius*: Togo > N.Carolina > Senegal;
- for *S.granulatus*: N.Carolina > Senegal > Togo;
- for the bacteria: N.Carolina > Togo > Senegal.

These experimental approaches are simplified and do not integrate all the factors involved in the bioavailability of mineral elements in soils, but they can provide a valuable insight into mobilization processes.

Reaction of the Natural Norway Spruce Mycorrhizal Flora to Liming and Acid Irrigation

A.F.S. Taylor and F. Brand
Institut für Systematische Botanik, Menzinger Strasse 67, D-8000 München 19, Germany.

In 1984, as part of the EXMAN programme, a field experiment examining the effects of acid irrigation and liming on tree growth was established in mature Norway spruce at the Höglwald, near München, Germany. Forest plots were treated with combinations of normal (pH 5.5) or acid (pH 2.7) irrigation with or without the addition of lime. The first results from a detailed study on the effects of the treatments on the mycorrhizal flora are reported here.

Between September and November 1990, six 4.5-cm diameter cores were collected from each of the six treatment plots. Each core was separated as follows; the organic matter was first measured then divided into LF and OH material and the top 10 cm of mineral soil was divided into two portions each of 5 cm length. Each soil fraction was then stored separately in a polythene bag at 4° C until examined. Analysis of the mycorrhizas in each sample was carried out by first soaking the fractions in water for at least 1 h then by wet sieving using 1.8-mm and 0.8-mm sieves. This method proved to be fast and effective and the percentage of root tips lost during processing was very small, <1%. The identity of the mycorrhizas was determined, where possible, from published descriptions. The numbers of each mycorrhizal type and the number of non-infected root tips was recorded.

A total of 25 mycorrhizal types were found, of which 11 were identified to the species level. There was consistently more types recovered from the plot which received both lime and normal irrigation, but this was not significantly different from the control. The most widespread mycorrhiza was that formed by *Russula ochroleuca*. The numbers of this type were not significantly affected by treatment but the percentage contribution was significantly reduced in the limed plots due to the presence of the three lime-dependent types, 'Piceirhiza nigra', *Amphinema byssoides* and *Tuber* sp. 'P.nigra', in particular, showed a very distinct distribution with large numbers occurring in all limed plots, making in the commonest mycorrhizal type recorded. In addition to the types noted above, two others formed by *Tylospora* sp. and *Elaphomyces* sp. were also common and were found in all the treatment plots. The latter type and that formed by *R. ochroleuca* were consistently more frequent in the mineral soil, whereas the lime dependent types were largely confined to the organic fractions.

Differential Effects of Fungicides on VA Fungal Viability and Efficiency

A. Trouvelot, G.M. Abdel-Fattah, S. Gianinazzi and V. Gianinazzi-Pearson
Laboratoire de Phytoparasitologie, INRA-CNRS, Station de Génétique et d'Amélioration des Plantes, INRA, 21034 Dijon, France.

Fungicide application is a common and often obligatory practice in plant production. Soil applications can directly or indirectly affect VA fungi but reported effects have been variable in previous studies, depending on application rates and procedures. We have been investigating the effects of different soil-applied fungicides on VA mycorrhiza formation and vitality in pot and field trials.

The influence of five soil fungicides (fosetyl-Al, benomyl, captan, mancozeb and quintozene), applied at recommended field application rates (40, 50, 20.7, 40 and 21 mg kg^{-1} of soil respectively), on plant growth and VA mycorrhiza infection were compared in pot grown soyabean plants inoculated or not with *Glomus mosseae*. In a field trial, four actinomycete fungicides (fosetyl-Al, etridiazole, furalaxyl and prothiocarb) were studied for their effects on plant growth and on infection establishment by native or introduced VA fungi; these fungicides were applied at field rates (15, 3.5, 5 and 8.7 g m^{-2} respectively) by soil drenching after sowing of leek seeds in nondisinfected or VA-inoculated (*G. mosseae*, *G. intraradix*) disinfected plots.

In pot trials, the growth of non-mycorrhizal plants was unaffected by fungicide applications. Two fungicides, benomyl and quintozene, eliminated mycorrhizal effects on shoot growth due to depressive effects on infection (trypan blue staining) and fungal viability (succinate dehydrogenase and alkaline phosphatase activities). Both fungicides reduced the infection potential of inoculum (-82%, -83%) but only benomyl inhibited subsequent colonization of root tissues (-68%). Fosetyl-Al and mancozeb stimulated VA mycorrhiza development ($+12\%$, $+13\%$) while captan was slightly inhibitory (-10%), but none affected the mycorrhizal effect on plant growth ($+69\%$ shoot dry mass).

In the field trial, early depressive effects of furalaxyl and fosetyl-Al on shoot growth were observed in non-disinfected soil, but these disappeared at harvest in the case of fosetyl-Al. VA mycorrhiza colonization was generally low (17–25%) and no significant changes occurred in any treatment. Furalaxyl also tended to decrease plant growth in the disinfected, reinoculated plots (-31%) while etridiazole enhanced production ($+27\%$). VA mycorrhiza colonization of roots was 40–66%, higher than in non-disinfected plots. It was significantly increased following treatment with etridiazole ($+65\%$), furalaxyl ($+49\%$) or prothiocarb ($+55\%$). The VA mycorrhiza potential (MPN estimation) after harvest was similar in fungicide treated plots, as compared with corresponding untreated controls (10 000 propagules kg^{-1} soil).

Moderate soil applications of fungicides can be compatible with VA mycorrhiza systems and in some cases of low application rates, they may even enhance the mycorrhizal effect.

Are Hydrophobic Ectomycorrhizas Important for Microbial Activity in the Forest Soil?

T. Unestam
Department of Forest Mycology and Pathology, Swedish University of Agricultural Sciences, Box 7026, 75007 Uppsala, Sweden.

Ectomycorrhizal short roots, mycelia, rhizomorphs, and mats from conifer soil were examined in relation to their hydrophobic properties. In some cases, connected fruit bodies were included in the study. Mycorrhizal soils gathered from the forest and/or colonized in a laboratory rhizoscope were studied, as were mycelia in pure culture. Most forest-derived species were hydrophobic. The drought resistant *Cenococcum geophilum* and the more ruderal and moisture dependent *Thelephora terrestris* were both strongly hydrophilic. The hydrophobic mycelium seemed solely responsible for the water-repellent properties, and adjacent soil and plant debris remained unaffected, and still hydrophilic. In hydrophobic fungi, mat formation was induced in the rhizoscope by hyphal contact with red alder litter leaves. This stimulating effect was not found when the leaves were covered by a water film or by using fresh, green alder leaves. *T. terrestris* did not form such mats *in vitro* and spread sparsely in air pockets as well as in the adjacent water film. It is discussed whether many mycorrhizal fungi in the forest may, in part, control their soil environment via aeration created by their hydrophobia.

The Effects of Cu and Ni on the Axenic Growth and on the Element Composition of *Cenococcum geophilum* and *Suillus variegatus*

H. Väre

Department of Botany, University of Oulu, SF-90570 Oulu, Finland.

The effects of Cu and Ni on the growth and on the element composition of *Cenococcum geophilum* and *Suillus variegatus* was studied in order to determine if Cu and Ni are chelated by mechanisms similar to those seen in Al-polyphosphate binding on *S. variegatus* hyphae.

Pure mycelial cultures of *C. geophilum* and *S. variegatus* were grown on Petri plates containing MMN media and various concentrations of copper or nickel (0, 50, 100 or 500 p.p.m.) in the form of $CuCl_2$ and $NiCl_3.6H_2O$.

The growth of fungal mycelium was measured after 60 days. The composition (K, Cl, P, S) of hyphae was analysed by SEM-EDS (scanning electron microscopy energy dispersion system) or STEM-EDS (scanning transmission electron microscopy energy dispersion system) (Cu, Ni). The results are qualitative and expressed as relative percentage of all elements detected. *S. variegatus* showed growth only on one medium containing Cu (50 p.p.m.). Therefore it was not possible to investigate *S. variegatus* by SEM-EDS or STEM-EDS.

Cenococcum showed growth on 50 or 100 p.p.m. Cu or Ni. Copper could not be detected in *Cenococcum* grown on Cu media. The low concentration of Cu in the fungal hyphae may be due to an active exclusion mechanism or to reactions between Cu and other elements in the media. The concentrations of Cl and S in *Cenococcum* hyphae decreased in higher concentrations of Cu.

Nickel was detected in *Cenococcum* hyphae growing in 50 or 100 p.p.m. Ni. Phosphorus concentrations increased with those of Ni, and Cl concentrations decreased.

An *S. variegatus* strain of high Al tolerance was sensitive to Cu and Ni. This indicates different sensitivities to different metals.

As a conclusion it is hypothesized that heavy metal depositions select for tolerant fungal strains and reduce the number of fungal species in nature resulting in a less diverse ecosystem. Further, Cu and Ni may chelate essential elements such as P in soil and on mycorrhizal and fungal hyphae.

The Influence of Organic and Inorganic Fertilization on Development of Indigenous VA Fungi in Roots of Red Clover

H. Vejsadová

Institute of Microbiology, Czechoslovak Academy of Sciences, Prague, Czechoslovakia.

Before any inoculation programme is initiated the interactions between native VA fungal populations and fertilization must be studied for each plant and soil. The extent of infec-

Table 1. The effect of fertilization on VA infection of red clover.

N	P	K	Manure	Frequency	Intensity	Arbuscule
	(kg ha^{-1})		(t ha^{-1})	(%)	(%)	level
0	0	0	0	35[b]	10[a]	1.3[a]
0	0	0	40	71[c]	38[b]	6.1[b]
80	200	200	40	3[d]	0[c]	0.0[d]
80	200	0	40	14[a]	3[a]	2.8[a]
80	0	200	40	88[c]	45[b]	11.9[c]

Means marked by different letters are significantly different on the level $P = 0.05$ (Duncan's test, $n = 20$).

tion by native VA fungi in roots of red clover was measured during a long-term field experiment involving application of manure and N, P, K fertilization (Table 1). The manure significantly stimulated VA frequency and intensity as compared with the control treatment (limed only). N, P, K applied together with manure caused the most dramatic decrease of infection in comparison with other treatments. The results confirmed the negative influence of P fertilization on the VA infection in clover. Additional N fertilization had a positive effect on mycorrhiza. The positive influence of organic manuring on VA infection probably depends on culture conditions. The present findings provide a better insight into mutual relationship between VA fungal populations and fertilizers.

Propagule Production by VA Fungi in Red Clover Plants Subjected to Periodic Removal of the Aerial Parts

A. Vilariño[1], J. Arines[1] and H. Schüepp[2]
[1]*Instituto de Investigaciones Agrobiologicas de Galicia, CSIC, 15080-Santiago de Compostela, Spain: [2]Swiss Federal Research Station, CH-8820 Wädenswil, Switzerland.*

Experiments carried out in our laboratory have shown that the removal of the aerial part of red clover plants colonized by different VA fungi, enhances inoculum potential in soil. Because of the implication of these results for inoculum production, we have studied the response of the VA fungi *Glomus mosseae*, and *Glomus epigaeum* to this treatment.

The results showed that independently of VA mycorrhiza colonization, the highest arbuscule abundance was found in root of plants subjected to removal of the aerial part. The effect of clipping on spore production and extraradicular mycelium growth was also different between inocula. *Glomus mosseae* increased sporulation in response to the clipping, without significant differences in mycelium growth. On the contrary, *Glomus epigaeum* produced more mycelium, and reduced its spore formation.

There was not a significant difference in MPN (most probable number) for extraradicular mycelium between clipped and control plants for both inocula, but clipping slightly increased this feature. These results suggest that the removal of the shoots of colonized red clover plants may be an effective method for increasing VA mycorrhiza inoculum potential in soil. Nevertheless, environmental conditions, clipping frequency, and time allowed for plant resprout are likely to play an important role in the VA endophyte response.

Acceleration of VA Mycorrhiza Development by Bacteria or Fungicides

H. von Alten and A. Lindemann
Institut für Pflanzenkrankheiten und Pflanzenschutz, Universität Hannover, Herrenhäuser Str. 2; D-3000 Hannover 21, Germany.

A controlled improvement of VA mycorrhiza development especially in young plants may increase plant stress tolerance and performance. It has recently been shown that a *Bacillus cereus* var. *mycoides* isolate increased the growth of several VA mycorrhizal isolates in a number of plants, measured as percentage mycorrhizal root length per plant. In this experiment mycorrhizal inoculum growing in expanded clay particles was mixed into the culture medium of the bacterium *B. cereus*. Then plants were inoculated with the two microorganisms by mixing 5% (vol/vol) into their growth medium. Only the early stages of VA development were enhanced. The effect depended on the type of soil used.

The increases of VA infection were not only seen as a faster colonization of the root systems but also as an increase in sporulation following removal of the host shoots 4 weeks after planting. This sporulation may be a reflection of the vigour of the symbiosis at the time of shoot removal. When VA fungi were inoculated together with the bacteria they produced twice as many spores as they did in the absence of *Bacillus*.

In addition to bacteria, several fungicides were tested for their ability to improve VA mycorrhizal activity after leaf application. Two fungicides known to be able to accelerate VA mycorrhiza formation when applied to barley leaves, and two active bacteria (*B. cereus* and *Pseudomonas fluorescens*, applied to the mycorrhizal expanded clay inoculum) were used in combination. The aim was to determine whether additive, synergistic, or antagonistic interactions of the two VA mycorrhiza improving factors occurred. The results show, that mycorrhizal colonization of the fungicide + bacteria treated plants was relatively poor in most combinations. Obviously *Bacillus*, *Pseudomonas*, and the fungicides improve VA development by different mechanisms. This assumption is supported by measurements of the exudation of mycorrhizal roots which was increased by bacterial applications, but not by the leaf fungicides. A greater root exudation after the bacterial treatment might attract the mycorrhizal hyphae and so lead to a faster colonization. Further indirect evidence for the view that *B. cereus* acts via the host plant and not directly by way of an improvement of inoculum infectivity is that in some unsterile soils the bacteria added to the inoculum particles could accelerate the development both of the indigenous and the introduced VA fungi.

Influence of water status on VA infection and growth of *Festuca rubra*

M. Vosatka, and V. Hadincova
Botanical Institute, Czechoslovak Academy of Science, 252–43 Pruhonice Near Prague, Czechoslovakia.

The aim of this study was to investigate seasonal changes of VA infection of *Festuca rubra* clones in meadow communities on sites of differing soil water status and the mycorrhizal symbiosis in two clones grown under four water regimes.

The field study was conducted in four communities: (1) *Angelico–Cirsietum palustris Scirpetum sylvatici* (A–Cp); (2) *Carici rostratae–Sphagnetum apiculati* (Cr–Sa); (3) *Nardo–Festucetum capillatae* (N–Fc); (4) *Sanquisorbo–Festucetum commutatae* (SFc). Sites 1, 2 and 4 were permanently flooded, and 3 was periodically flooded by water. Phytocenological observations and measurement of seasonal dynamics of underground water were made over a period of 3 years. Ten soil samples from each plot were collected in 1989 from two depths (0–5 cm and 5–10 cm). 150 × 1-cm root segments were analysed. No difference in VA infection between roots from two depths (0–5 cm and 5–10 cm) was found. Seasonal dynamics of infection were similar at all wet localities. Plants from the driest site (N–Fc) exhibited steady increase of infection during a growing season. Fluctuation of underground water is probably more favourable for infection development than steady wet conditions in accordance with previous reported results.

Two clones of *F. rubra* selected from N–Fc and Cr–Sa communities were used in the pot experiment. Plants were grown from tillers in a 1.5 litre pot filled with soil and inserted into a larger 2 litre pot to supply water. Sandy loam with a high population of *Glomus mosseae*, *G. fasciculatum* was used – numbers of spores and infective propagules were 149 + 82 and 506 + 65 per 100 g of soil respectively. Plants were grown from September to April under the same conditions and then 6 months under four water regimes: dry (no additional irrigation); moist (irrigated to the soil water capacity); moist–wet (when water in outer pot evaporated it was refilled) and wet (outer pot always full of water). Plant biomass and number of tillers were the highest in moist treatment in the N–Fc clone probably due to its better adaptation to similar conditions in the field. Cr–Sa plants showed better adaptability to wet soil in that they had the highest infection in the wet treatment. *F. rubra* is a facultative symbiont which showed considerable clonal plasticity, the VA responses of each of which are probably adapted to the water status of their environments.

VA Mycorrhizal Colonization of Maize in an Industrially Polluted Soil and Heavy Metal Transfer to the Plant

I. Weissenhorn, C Leyval and J. Berthelin
Centre de Pédologie Biologique, UPR 6831 du CNRS, associé à l'Université Nancy I, B.P.5–54501, Vandoeuvre-les-Nancy Cedex, France.

Heavy metals in soils are reported to decrease mycorrhizal infection. However, metal polluted soils contain mycorrhizal fungi which may be tolerant strains. Mycorrhizal effects on metal uptake by plants in polluted soils may involve enhancement of uptake combined with plant growth decrease and increased tolerance of mycorrhizal plants to toxic metal concentrations in soil. In field and pot trials on an industrially polluted soil we have studied (1) the effect of heavy metals (Cd, Pb, Zn, Cu) on the mycorrhizal colonization of maize roots and (2) the mycorrhizal effect on the transfer of metals to these plants.

Maize was grown in field plots or in pots in a growth chamber on a soil polluted by heavy metals via atmospheric deposition derived from a metal smelter (Metaleurop, Nord-Pas-de-Calais). Roots were cleared and stained with trypan blue to score the frequency (% of infected fragments) and the intensity (% of infected root cortex) of mycorrhizal infection. Metal concentrations in shoot tissue were analysed after digestion in nitric acid by plasma emission spectrometry.

The frequency and particularly the intensity of mycorrhizal infection of maize roots was lower on the metal polluted plot (87.5% and 30% respectively) than on the control (100% and 48.2%). In a pot assay we found after 4, 7 and 12 weeks of cultivation an even stronger decrease of mycorrhizal infection in the polluted soil compared with the control (90% to 75% frequency and 28% to 11% intensity). This indicates a delay of mycorrhizal development in the polluted soil.

In maize plants grown for 9 weeks in pots on the polluted soil (sterilized by irradiation or untreated) we observed a significantly lower Cd, Zn ($P < 0.01$) and Cu ($P < 0.05$) concentration (16.321 and 13.5 p.p.m. respectively) in the shoots of mycorrhizal (24.5% infection intensity) than in non-mycorrhizal plants (29.446 and 15.7 p.p.m. respectively).

These preliminary results show: (1) A negative effect of metals on the development of VA mycorrhiza. However considerable mycorrhizal infection was found in the polluted soil, suggesting the presence of tolerant strains. In further experiments strains isolated from the polluted soil will be compared with reference strains. (2) A possible mycorrhizal control of metal uptake into plants from polluted soils decreasing their phytotoxic effects.

The Possible Application of Enzyme-linked Immunosorbent Assay (ELISA) for Detection of *Tuber albidum* Ectomycorrhizas

A. Zambonelli[1], L. Giunchedi[2] and C. Poggi Pollini[2]
[1]Dipartimento di Protezione e Valorizzazione Agroalimentare; [2]Instituto di Patologia Vegetale, Università degli Studi di Bologna, Italy.

Seedling mycorrhizal inoculation with *Tuber* spp. is extensively used in Italy for truffle cultivation. Mycorrhiza formation is usually assessed by morphological methods. However, only a specialist can recognize distinctive *Tuber* sp. mycorrhiza. In particular *Tuber albidum* and *Tuber magnatum* mycorrhiza are very similar in colour, size, shape and mantle ornamentation and it is possible to confuse them.

The aim of this study was to determine whether the immunoenzymatic technique ELISA is suitable for characterization and detection *T. albidum* ectomycorrhiza.

An antiserum was prepared in rabbit against an acetone precipitate of a culture fluid of *T. albidum*. F(ab^1)$_2$ indirect ELISA technique was used to detect the presence of *T. albidum* in root tips of *Quercus pubescens* and *Pinus pinea* seedlings inoculated under controlled conditions. Samples were grown in phosphate-buffered saline (PBS) Tween containing 2% polyvinyl polypyrrolidone (PVP) 25K and 0.2% ovoalbumin. The ELISA tests were conducted in microtitre plates (Dynatech) coated with $F(ab^1)_2$ fragments of IgG, diluted 1:400 in carbonate buffer. Trapped antigens were detected with whole immunoglobulin (as the second antibody), diluted 1:600 in PBS Tween, and Protein A alkaline phosphatese conjugate (Sigma), diluted 1:1000 in the same buffer. Enzymatic activity was expressed spectrophotometrically as absorptivity at 405 nm.

Our results show that the $F(ab^1)_2$ indirect ELISA test is suitable, and highly specific, for discriminating amongst *T. albidum* ectomycorrhiza and other ectomycorrhiza on both *Q. pubescens* and *P. pinea* seedlings. In particular this ELISA method can differentiate *T. albidum* and *T. magnatum* mycorrhiza. Moreover, it is able to detect *T. albidum* ectomycorrhiza at low concentrations of infected root tips.

The variability of the absorbance values of *T. albidum* mycorrhiza is due to the different ages and sizes in the ectomycorrhiza tested. The young, branched ectomycorrhiza with a well-developed fungal mantle are the most reactive.

In general, *P. pinea* mycorrhiza are better developed than those of *Q. pubescens* and thus they can be detected more readily by ELISA.

Index